Bridges

Baidar Bakht • Aftab Mufti

Bridges

Analysis, Design, Structural Health Monitoring, and Rehabilitation

Second Edition

Springer

Baidar Bakht
Department of Civil Engineering
University of Manitoba
Winnipeg, MB, Canada

Aftab Mufti
Department of Civil Engineering
University of Manitoba
Winnipeg, MB, Canada

Additional material to this book can be downloaded from http://extras.springer.com
All photos are provided courtesy of JMBT Structures Research Inc.

ISBN 978-3-319-17842-4 ISBN 978-3-319-17843-1 (eBook)
DOI 10.1007/978-3-319-17843-1

Library of Congress Control Number: 2015951973

Springer Cham Heidelberg New York Dordrecht London

Printed on acid-free paper

Springer International Publishing AG Switzerland is part of Springer Science+Business Media (www.springer.com)

This book is dedicated to the memory of Leslie G. Jaeger CM, DSc, FRSE, P.Eng. (1926–2013),
An outstanding teacher, a cherished research colleague, a dear friend, and a mentor to authors

Foreword

The authors, sometimes accompanied by Professor Leslie Jaeger (1926–2013), have given courses on bridge engineering in various countries of the world since early 1990s; these courses dealt with the aspects of bridge engineering in which the authors not only had a hands-on experience, but had also published technical papers. The course notes, initially comprising copies of published papers, progressed into a seminal document. The latest and most up-to-date version of this document is now in your hands in the form of a self-contained book.

It is the hope of the authors that both practicing bridge engineers and graduate students of structural engineering will find this book practical and useful. A number of computer programs, which can be downloaded from http://extras.springer.com, are expected to provide solutions to everyday problems facing bridge engineers. Some topics, such as arching in deck slabs, fibre reinforced polymers, and structural health monitoring, might not be covered by the design codes of many countries. Hopefully, the chapters on these topics that provide copious references and case histories will pave the way for their adoption in these countries.

Our outstanding graduate students and research associates have been acknowledged through the references cited in the books. We would also like to express our gratitude to the following persons, some of whom are not acknowledged through the references: Gamil Tadros for his friendship and collaboration on many engineering projects; Evangeline Murison for her help in many SHM projects; Mike Wilson of Atlantic Industries Limited, Canada, for his permission to use the details of the Whitehorse Creek soil-steel bridge. Also, we like to thank our various secretaries at Dalhousie University and University of Manitoba, who have assisted us in producing the book.

It is acknowledged that the CHBDC (2006) design provisions, described in Chaps. 8 and 9 were formulated by the Technical Subcommittee, which comprised B. Bakht, N. Banthia, B. Benmokrane, G. Desgagné, R. Eden, M.-A. Erki,

V. Karbhari, J. Kroman, D. Lai, A. Machida, A.A. Mufti, K. Neale, G. Tadros, B. Täljsten, with A.A. Mufti being the chair. The current chair of the CHBDC Technical Subcommittee for Fibre Reinforced Structures is S. Sheikh.

Winnipeg, MB, Canada
February 21, 2015

Baidar Bakht C.M.
Aftab Mufti C.M.

Contents

Chapter 1
Loads and Codes

1.1 Introduction

Although not generally appreciated by lay people, it is not possible to design and construct a structure that will remain safe against failure under all conditions and at all times.

The several reasons for a structure being prone to failures include: (a) the strength of the various components of the structure cannot be assessed with full certainty; (b) the loads that a structure will be called upon to sustain also cannot be predicted with certainty; and (c) the condition of a structure may deteriorate with time due to the effects of the environment, causing it to lose strength. Because of these factors, there exists a probability that the strength of a structure will at some time be exceeded by the loads that it has to sustain, resulting in the failure of the structure. As noted in Sect. 1.3.2, the term failure is being used here not only to signify the collapse of the whole structure, but also to include the situation of the structure not being able to fulfil one or more of its intended functions.

The probability of failure of a structure can be reduced by increasing its design strength, which invariably leads to a higher first cost. The role of the structural engineer is to strike a socially acceptable balance between the risk of failure and the cost of the structure. For example, a bridge can indeed be built to have the same probability of failure as the pyramids of Giza, shown in Fig. 1.1. The cost of such a bridge, however, is likely to be so high that society may not be prepared to pay for it. By contrast, society may not be prepared to accept in a bridge the same high frequency of failure as in an automobile.

It is sometimes argued that a good engineer can strike a balance intuitively between the cost and safety of a structure, and that design codes tend to restrict the creative ability of the designer. The ideal criteria for structural design, it is argued, are those which merely require that a structure remain safe while fulfilling its intended functions. Examples of the world's most spectacular bridges, which

B. Bakht, A. Mufti, *Bridges*, DOI 10.1007/978-3-319-17843-1_1

Fig. 1.1 Pyramids of Giza in Egypt, examples of structures with low probability of failure

have very long spans and for which there existed no design codes until recently, are given in defence of the argument for having no design code at all.

It can be demonstrated readily that, due to the lack of a set of comprehensive design criteria, different structures designed by different designers are likely to have different probabilities of failure. This situation is particularly undesirable for bridges on the same roadway system. Since all such bridges are likely to be subjected to nearly the same maximum vehicle and environmental loads, the bridge with the highest probability of failure will govern the capacity of the road; in this case, it can be readily appreciated that the resources put into making the rest of the bridges extra-safe are not being expended wisely.

Since the designs of short and medium span highway bridges are governed mainly by vehicle weights, the design live loads constitute a very important part of the design criteria. It is surprising that little attempt is usually made to ensure a realistic correspondence between the actual vehicle weights in a jurisdiction and the design live loads for its bridges.

This chapter presents a method using which any number of vehicles can be compared with each other with respect to the maximum load effects they induce on bridges; this method can also be used to formulate one or more design vehicles corresponding to a given population of vehicles. The chapter also provides the basics of the probabilistic methods, which are used to quantify safety in modern design codes.

1.2 Vehicle Loads

Designs of most short and medium span highway bridges are governed predominantly by longitudinal moments and shears. The live load components of these responses are caused by heavy commercial vehicles and are governed by the spacing and weights of their axles. The task of quantifying the commercial vehicles

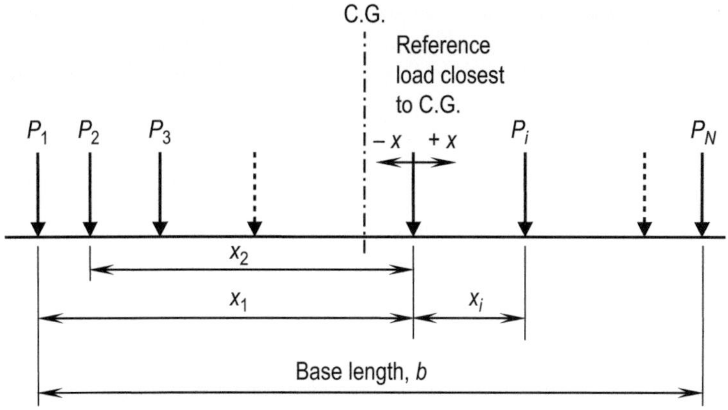

Fig. 1.2 Notation for a series of point loads and their spacing

with respect to the load effects which they induce in bridges is made difficult by the very large number of axle weight and spacing combinations that are encountered in practice.

With the help of the method described in Sect. 1.2.1, a set of discrete loads can be reduced to an equivalent uniformly distributed load (UDL) which gives very closely the same maximum moments and shears in one-dimensional beams as the discrete loads. The equivalent UDL, as explained later, is useful in comparing the effects of different vehicles in all bridges.

1.2.1 Equivalent Base Length

It has been shown by Csagoly and Dorton (1978) that N discrete loads, with a total weight of W, on a beam can be replaced by a UDL which is also of total weight W, and has a length B_m so that the moment envelope along the beam due to the UDL is very nearly the same to the moment envelope due to the set of discrete loads. The length B_m, which is referred to as the *equivalent base length*, is given by the following equation:

$$B_m = \frac{4}{W}\sum_{i=1}^{N} |P_i x_i| - \frac{2(N-1)}{bNW^2}\left\{\sum_{i=1}^{N} (P_i x_i)\right\}^2 \tag{1.1}$$

where N is the total number of discrete loads and other notation is as illustrated in Fig. 1.2. The load closest to the centre of gravity of the set of loads is taken as the reference load and distances of other loads x_i, are measured with reference to this load.

It can be seen that Eq. (1.1) is independent of the span length of the beam; it gives only approximate values of B_m which, as shown later, are accurate enough for most practical purposes. Equation (1.1) is adapted from the following more accurate expression which incorporates the span length, L, of the beam and which is reported by Jung and Witecki (1971).

$$B_m = \frac{4}{W}\sum_{i=1}^{N} |P_i x_i| - \frac{2}{LW^2}\left\{\sum_{i=1}^{N} (P_i x_i)\right\}^2 \tag{1.2}$$

1.2.1.1 Accuracy

The percentage of error incurred in the determination of beam moments through the simplified approach of equivalent base length defined by Eq. (1.1) is denoted by Δ and quantified by:

$$\Delta = \left\{\left(\frac{M_B}{M}\right) - 1\right\} \times 100 \tag{1.3}$$

where M_B is the maximum beam moment at a reference point due to the uniformly distributed load of length B_m obtained by Eq. (1.1), and M is the corresponding maximum moment due to the given set of discrete loads.

Values of Δ are plotted in Fig. 1.3a against span length for moments in simply supported beams due to a truck with five axles. It can be seen in this illustrative example that the degree of error is within +1 % and −8 % for all reference points considered. Values of Δ are large only where the magnitude of moment is small and hence the magnitude of Δ is irrelevant.

Although Eq. (1.1) was developed for moments in simply supported beams, it is also valid for shears and for continuous beams. In Fig. 1.3b, values of Δ are plotted against the span length of a two-span continuous beam for maximum moments at different points also due to a truck with five axles. It will be noted that the values of Δ are somewhat larger than those of their counterparts in the simply supported beam, but are still small, being within +1 % and −10 %. In both beams, Δ reduces with the increase in span length.

The actual envelope of maximum moments in a simply supported beam with a span of 10.67 m span to a five-axle truck is compared in Fig. 1.4 with the envelope of maximum moments due to the UDL of length B_m obtained by Eq. (1.1). The closeness of the two envelopes is striking. The figure also shows the variation of Δ along the span. It can be seen that the values of Δ are very small in the middle half of the bridge, where moments are usually considered in design, being within ± 5 % in this region.

Near the supports, where the magnitudes of moments are small, Δ becomes as high as about −13 % but is not of concern in design.

It has been shown by Csagoly and Dorton (1973) that the value of Δ increases as the number of discrete loads is reduced. However, even for a set of three

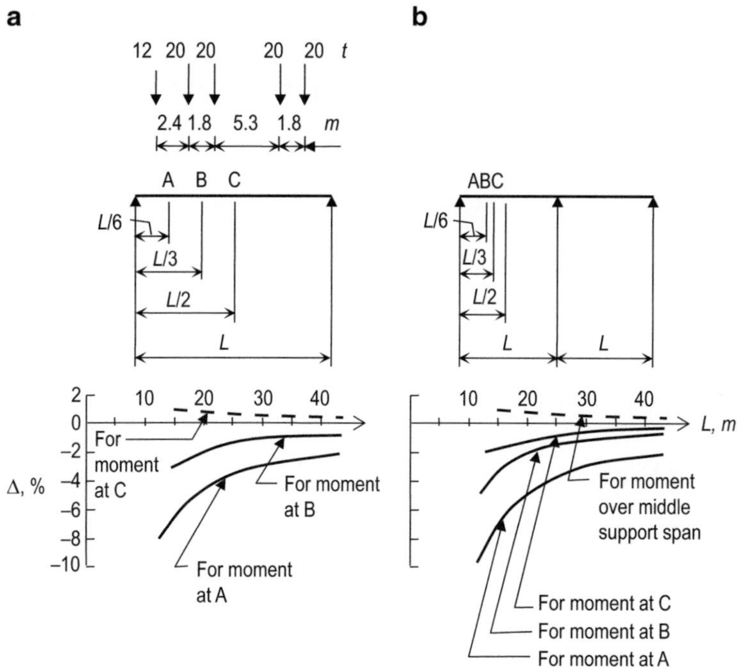

Fig. 1.3 Δ plotted against span length: (**a**) simply supported beam; (**b**) two-span continuous beam

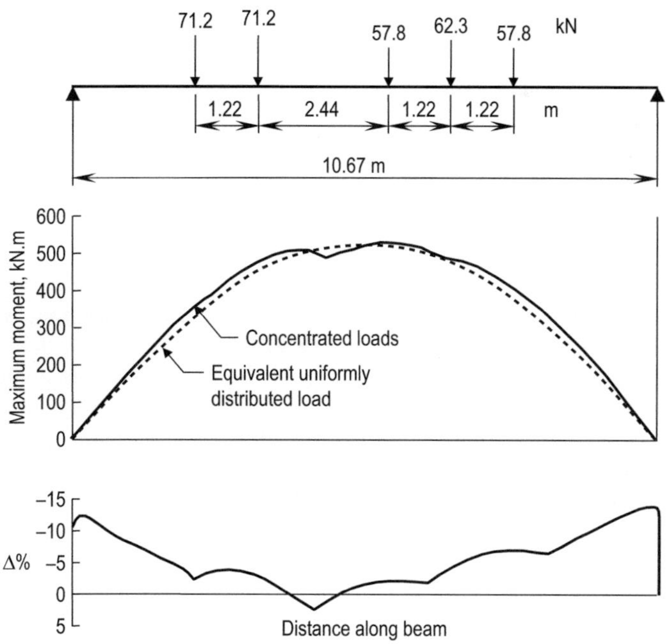

Fig. 1.4 Comparison of moments due to a series of point loads and the equivalent uniformly distributed load

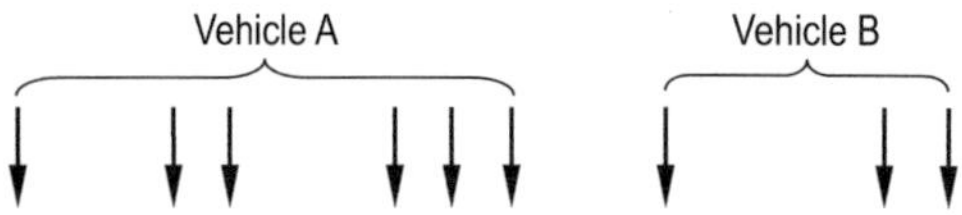

Fig. 1.5 Axles of two different vehicles

concentrated loads, Δ is found to be within +3 % and −11 % for simply supported beams, and +3.5 % and −13 % for two-span continuous beams. The above ranges cited above are for moments near end supports. For moments in the middle regions of beams, these ranges are much narrower.

It is concluded that a set of discrete loads can be realistically transformed as a uniformly distributed load without taking account of the beam span.

1.2.1.2 W-B_m Space

The elimination of the span length from consideration simplifies the task of comparing two sets of discrete loads with regard to their load effects, namely moments and shears, in beams of different spans.

For example, two sets of loads, identified as Vehicles A and B in Fig. 1.5, are compared by using the maximum moments they induce in any beam as the basis. If the two vehicles have the same total weight, W, then clearly the vehicle with shorter equivalent base length, B_m, will induce higher maximum moments. Alternatively, if the two vehicles have the same B_m, then the vehicle with larger W will induce higher maximum moments. The total longitudinal moment, or shear, across the cross-section of a right bridge due to a single vehicle is used as a basis for comparison and is hereafter referred to simply as the moment, or shear. Thus, the bridge is reduced to a beam and each axle load to a discrete load.

Each set of axles of a given vehicle leads to one value each of W and B_m. Therefore, on a two-dimensional surface, with W and B_m as the orthogonal axes, each combination of axles is represented as a single point. Obviously, the W-B_m space, as this surface is called herein, can accommodate any number of points.

Truck survey data involving a very large number of vehicles can now be condensed on a single sheet of paper. Schematically, such a diagram would be as shown in Fig. 1.6. An upper-bound envelope of the type shown in this figure, or a curve parallel to it, can then be used as the basis for vehicle weight control. Such a curve of permissible vehicles can be established after superimposing an envelope of bridge live load and capacity of existing bridges and making allowance for a compliance factor, as illustrated in Fig. 1.6. A point outside this curve would indicate a vehicle, which would produce higher moments and shears in all bridges than vehicles of permissible weights. Agarwal (1978) has used the W-B_m space as a convenient device for comparing vehicle weight regulations across Canada.

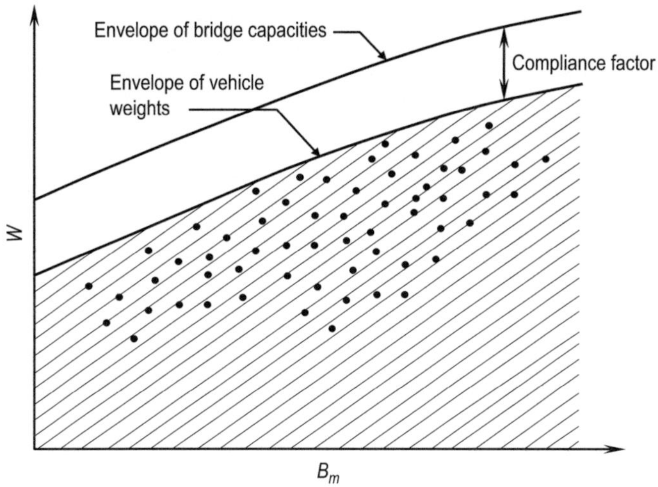

Fig. 1.6 Vehicle weight data plotted on W-B_m axes

1.2.2 Formulation of Design Live Loads

It is not feasible to design a bridge individually for each of the millions of vehicles that are likely to cross it during its lifetime. Accordingly, bridge design codes specify a limited number of design live loads, which are representative of the expected traffic. The design live loads, which usually comprise discrete point loads and/or UDLs, are formulated in such a way that the load effects induced by them in any bridge component constitute, with a known degree of certainty, an upper-bound of the corresponding load effects caused by all vehicles that are expected to cross the bridge.

Typically, design loadings are developed on a per-lane basis, with the multiple presences of vehicles in more than one lane being accounted for by means of reduction factors. Older bridge design codes contain simplified design loadings which are in the form of UDLs and knife-edge loads.

Such simplifications were necessary in order to keep the analytical calculations to the minimum. The trend in modern design codes has been to keep the design loading as close in configuration to the actual vehicles as possible. The reason for this preference is explained in the following.

Bridge design live loads are formulated, almost without exception, on the basis of maximum bending moments and shears in simply supported beams. The design live loads thus formulated are usually adequate for longitudinal components in conventional bridges in which the governing load effects are related to beam moments and shears. As shown conceptually in Fig. 1.4, a vehicle on a simply supported beam can be represented realistically by a UDL in such a way that the maximum bending moments, or shears, induced by the two loads are the same. However, such a UDL may not be able to represent adequately the vehicle loading on a spandrel-filled arch bridge or a soil-steel bridge, the latter being dealt with in Chap. 7. The lack of correspondence between the vehicle loads and the distributed

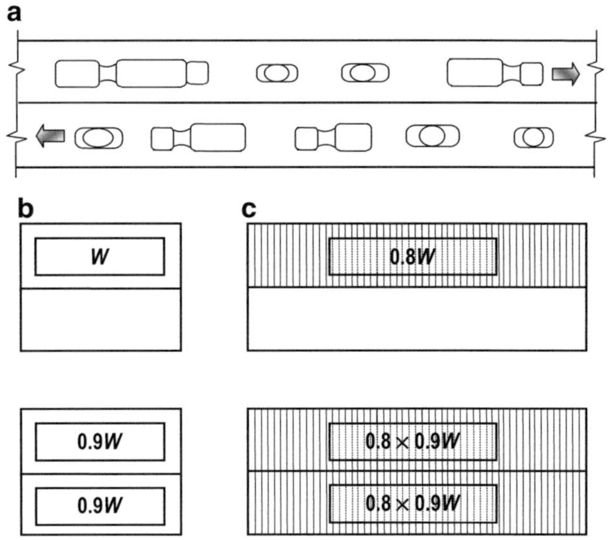

Fig. 1.7 Vehicles on a two-lane road and their representation by bridge design loads: (**a**) an example of actual vehicle traffic; (**b**) design loads on a short span bridge; (**c**) design loads on a medium span bridge

load is because the concentrated loads of the vehicle disperse through the fill in such a manner that the equivalence between the actual and idealized loadings, which is valid for beam-type bridges, is no longer maintained for arch-type bridges.

A UDL, even when accompanied by a knife-edge load, can fail to represent vehicle loads adequately for some bridge components, notably those which span generally in the transverse direction. The cantilever deck slab overhang dealt with in Chap. 5 is an example of such a component.

In modern design codes, for example the Canadian Highway Bridge Design Code (CHBDC, 2006), the design live loading per lane comprises two alternatives, one of which is a *design vehicle* with several axles, with the centres of the wheels in each axle being, more or less, at the same distance as that in typical commercial vehicles. For the other alternative, the loading comprises the design vehicle reduced by a prescribed fraction together with a UDL whose intensity per unit length of the span does not change with the span length of the bridge. The various components of a bridge are required to be designed for the higher of the load effects induced by the two alternatives.

The design vehicle is formulated to represent all the actual or foreseen vehicles individually. Accordingly, it governs the design of short span transverse components, and the longitudinal components of those bridges which have spans of less than about 20 m. The main components of medium span bridges, which have spans between 20 and 125 m, are generally governed by the second alternative loading. It can be appreciated that the UDL component of this alternative accounts for more than one vehicle in the loaded length of one lane.

The representation of the actual vehicle loads on a road by design live loads is illustrated by the various sketches in Fig. 1.7. The sketch in Fig. 1.7a shows in plan the mix of vehicles on a two-lane highway. As expected, the traffic on the road consists of large and small trucks interspersed with much lighter passenger cars. A short span bridge on the two-lane highway is shown in plan in Fig. 1.7b. For convenience of illustration, the bridge has a larger span than the overall length of the design vehicle. Such a bridge will typically be investigated for two load cases involving the design live loads. In one load case, the bridge will be subjected to the full design vehicle, with a total load W, placed strategically in only one lane. In the other load case, each of the two lanes will carry a design vehicle the weight of which is multiplied by a factor which is smaller than 1.0, say 0.9; this factor, called the multi-presence reduction factor, is discussed later.

A medium span bridge on the same two-lane highway is now considered. Since it is long enough to contain more than one truck on one lane, the design of its longitudinal components is likely to be governed by one or the other of the two load cases shown in Fig. 1.7c.

In one load case, one lane of the bridge is subjected to the specified uniformly distributed load of w/unit length and the design vehicle whose weight is multiplied by a factor smaller than 1.0, say 0.8. In the other loading case, each of the two lanes carry the same loading as in the first load case except that all the loads are proportionally reduced by multiplying them by a multi presence reduction factor, which is smaller than by 1.0, say 0.9.

1.2.2.1 Design Vehicle

From the preceding discussion it can be seen that one component of design live loading should ideally resemble actual trucks; this component is termed herein as the *design vehicle*. As explained in the following, the W-B_m space, discussed in Sect. 1.2.1, can be used conveniently to develop the design vehicle.

In order to explain the technique for developing a design vehicle through the W-B_m space, three fictitious heaviest vehicles are chosen which are to be represented by a single design vehicle. As shown in Fig. 1.8, one of the heaviest vehicles has three axles, another has five axles, and the last one has six axles. The weights and spacing of the axles of the three vehicles are also shown in this figure, together with all the relevant sub-configurations of these vehicles involving one to six axles. It can be seen that the three heaviest vehicles lead to 22 relevant individual combinations of axles which should be considered with respect to their effects on bridges.

Points representing W and B_m values for most of the 22 combinations of axles are plotted on the W-B_m space of Fig. 1.9, which also shows an upper-bound curve enveloping all the 22 points.

For a single concentrated load, $B_m = 0.0$. For uniformly-spaced equal point loads, B_m can be obtained from the following equation.

$$B_m = Kb \tag{1.4}$$

where the value of K is plotted in Fig. 1.10 against the total number of points N. As expected, K converges to 1.0 as N becomes large.

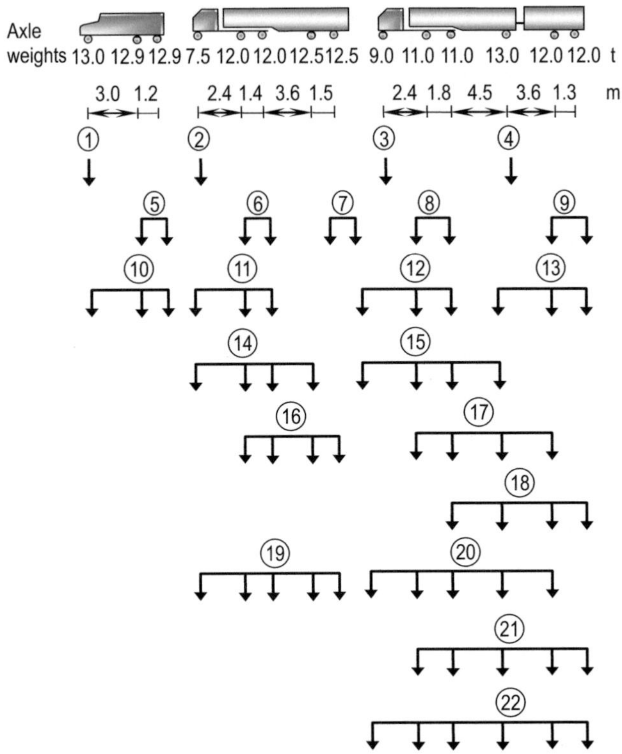

Fig. 1.8 Axle weights and spacing of three trucks and their sub-configurations

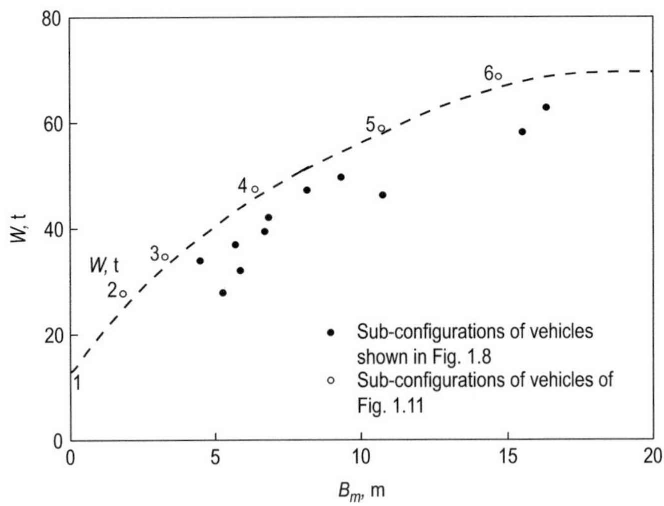

Fig. 1.9 W-B_m data corresponding to the axle load combinations shown in Fig. 1.8

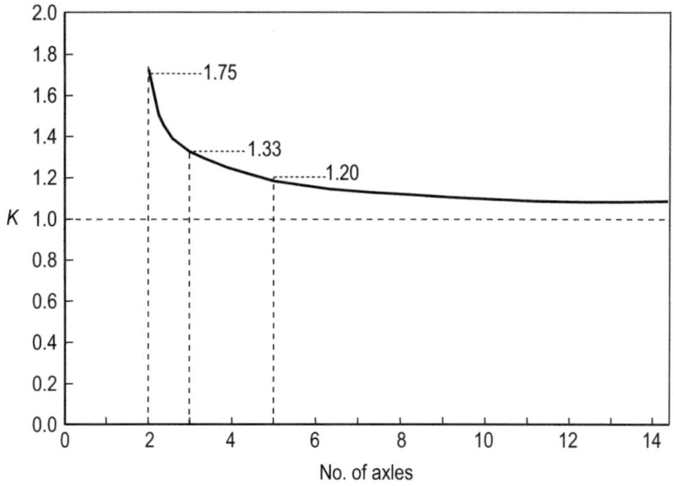

Fig. 1.10 *K* plotted against the no. of equal point loads or axles

To formulate a design vehicle, the following considerations have been found helpful:

(a) The design vehicle should have at least one axle representing the heaviest axle of the given vehicles.
(b) The design vehicle should have a group of closely-spaced axles representing closely-spaced axle groups of the given vehicles.
(c) The length of the design vehicle, i.e. its base length, should be as close to the longest vehicle as possible.
(d) The number of axles of the design vehicle should be neither too small nor too large.

Using the foregoing considerations as constraints, a six-axle design vehicle is selected in which the following components are pre-selected.

(a) Axle number 5 is chosen to represent the heaviest single axle, being 13.0 t.
(b) Axles numbers 2 and 3 are spaced at 1.0 m, with each carrying 12.0 t; these two axles can be shown to represent the heaviest of the closely-spaced axles of the given vehicles.

The weights of the other three axles and the remaining inter-axle spacing are now determined by an iterative process. It can be appreciated that this problem has an infinite number of solutions only a few of which will, however, appeal to engineering judgement. A proposed formulation of the resulting design vehicles is shown in Fig. 1.11. The W-B_m points for the various sub-configurations of the design vehicle are plotted in Fig. 1.9, in which it can be seen that these points closely cling to the upper-bound curve corresponding to the three given vehicles.

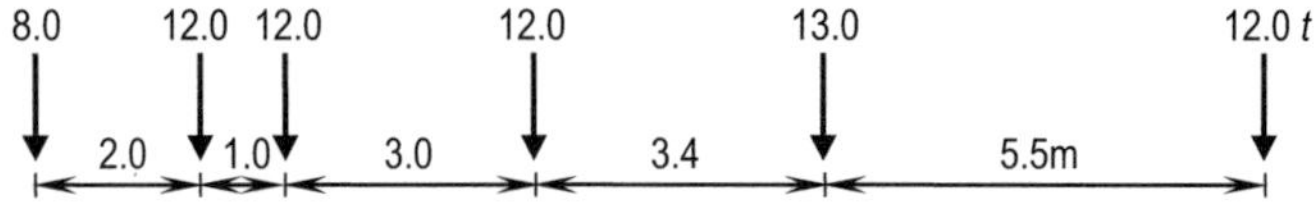

Fig. 1.11 Example of a design vehicle

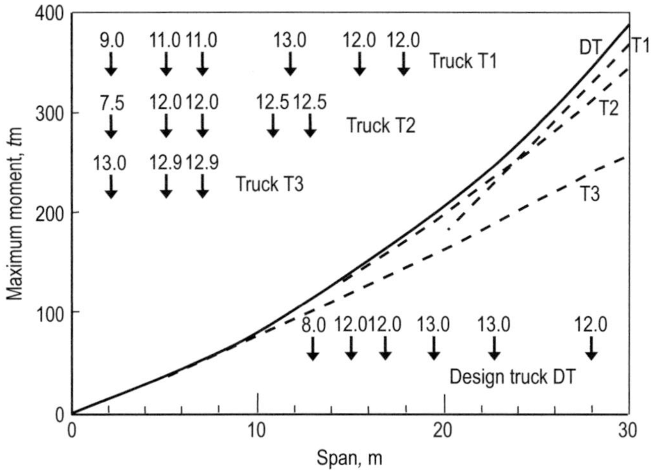

Fig. 1.12 Maximum bending moments due to actual and proposed design vehicles

A more-readily understandable measure of the effectiveness of the design vehicle is provided in Fig. 1.12, in which maximum moments in simply supported beams due to the three given vehicles and the proposed design vehicle are plotted against the span length of the beams. It can be seen in this figure that the envelope of maximum moments by the design vehicle represents well the envelope of maximum moments due to the given vehicles. The same will be found to be the case for maximum beam shears.

The technique of formulating the design vehicle described above is valid only when the transverse centre-to-centre distance between the longitudinal lines of wheels of the design vehicle is nearly the same as the corresponding distance in actual vehicles. This distance is about 1.8 m in most heavy commercial vehicles.

It will be appreciated that a design vehicle can represent the upper-bound effects of either the permissible (legal) vehicle weights or the maximum observed loads (MOL). As shown in Fig. 1.13, the design vehicle of the Ontario Highway Bridge Design Code (OHBDC, 1992) is based on MOL. The design loading of the CHBDC (2006), on the other hand, is based on permissible weights. Both these design vehicles are based on results of extensive vehicle weight surveys, which can be conducted by using bridge as weighing-in-motion devices, as explained in Chap. 11. The differences in the bases of the design loadings are accounted for explicitly by the load factors, which are discussed in Sect. 1.3.2.

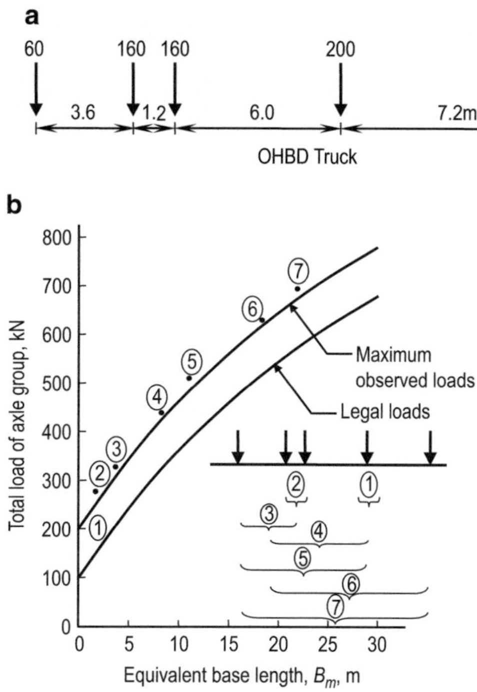

Fig. 1.13 The OHBDC design vehicle (**a**) on the W-B_m space (**b**)

Unlike the OHBDC design loading, the magnitude of the CHBDC design live loads is not fixed; the truck loading is designated as CL-W Truck, and the lane loading as CL-W Lane Load. 'W' is a variable, defining the total load of the truck, which a jurisdiction can adopt according to its own vehicle weight regulations and degree of its enforcement. It is noted that in Canada, regulations pertaining to highway transportation, falling under provincial jurisdiction, vary from province to province. The CHBDC recommends that bridges used for inter-provincial transportation be designed for at least CL-625 Loading, in which $W = 625$ kN; this loading is based on a set of regulations for inter-provincial transportation contained in the Memorandum of Understanding (MOU) on Vehicle Weights and Dimensions signed by representatives of all Canadian provinces, initially in 1988, and amended in 1991 (TAC 1991).

Details of the CL-625 Truck are shown in Fig. 1.14, and those of the CL-625 Lane Load in Fig. 1.15.

1.2.2.2 Computer Program

A Fortran program 'TRUCK' can be downloaded from http://extras.springer.com; this program uses information about a set of axle weights and the inter-axle spacing, and calculates the W and B_m values for all the sub-configurations of successive

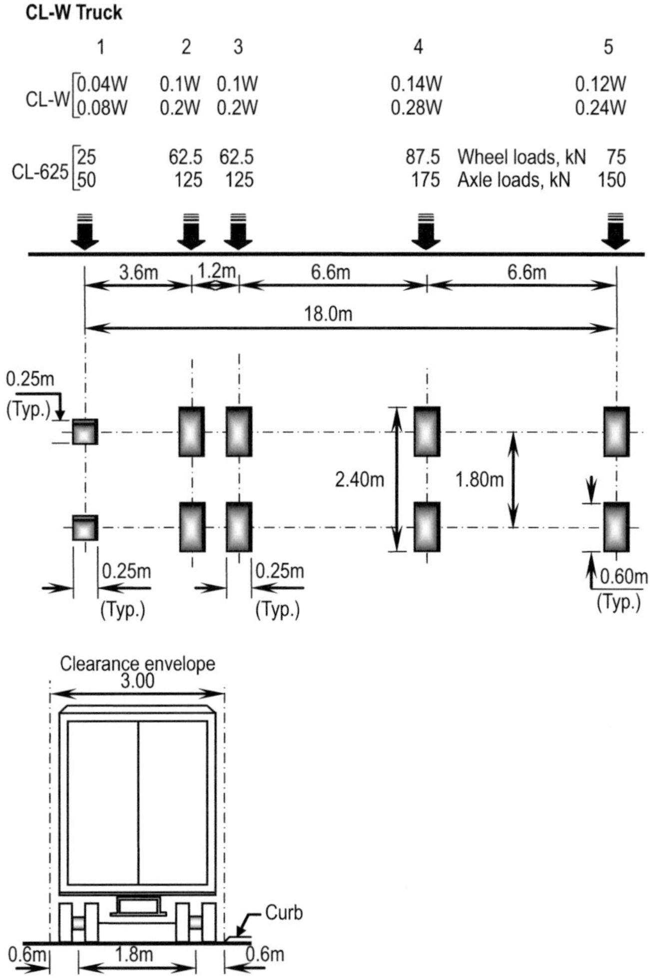

Fig. 1.14 The CL-W and CL-625 trucks of the CHBDC

axles. It is hoped that this program will be found useful not only in the formulation of a design truck but also in the processing of data from vehicle weight surveys.

The program DTRUCK uses Eq. (1.1) to calculate the values W and B_m for all the sub-configurations of a given set of point loads.

Data Input

The program requires the input as defined below, it being noted that each set constitutes one line of data with the entries being separated by commas or spaces.

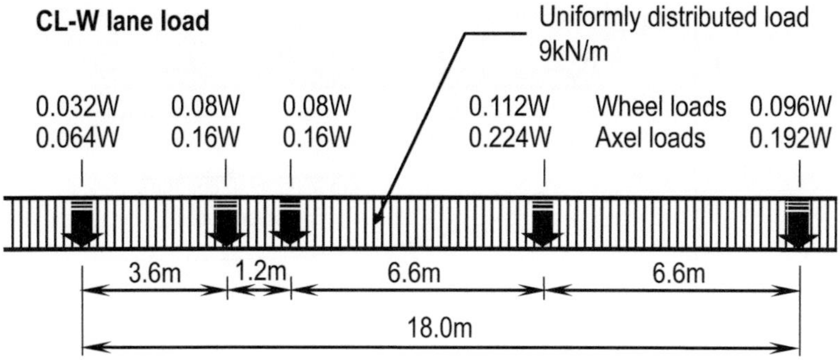

Fig. 1.15 CL-625 lane load of the CHBDC

Set 1	Title (maximum 52 characters)
Set 2	Number of trucks, *NV*
Set 3	Number of axles, *NA*
Set 4	Weights of axles, starting from left [*NA* entries]
Set 5	Inter-axle spacings, starting from left [(*NA* – 1) entries]

Data sets 3–5 are repeated *NV* times.

Running of Program

To run the program, the data are stored in a file named TRUCK.DAT. The user should simply click on the TRUCK icon to run the program, and the program will proceed to execute the stored data.

Reviewing Results

After it has completed the computations, the program stores the results in a file named TRUCK.RES. This file can reviewed by using any text editor for results, which are labelled adequately enough to be self-explanatory.

1.2.2.3 Multi-Presence in One Lane

After loads have been multiplied by the appropriate load factor, both the OHBDC and CHBDC design trucks represent the heaviest expected vehicle. It is not reasonable to expect with virtual certainty that a lane of a bridge will carry a train of these heaviest vehicles at a fixed distance from each other. It is intuitively obvious that a bridge lane, which can accommodate more than one design vehicle, should contain one design vehicle and other lighter vehicles, the weight and

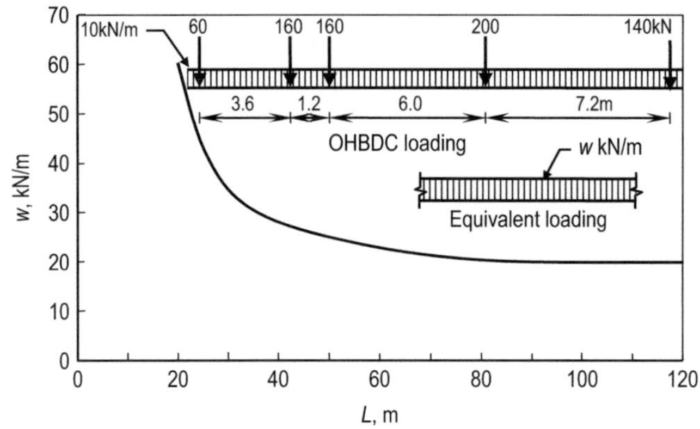

Fig. 1.16 Variable uniformly distributed design load

frequency of which should reduce with increase in the loaded length. The mix of vehicles, which should be considered in the longer loaded lengths for the development of the bridge design loading, is clearly a statistical problem which cannot be solved by deterministic means.

The most common, and scientifically-defensible, method of developing bridge design loading for loaded lengths greater than the length of actual trucks is that of computer-based simulations. Such simulations are typically conducted by using known vehicle weights and either observed or assumed distances between the vehicles. Two examples of single-lane design loading developed from independent computer simulations are those adopted for medium span bridges by the OHBDC (1992), and that proposed for long span bridges by Buckland and Sexsmith (1981), developed for a technical committee of the American Society of Civil Engineering and known informally as the ASCE loading. The OHBDC loading comprises 70 % of the OHBDC design vehicle superimposed centrally over a UDL of 10 kN/m length of the lane. The ASCE loading is a combination of a knife-edge load and a uniformly distributed load, the intensities of both of which vary with the span of the bridge. Whereas the intensity of the UDL decreases with increase in span length, the magnitude of the knife-edge load increases with span length. Despite the fact that the intensity of the uniformly distributed load of the OHBDC loading is constant, the net effect of the total design load is to reflect the overall reduction of loads with increase in span length. This observation can be confirmed quantitatively by replacing the vehicle load and the uniformly distributed load of the OHBDC design loading by a uniformly distributed load w, which will lead to the same maximum bending moment in simply-supported beams as the former loading. As can be seen in Fig. 1.16, w decreases with increase in L.

As described in the commentary (2006) to the CHBDC, the CL-W Lane Load is derived from the ASCE loading; it comprises 80 % of the Truck load superimposed on a UDL of 9 kN/m length of a lane.

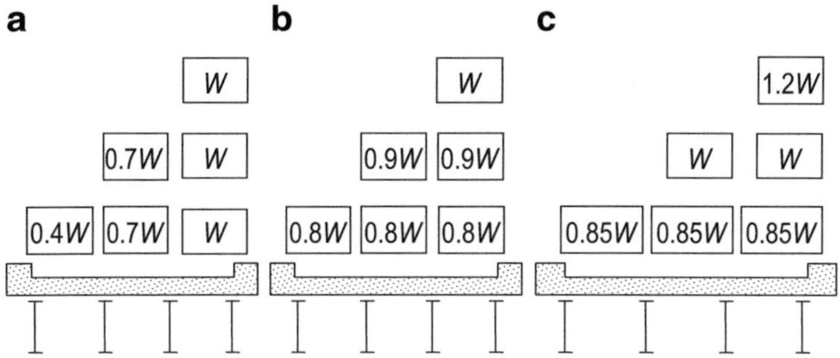

Fig. 1.17 Modification factors for multi-lane loading: (**a**) traditional factors for long span bridges; (**b**) factors specified in CHBDC (CHBDC 2006); (**c**) factors specified in AASHTO LRFD (2010)

1.2.2.4 Multi-Presence in Several Lanes

As noted earlier, the design live loads on highway bridges are usually specified on a per-lane basis with the loading in a single lane being related to the heaviest vehicles expected to cross the bridge during its lifetime.

For multi-lane loading, the loads per lane are reduced so as to account for the low probability of two or more lanes being simultaneous loaded by the heaviest vehicles. In long span bridges, it had been found convenient to achieve such reduction by loading one lane with the maximum loading in one lane and then gradually reducing the load in successive lanes (e.g. see Buckland and Sexsmith, 1981). This process is shown schematically in Fig. 1.17a, it being noted that W in this figure represents the load that would be specified if only one lane of the bridge were loaded. For multi-lane loading in short and medium span bridges, it is customary to use a scaled-down version of the single lane loading in each lane. Thus the reduced loading for each lane is obtained by multiplying each component of single lane loading by a reduction factor, which is usually less than 1.0 and, which is often referred to as the multiple-presence reduction factor.

These reduction factors specified for design purposes in the OHBDC (1992) for two- three- and four-lane bridges are 0.90, 0.80 and 0.70, respectively; factors for one, two and three lanes are shown in Fig. 1.17b. Unlike previous practice, the CHBDC has specified the same multi-presence reduction factors for long span bridges as well.

The reduction factors specified in the LRFD (Load and Resistance Factor Design) bridge design specifications of the U.S.A. (AASHTO, 2010) for one, two, three and four lanes are 1.20, 1.00, 0.85 and 0.65, respectively, the earlier three of which are shown in Fig. 1.17c. It is interesting to note that the two-lane loading has been used in the AASHTO Specification as the reference loading. Both the CHBDC and AASHTO multi-presence reduction factors for design of new bridges are independent of the volume of traffic on the bridge.

It is intuitively obvious that the probability of two or more lanes of a bridge being simultaneously loaded by the heaviest vehicles is higher if the bridge is

Table 1.1 Criteria for the various classes of highways

Highway class	Average no. of trucks per lane per day	Average no. of all vehicles per lane per day	Criteria when traffic data are not available
A	>1000	>4000	Highways primarily for through traffic
B	>250 and ≤1000	>1000 and ≤4000	Roads primarily for property access which carry moderate commercial traffic
C	≤250	≤1000	Roads which carry little or no commercial traffic

Table 1.2 Values of multiple presence reduction factor m_f

	Number of loaded lanes			
Highway class	2	3	4	5
A	0.90	0.80	0.75	0.70
B	0.90	0.80	0.70	[a]
C	0.85	0.70	[a]	[a]

[a]These cases need not be considered at all

located on a busy highway than if it is on a secluded lane. The concept of a reduction factor based on the volume of traffic can be used with advantage in the evaluation of the load carrying capacity of existing bridges. It can be appreciated that accounting for the volume of traffic through multiple-presence reduction factors may not be advisable in the design of new bridges for which the prediction of future traffic volumes is usually fraught with a high degree of uncertainty.

Jaeger and Bakht (1987) have shown that the multiple-presence reduction factor, for static vehicle loads m_f, depends not only on the volume of traffic but also on a factor $BT^{(N-1)}$, in which B is the life of bridge over which the projected volume of traffic is expected to be maintained, T, is the time required for a vehicle to cross the middle one-third of the span under consideration, and N, is the number of traffic lanes on the bridge.

From the consideration of the volume of traffic, highways can be divided into three categories, being A, B and C, with class A highway carrying the densest traffic and class C highway the lightest. The criteria for these three classes of highways are listed in Table 1.1.

Jaeger and Bakht (1987) have further shown that the multiple presence reduction factor, m_f, used in design or evaluation is a combination of m_{fs} and m_{fd} with the former being related to static vehicle weights and the latter factor to dynamic amplification of loads. The values of m_f that they have proposed for different classes of highways are listed in Table 1.2.

As discussed by Jaeger and Bakht (1987), the values listed in Table 1.2 are safe-side estimates of the reduction factors which are applicable to both the AASHTO and OHBDC.

There is no reason to believe that these factors are not applicable to other codes. The reduction factors listed in Table 1.2 have been adopted by both the OHBDC (1992) and CHBDC (2006) for the evaluation of the load carrying capacity of existing bridges.

1.2.3 Accounting for Dynamic Loads

Vehicles that are expected to cross a bridge during its lifetime are accounted for in the design or evaluation of the bridge through a static design loading and a certain prescribed fraction which is traditionally referred to as the impact factor. The static design loading is a tangible entity which, as described earlier, can be formulated from the static weights of actual and foreseen vehicles. The impact factor, on the other hand, is an abstract entity, which is supposed to account for the magnification of load effects in a bridge caused by the dynamic interaction of the bridge and moving vehicles.

Despite its abstract nature, the impact factor has been used in the design of bridges for several decades.

There have been numerous attempts to measure this elusive entity in bridges through dynamic testing of bridges. Bakht and Pinjarkar (1990), through an extensive survey of technical literature dealing with the dynamic testing of bridges, have shown that there is a general lack of consistency in the manner in which the test data are interpreted to obtain representative values of the impact factor. They have shown that the same bridge test data can give impact factors varying between 0.2 and 0.5 depending upon the various definitions used for the impact factor. Bakht et al. (1992) have shown that the very large values of the impact factors calculated from test data and sometimes reported in the technical literatures are the result of misinterpretation of the data. A case against the use of impact factors in bridge design and evaluation is made by Bakht et al. (2003).

Notwithstanding the above comments, since it is specified in the design codes the impact factor has to be included in calculations for design and evaluation. The OHBDC first introduced a new concept of the impact factor, which was renamed to dynamic load allowance (*DLA*) to reflect the fact that the dynamic amplification of load effects is not always due to the impactive action of a wheel. After a careful study of data from dynamic tests on a large number of bridges, the OHBDC (1979) decided to prescribe *DLA* as a function of the first flexural frequency, *f*, of the component under consideration. The *DLA* was required to be obtained from a chart, which is reproduced in Fig. 1.18.

Recognizing that the calculation of *f* is not always simple, the third edition of the OHBDC (1992) specified the values of *DLA* to depend upon the number of axles. The CHBDC (2006) values of the *DLA*, modified slightly from the OHBDC values, are given in Table 1.3. The values listed in this table are not significantly different from those obtained from Fig. 1.18, and could be adopted directly by other design

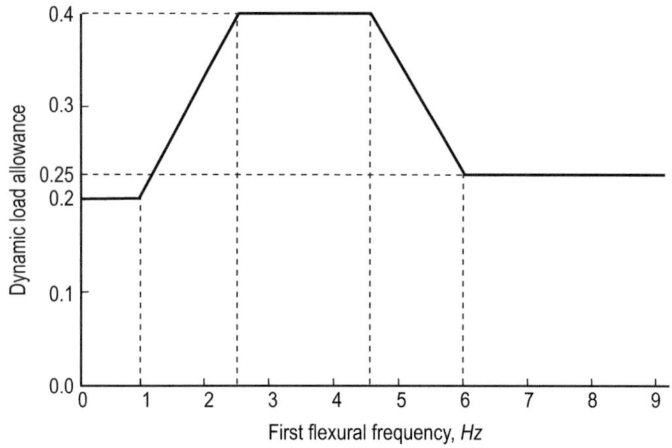

Fig. 1.18 *DLA* specified as a function of *f* in earlier editions of OHBDC

Table 1.3 Values of *DLA* prescribed by CHBDC (2006)

No. of axles	*DLA*	Notes
1	0.40	Except for deck joints, for which *DLA* = 0.50
2	0.30	Also for axle Nos. 1, 2 or 3 of CL-W Truck
3 or more	0.25	Except for axle Nos. 1, 2, 3 of CL-W Truck

codes. It may be noted that the values of *DLA* in Table 1.3 are subject to the same multi-presence reduction factors as the static design loads.

1.3 Design Philosophy

1.3.1 Probabilistic Mechanics

It is well known that the resistance of a structural component, regardless of its material composition is not deterministic because it can vary from sample to sample even if the components are of the same nominal size. As a tool for discussion, a large number of similar tension ties are considered. If all the ties were tested to failure, the histogram of the tensile strength of the ties *R* may be as shown in Fig. 1.19a. It is well known that the histogram of the kind shown in this figure can be represented in the limit by a continuous distribution curve shown in Fig. 1.19b. For the purpose of discussion, it is assumed that the distribution of the tensile strengths of the ties under consideration is normal, and that minimum and maximum observed tensile strengths are 750 and 1450 units of force. This distribution can be characterized adequately by two parameters, being mean μ_R and standard

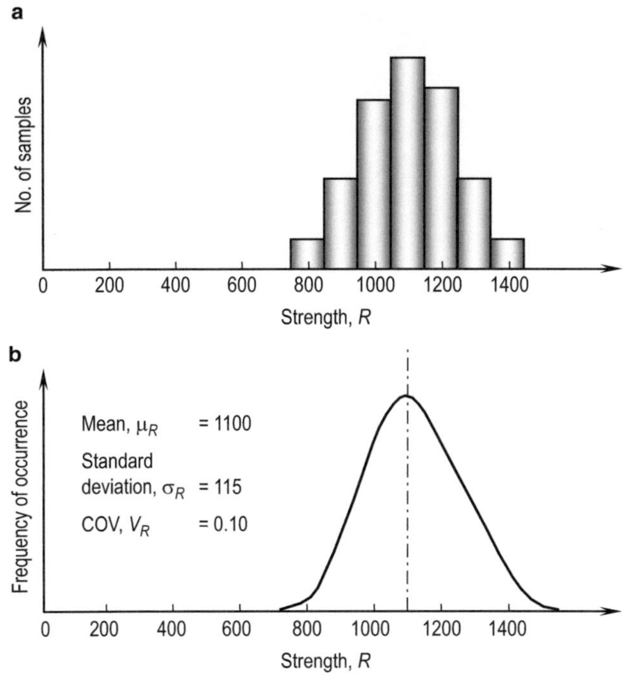

Fig. 1.19 Distribution of tensile strengths of nominally similar ties: (**a**) histogram of strengths; (**b**) probability distribution

deviation σ_R, with the latter relating to the variability. For the case shown in Fig. 1.19b, $\mu_R = 1100$ units and $\sigma_R = 115$ units.

The variability of the statistical distribution is defined by the coefficient of variation V_R which is the ratio of standard deviation and mean, so that:

$$V_R = \sigma_R / \mu_R \tag{1.5}$$

As noted in Fig. 1.19b, $V_R = 0.10$ for the data under consideration.

It is assumed that there are N ties with strengths R_1, R_2, R_N where N is a suitably large number, say 100. It is further assumed, that as shown in Fig. 1.20, these ties respectively support loads W_1, W_2, W_N inducing tensile forces S_1, S_2, S_N, respectively. The loads supported by the ties, hence their tensile forces, have the same nominal value but are subject to small variations. The smallest and largest values of the tensile forces in the 100 ties are assumed to be 700 and 1120 force units, respectively. The statistical distribution of the tensile forces, which is also assumed to be normal, is shown in Fig. 1.21 The mean, μ_s, and standard deviation, σ_s, of the tensile forces are 900 and 55 force units respectively, giving the coefficient of variation, $V_s = 0.06$.

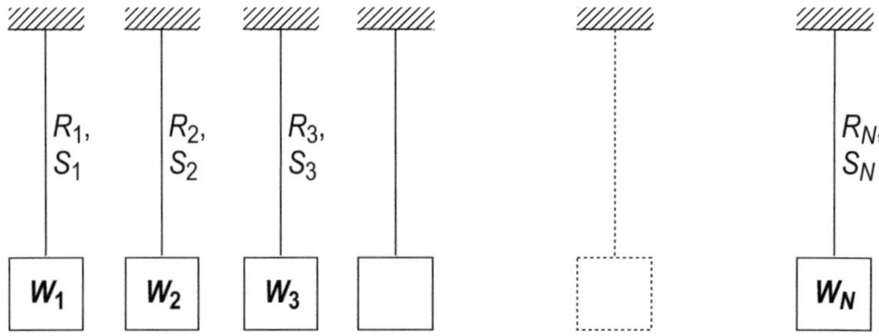

Fig. 1.20 A number of similar ties supporting similar loads

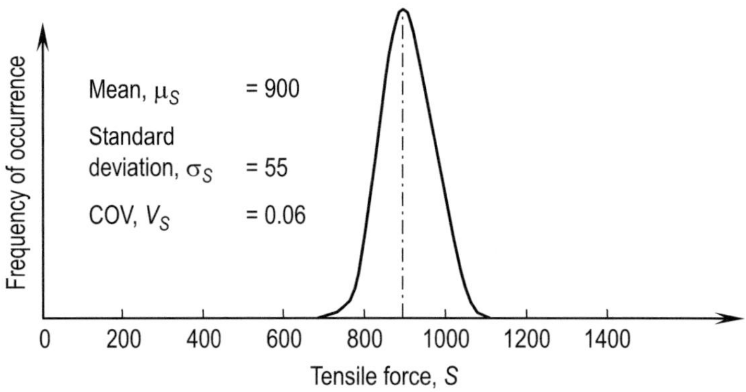

Fig. 1.21 Distribution of tensile forces in the ties

Conventionally, the safety margin in a component is defined by the safety factor, which is the ratio of the nominal resistance and load effect. For the ties under consideration, the central safety factor is equal to 1100/900, or 1.22. Such a small factor of safety is likely to be deemed unsatisfactory by most codes simply because it is too close to 1.0, and not because it represents an unacceptably high probability of failure.

If one examines carefully the concept of conventional factor of safety, one can see that this concept is based incorrectly upon absolute limits of uncertain quantities. For different structures, the same value of the factor of safety can represent different probabilities of failure. It can, therefore, be seen that the actual margin of safety in a structure cannot be represented realistically by the conventional factor of safety despite the fact that this entity had been extremely useful to engineers who are called upon to design a "safe" structure, notwithstanding the uncertainties involved.

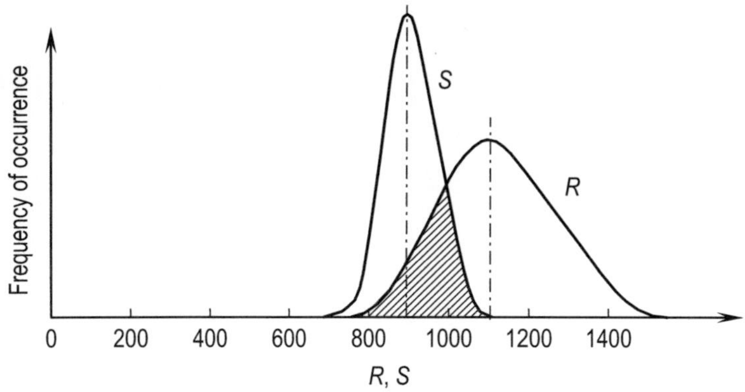

Fig. 1.22 Distributions of tensile strengths and forces plotted on the same chart

1.3.1.1 Safety Index

Figure 1.22 shows both strengths *R* and tensile forces *S* plotted on the same space. It can be seen that a considerable portion of the distribution of strength *R* overlaps with the distribution of the load effects, i.e. tensile forces *S*, indicating that there is a significant likelihood that the strength of the ties may be exceeded by the load effects, thus causing failure. The overlapping area of the two distributions, which is shown shaded in this figure does not, however, give a quantitative assessment of the likelihood of failure. The probability of the failure of the bar can be obtained by studying the distribution of the quantity ($R-S$) which is denoted as g, so that:

$$g = R - S \tag{1.6}$$

Since both *R* and *S* are assumed to be normally distributed, it follows that the quantity g is also normally distributed, so that according to elementary principles of statistics, its mean μ_g and standard deviation σ_g are given by the following equations:

$$\mu_g = \mu_R - \mu_S \tag{1.7}$$

and

$$\sigma_g = \left(\sigma_R^2 + \sigma_S^2\right)^{0.5} \tag{1.8}$$

From the specific values μ_R, μ_S, σ_R and σ_S, given earlier, it can be shown that $\mu_g = 200$ units and $\sigma_g = 127.5$ units. Using these values of μ_g and σ_g, the distribution of g can be plotted as shown in Fig. 1.23. It can be seen in this figure that a portion of the distribution diagram of g, shown shaded, lies in the negative range. This area, which represents the probability of *S* exceeding *R*, or the probability of failure, is bounded on the right by the line $g = 0$, which is β times the standard deviation

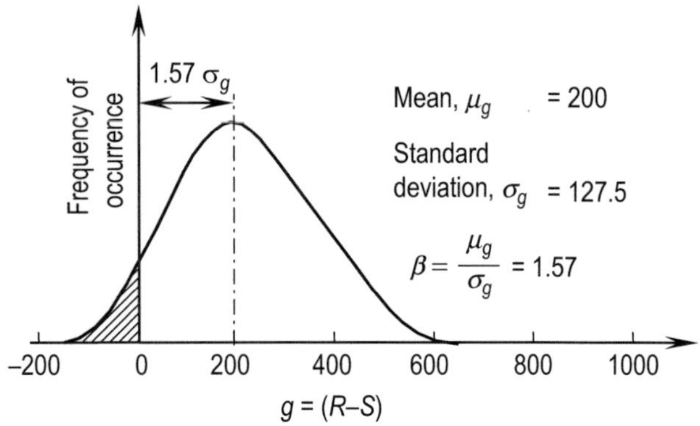

Fig. 1.23 Distribution of quantity g

below the mean. It can be readily shown that for the specific case under consideration, $\beta = 1.57$ and that the shaded area of the distribution diagram is about 6 % of the total area under the distribution curve. The shaded area represents a probability of failure of 0.06, implying that if a hundred similar ties were constructed corresponding to the data of Figs. 1.19b and 1.21, it is virtually certain that six of those ties will have failed.

The quantity which is denoted as β (beta) above is a first order measure of the reliability of the component; it is known as the *safety index*. This index is also called the second moment reliability index because it is based on the assumption that uncertainties relating to the reliability of a component can be expressed solely by the mean and standard deviation of resistances and load effects. The safety index β is defined by the following equation:

$$\beta = \frac{\mu_g}{\sigma_g} \tag{1.9}$$

If it is assumed that the tie fails when R is exceeded by S, then it can be shown that for normal distribution of R and S, which are assumed to be unrelated, the expression for β is as follows:

$$\beta = \frac{\mu_R - \mu_S}{\left(\sigma_R^2 + \sigma_S^2\right)^{0.5}} \tag{1.10}$$

This operation indeed follows directly from the principles that led to the calculation of $\beta = 1.57$ for the specific case of Fig. 1.23. Equation (1.10) can also be written as the following equation.

$$\beta = \frac{\theta - 1}{\left(\theta^2 V_R^2 + V_S^2\right)^{0.5}} \tag{1.11}$$

where $\theta = \mu_R/\mu_S$, and V_R and V_S are the coefficients of variation of R and S, respectively. The factor θ is usually referred to as the central safety factor.

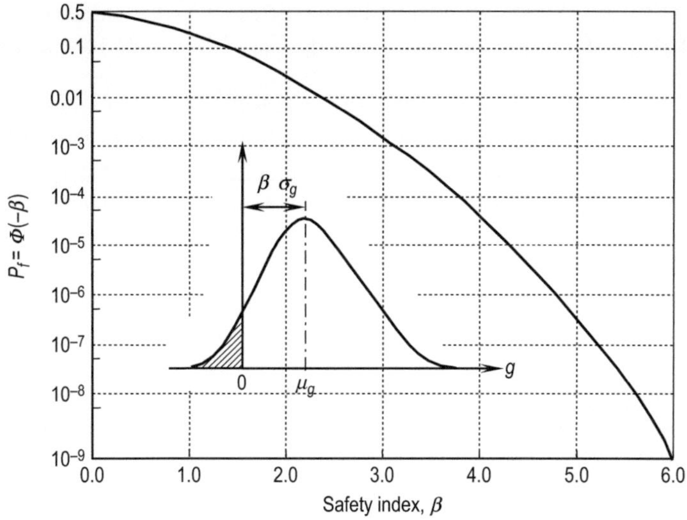

Fig. 1.24 Properties of normal distribution

The function which relates β to the probability of failure P_f is denoted as Φ (- β), thus:

$$P_f = \Phi(-\beta) \tag{1.12}$$

Values of P_f for different values of β can be obtained from the properties of normal distribution, which are plotted in Fig. 1.24.

In the development of Eq. (1.10), it was assumed that failure takes place when R is exceeded by S. This criterion is not unique when normal distributions are assumed for R and S. This is because for normal distributions, both R and S can theoretically have negative values. An alternative to the previous approach is to assume that failure takes place when log R is exceeded by log S. In this case, the expression for β can be written as follows (e.g. see Madsen et al., 1986).

$$\beta = \frac{\ell n\mu_R - \ell n\mu_S}{\left(\sigma^2_{\ell nR} + \sigma^2_{\ell nS}\right)}^{0.5} \tag{1.13}$$

Where $\sigma_{\ell nR}$ and $\sigma_{\ell nS}$ are the standard directions of the logs of R and S respectively.

When the coefficients of variation are small, β can be shown to be given approximately by the following equation:

$$\beta = \frac{\ell n\theta}{\left(V_R^2 + V_S^2\right)^{0.5}} \tag{1.14}$$

Equations (1.9) and (1.13) will be found to be different for the same sets of R and S, showing that the definition of β as given by Eq. (1.9) is not independent of the distribution.

The first generation of probabilistic-based design codes employ the kind of equations given above and are usually calibrated to $\beta = 3.5$. It should be emphasized that the value of β obtained from Eq. (1.11) or Eq. (1.14) responds to the theoretical probability of failure of single components. The probability of failure of the full structures, which are assemblies of individual components, is mercifully much smaller. The safety index assumed in the first generation design codes should only regarded as a relative measure of safety the use of which ensures that all the components of a structure have similar margins of safety against failure.

Probabilistic mechanics of systems of components is too complex to be dealt with in an introductory chapter. For further reading on the subject, reference should be made to standard text books on the subject, e.g. Madsen et al. (1986), Nowak and Collins (2000).

It may be noted that for the system of ties shown in Fig. 1.20, the probability of failure will not be zero even if the smallest observed failure load of the 100 ties was larger than the heaviest expected load. This is because the smallest strengths and heaviest loads cannot be predicted with certainty. It is quite possible that one tie in a million will come with a defect that reduces its strength significantly and hence causes failure.

1.3.1.2 Maximum Load Effects

The distribution of tensile forces in the ties, shown in Fig. 1.21, was based on the assumption that the loads are permanently attached to the ties. In this case, the statistics of individual loads will be the same as the statistics of the load effects in the ties. When the ties are temporary and are used several times for different loads, as for examples hangers in a warehouse, then for obtaining the safety index, one would have to determine the statistics of the maximum loads that each tie will sustain during its lifetime. The distribution of maximum tensile forces in ties can be determined either by numerical simulations or by physical measurements.

1.3.1.3 Analogy Between Ties and Bridges

In order to draw the analogy between the ties of Fig. 1.20, a 3-lane bridge with four simply supported spans, and a total of 28 reinforced concrete girders, is considered (see Fig. 1.25). Moment capacity of the girders at their mid-spans are examined. It is well known that if all the girders were tested to failure, it is likely to find that each girder has a somewhat different moment of resistance than those of the others. A few girders are likely to have very small moment capacity, while a few others would have a very high moment capacity. The majority of girders would perhaps have their moments of resistance within a narrow range. The histogram of the

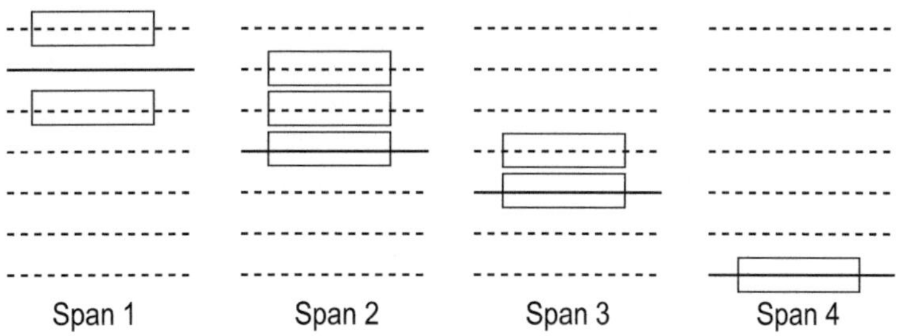

Fig. 1.25 Comparison of factored moments

moments of resistance of the girders can be represented by a continuous distribution curve, similar to Fig. 1.19b. Although the failure tests on the 28 girders may not show zero, or very small, strengths, the R curve with normal distribution implies the presence of even zero strengths.

During the lifetime of the bridge, each girder is likely to be subjected to a different maximum live load moment. For example, as illustrated in Fig. 1.25, a girder in Span 1 shown in solid line, might receive its maximum live load moment from two exceptionally heavy trucks, present on the bridge at the same time. Similarly, a girder in Span 2, also shown in solid line in Fig. 1.25, might experience its maximum moment under three different heavy trucks, simultaneously present on the span and so on. A distribution curve representing these maximum moments that the girders are expected to experience during their lifetime can be similar to the curve in Fig. 1.21 representing the tensile forces in the ties of Fig. 1.20.

To the authors' knowledge, the statistics of maximum load effects in bridge components have always been obtained by numerical simulations. There is clearly a need to verify the outcome of these simulations by physical measurements, i.e. by structural health monitoring, discussed in Chap. 10.

1.3.2 Limit States Design

As noted, for example, by Kennedy (1974), a structure in serving its intended purpose must satisfy two basic requirements: (a) it must remain functional with an acceptably high degree of certainty, and (b) the probability of its collapse must be sufficiently low. The states of 'unserviceability' and collapse are called the limit states. The limit states design is that method of design in which the performance of the structure is checked with respect to the various limit states at appropriate levels of load.

There are three limit states currently employed in bridge design codes: (a) the ultimate limit state (ULS) relating to the collapse of the structure, or more appropriately, a component; (b) the serviceability limit state (SLS); and (c) the fatigue limit state (FLS), which used to be grouped with SLS in earlier design codes. The SLS includes such responses of the structure as deflection, cracking and vibrations.

In limit states design, the satisfactory performance of a structure, or component, is defined by the requirement given below in the form of an inequality, it being noted that this particular condition, which is given only as an example, relates to one of ULS loading combinations.

$$\phi R_n \geq \sum_{i=l}^{N} \alpha_{Di} D_{ni} + \alpha_L (l + DLA) L_n \tag{1.15}$$

where

ϕ = resistance factor, which is ≤ 1.0 and which is supposed to account for the variability of resistance, as defined for example by V_R in the example of Fig. 1.19b.

R_n = the nominal failure strength corresponding to a given load effect, which may be either the mean or any identified level of other strength, the most commonly adopted one being the 5th percentile strength.

N = the number of dead loads, e.g. factory-made components, cast-in-place concrete components, and wearing course, etc.

α_{Di} = the dead load factor for dead load number i, which is ≥ 1.0 and accounts for the variability of both the load itself and the prediction of its load effect.

D_{ni} = the load effect due to nominal dead load number i, which is typically the mean load but can be assumed to be any other.

α_L = the load factor for live loads, which is ≥ 1.0 and accounts for the variability of both live loads and the prediction of their load effects.

DLA = the dynamic load allowance.

L_n = the load effect due to the nominal design live loading.

The inequality in Eq. (1.15) is sometimes expressed as "factored resistance must exceed factored loads." It can be appreciated that the various factors are so interlinked with each other and the nominal loads and resistances that their values cannot be determined in isolation. Determination of the values of the various factors is done through an iterative process known as code calibration. The calibration report for the CHBDC (2000) is appended to the Commentary of the code (Commentary, 2001).

1.3.3 Safety Factor

1.3.3.1 Comparison of Different Codes

For the reasons discussed immediately above, a quantitative comparison of two different codes is feasible only if they employ the same format of design condition.

Table 1.4 Comparison of factors which are used to reduce nominal strength

Material	Reciprocal of partial factors for material strength BD21/93	Resistance factors, OHBDC, (1992)
Steel	0.77–0.95	0.90–0.95
Concrete	0.67	0.75
Reinforcing steel	0.87	0.85

However, even for codes similar in format, it is not appropriate to compare their design, or assessment, loads or factors without taking account of all the other quantities on both sides of the design inequality.

Bakht in an as yet unreported paper had attempted to compare the bridge assessment loads of the UK (BD21/93) with the corresponding evaluation loads of the OHBDC (1992), it being noted that both codes are based on limit states format. Relevant segments of this paper are summarized in the following as a case study which may be useful in explaining the point raised above.

1.3.3.2 Vehicle Weights

The first step in a quantitative comparison of the design or evaluation loads of two jurisdictions appears to be a comparison of their respective vehicle weight regulations. By comparing these regulations for the UK and the Canadian province Ontario, it was concluded that the total vehicle weights permitted in Ontario were always higher than those permitted in the UK, with the increase being about 60 % for vehicles with seven or more axles. For single axles and two-axle tandems, the UK-permitted weights are greater than the Ontario loads by up to 6 %.

There is no reason to believe that the incidence and magnitude of overloading in the UK is any higher than it is in Ontario. Accordingly, it is obvious that the factored assessment loads of the UK document should be somewhat lower than the factored evaluation loads of the OHBDC, provided of course that the various quantities in the design condition are similar and compatible; their correspondence is discussed in the following.

1.3.3.3 Resistance Factors

The UK document reduces the nominal strength of components by dividing them by *partial factors for material strength*, which are greater than 1.0 and which are the reciprocals of the resistance factors defined earlier. The reciprocals of the partial factors for material strength are compared with the resistance factors of OHBDC (1992) in Table 1.4 for different materials.

It can be seen from Table 1.4 that the factors specified in the two documents reduce the nominal strength by factors which are not significantly dissimilar. It is

Table 1.5 Net load factors for dead loads for ULS

Loads	BD 21/93	OHBDC (1992)
Factory produced components	1.15	1.10
Cast-in-place concrete; timber	1.26	1.20
Wearing course	1.92	1.50
Earth fill	1.32	1.25

assumed that the nominal strengths specified in the two documents are based on 5th percentile strengths and are, therefore, comparable.

1.3.3.4 Dead Load Factors

BD 21/93 specifies a factor $\gamma_{f3} = 1.1$, by which the load effects due to factored loads are to be multiplied to obtain the right hand side of Condition (1.15). In effect, the net load factors are the product of 1.1 and the partial factors for loads. These net load factors are compared in Table 1.5 for various dead loads in bridges with those specified in the OHBDC.

It can be noted from Table 1.5 that the net dead load factors of the UK document are higher, but only slightly, than those of the OHBDC.

1.3.3.5 Comparison of Live Loads

Similarities between the philosophy of assessment, along with the near equality of the factors for material strength and dead loads, permit a direct comparison of the factored live loads of the UK and Ontario documents for bridge evaluation. Since the live loads specified in the two jurisdictions are of different configurations, the maximum bending moments induced respectively by the two loadings in simply supported beams are used as a basis for the comparison. The case of loading in only a single lane is considered first.

For BD21/93, the nominal HA loading, which is its assessment load, is multiplied by the following factors:

(a) partial factor for live loads = 1.5
(b) analysis factor $\gamma_{f3} = 1.1$
(c) lane factor $\beta = 1.0$ (for lane width ≥ 3.65 m) width
(d) reduction factor = 0.91 (for full range of vehicles)

The above factors combined together are equivalent to a multiplier of 1.50 to the nominal HA loading.

The OHBDC (1992) specifies a *DLA* of 0.25 and a live load factor of 1.40 for the primary members of multi-load path structures; for secondary members, the live load factor is reduced to 1.30. For the comparison at hand, the factor for primary members is adopted, giving a net multiplier of 1.75. For short spans, say less than

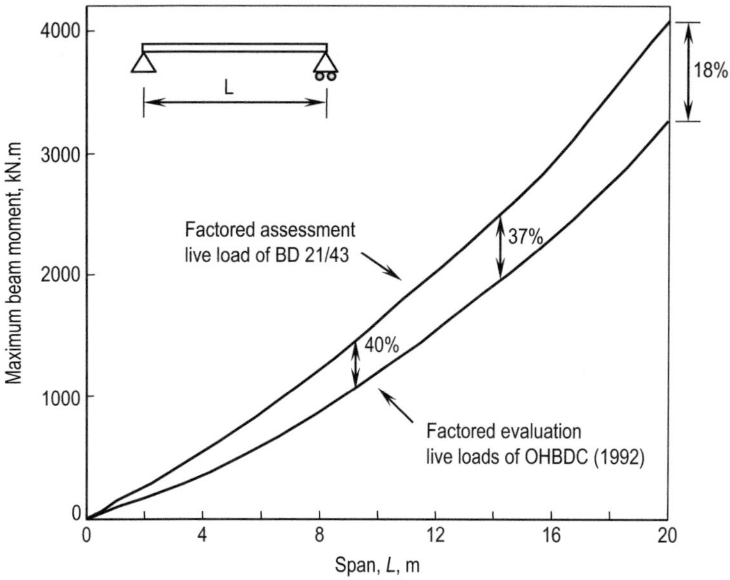

Fig. 1.26 Comparison of factored moments

20 m, the maximum longitudinal load effects are governed by the evaluation truck with a total wheel spread of 18.0 m and a gross weight of 740 kN. For larger spans, the truck loading is effectively supplemented by the UDL which accounts for the presence of more than one vehicle in the loaded length. In order to avoid the complication of factors such as the clear distance between vehicles, the comparison between the two assessment loadings is restricted to spans below 20.0 m.

Maximum moments in simply supported beams were calculated by the BD21/93 Bakht et al. (2003) and OHBDC (1992) assessment loads, each factored by the relevant multiplier identified above. The resulting maximum moments are compared in Fig. 1.26 for spans of up to 20 m. It can be seen in this figure that the moments due to the UK loading are always higher than those due to the Ontario loading. The difference between the two is very high at the lower span ranges and tapers off to about 18 % at the span of 20.0 m. This outcome is very surprising since the gross vehicle loads permitted in the UK are much lighter than those permitted in Ontario, as noted earlier. It is obvious that the assessment loads specified in BD21/93 are overly conservative in comparison to those of the OHBDC.

The degree of conservatism prescribed by the UK document is further increased when the bridge has more than one lane. For example, in the case of loading in two lanes, the live loads are not reduced at all by BD21/93. The OHBDC (1992), on the other hand, reduces the load in each of the two lanes by 10 % if the bridge is on a busy highway and by 15 % if it is on a lightly-travelled road.

To get a feel for the degree of conservatism implicit in BD21/93, it is helpful to consider two identical 20 m span two-lane bridges, one on a low volume road in Ontario and the other in the UK also on a lightly-travelled road. The former bridge is subjected to 59 % heavier loads than the latter. The total factored assessment live load for the Ontario bridge is, however, 28 % smaller than that for the bridge in the UK. It is obvious that the safety margin implicit in BD21/93 is excessively large in comparison with the safety margin in the OHBDC.

1.3.3.6 Adopting Codes of Other Countries

On the basis of only cursory consideration, it might at first appear appropriate for a jurisdiction to directly adopt the design code of another jurisdiction, provided that their respective vehicle populations are similar with respect to weights. However, it can be demonstrated easily that similarities in heavy vehicle weights and military hardware are not, on their own, sufficient reason to adopt codes of other countries. Although a bridge is designed to carry mainly traffic loads, the periodical damage that it experiences is mainly caused by environmental effects. A bridge should, therefore, be designed with its environment clearly in mind, so that both the long term and short term maintenance needs are minimized with respect to the total overall cost of the bridge. A code written for bridges in one set of climatic conditions may not be suitable for bridges in another environment.

It can be appreciated that the indiscriminate adoption of the design code of one jurisdiction by another is highly undesirable. Even more undesirable is the permission to adopt combinations of several codes without paying due regard to the consequences of the mixing of the various provisions. The authors are aware of several countries, which do not have design codes of their own and where the authorities having jurisdiction over bridges have permitted combinations of any of the "recognized" design codes. Such permission can lead to absurd results, as may be seen by considering a combination in which design live loads and live load factors are adopted from two different codes, say AASHTO and OHBDC.

The current AASHTO loading was developed several decades ago; because of the lack of direct correspondence between the actual vehicle weights and design live loading, AASHTO specifies a live load factor which is about 2.1. The OHBDC design live loading, on the other hand, has a direct correspondence with the maximum observed vehicle loads and has a much smaller live load factor being 1.4. The factored design live loads of both AASHTO and OHBDC are very similar to each other. It is assumed that the factored moment in a particular beam due to both AASHTO and OHBDC loads is $2.1\,M$. Thus the unfactored moments due to AASHTO and OHBDC loads (using live load factors of 2.1 and 1.4) are M and $1.5\,M$, respectively.

If the ASSHTO load factors and the OHBDC design loads are used, the maximum beam moment due to factored live loads will be $3.15\,M$. A combination of the OHBDC load factors and AASHTO design loading will give $1.4\,M$ for the same moment. It can be seen that one combination gives 50 % higher moments than the actual and the other 33 % lower. It is obvious that an indiscriminate combination of two design codes can lead to patently absurd results.

References

AASHTO (2010) American Association of State Highway and Transportation Officials. LRFD Design Specifications for Highway Bridges, Washington, DC

Agarwal A (1978) Vehicle weight regulations across Canada, a technical review with respect to the capacity of highway systems. Paper presented to RTAC Vehicle Weight and Dimensions Committee, Ministry of Transportation and Communications, Downsview

Bakht B, Pinjarkar SG (1990) Dynamic testing of highway bridges – a review, Transportation Research Record 1223. Transportation Research Board, Washington, DC

Bakht B, Billing JR, Agrawal AC (1992) Discussion on paper entitled "Wheel loads from highway bridge strains: field studies." ASCE J Struct Eng 118(6):93–100

Bakht B, Mufti AA, Clayton A, Saltzberg W, Klowak C (2003) Interpretation of bridge test data to determine dynamic load allowance and its influence on bridge design and evaluation. In: Proceedings of the international workshop on structural health monitoring of Bridges/Colloquium on bridge vibrations, Kitami Institute of Technology, Japan, pp 101–107

BD21/93 (1993) The assessment of highway bridges and structures Department of Transport, London

Buckland PG, Sexsmith RG (1981) A comparison of design loads for highway bridges. Can J Civ Eng 8(1):16–21

CHBDC (2006) Canadian highway bridge design code. CSA International, Toronto

CHBDC Commentary (2001) Canadian Highway Bridge Design Code. CSA International, Toronto

CSA (1988) Design of Highway Bridges, CAN/CSA-S6-88 Canadian Standards Association. Rexdale, Rexdale

Csagoly PF, Dorton RA (1973) Proposed Ontario bridge design load, research report 186. Ministry of Transportation and Communications, Downsview

Csagoly PF, Dorton RA (1978) Truck weights and bridge design loads in Canada, Structural research report, SRR-79-2. Ministry of Transportation and Communications, Downsview

Jaeger LG, Bakht B (1987) Multiple presence reduction factors of bridges. In: ASCE structures congress proceedings entitled "Bridges and Transmission Line Structures." Orlando, pp 47–59

Jung FW, Witecki AA (1971) Determining the maximum permissible weights of vehicles on bridges, Research report 175. Ministry of Transportation and Communications, Downsview, Ontario, Canada

Kennedy DJL (1974) Limit states design – an innovation in design standards for steel structures. Can J Civ Eng 1(1):1–13

Madsen HD, Krenk S, Lind NC (1986) Methods of structural safety. Prentice Hall, Eaglewood Cliffs

Nowak AS, Collins KR (2000) Reliability of structures. McGraw-Hill, New York

OHBDC (1979, 1992) Ontario highway bridge design code. Ministry of Transportation of Ontario, Downsview

TAC (1991) Memorandum of understanding respecting a federal-provincial-territorial agreement on vehicle weights and dimensions. Council of Ministers Responsible for Transportation and Highway Safety, Transportation Association of Canada, Ottawa

Chapter 2
Analysis by Manual Calculations

2.1 Introduction

In structural engineering, the term *analysis* usually refers to *force analysis* in which the distribution of force effects is determined in the various components of a structure. The responses of a structure such as deflections and bending moments are often referred to as *load effects*. Another infrequently used term in structural engineering is *strength analysis* which refers to the process of determining the strength of the whole structure or its components. The term *analysis* is used in this book only in the meaning of force analysis.

In bridge engineering the term *analysis* is also used for determining the effects of *load distribution* mainly in the longitudinal components of a bridge due to a vehicle. This chapter provides details of some methods which can be used to analyse a bridge for load distribution through manual calculations.

Notwithstanding the fact all calculations for bridge analysis are currently done with the help of computers, manual methods are important in permitting engineers to retain the physical feel of the distribution of load effects in a bridge, and confirming the results of computer analyses in a broad sense. The manual methods of calculations are also known as the 'simplified' methods. Bakht and Jaeger (1985) have written a book dealing with manual analysis of bridges, entitled as 'Bridge Analysis Simplified'.

It is very important to note that the specific manual methods of analysis presented in this chapter explicitly include: (a) the number of lines of wheels in the design vehicle and their transverse spacing, and (b) the reduction factors for multi-lane loadings specified by the design code. The manual methods of analysis presented in this chapter should be used only if the conditions of load placement are the same as those included in the methods.

B. Bakht, A. Mufti, *Bridges*, DOI 10.1007/978-3-319-17843-1_2

2.2 Distribution Coefficient Methods

The basis of many simplified/manual methods is the distribution coefficient methods prevalent in pre-computer days; the background to these methods is discussed in following. The distribution coefficient methods, e.g. Morice and Little (1956), were well known to bridge designers in the U.K. and in those Asian countries where bridge design practice is, or was, influenced significantly by the British practice. These methods can be applied manually to obtain the values of various load effects at any reference point on a transverse section of the bridge. For most of the load distribution coefficient methods, a right simply supported bridge is idealized as an orthotropic plate whose load distribution characteristics are governed by two dimensionless parameters α and θ, defined as follows:

$$\alpha = \frac{D_{xy} + D_{yx} + D_1 + D_2}{2(D_x D_y)^{0.5}} \tag{2.1}$$

$$\theta = \frac{b}{L}\left(\frac{D_x}{D_y}\right)^{0.25} \tag{2.2}$$

where the notation is as defined in the following:

x direction = the longitudinal direction, i.e., the direction of traffic flow
y direction = the transverse direction (perpendicular to the longitudinal direction)
D_x = the longitudinal flexural rigidity per unit width (corresponding to flexural rigidity EI in a longitudinal beam)
D_y = the transverse flexural rigidity per unit length (corresponding to flexural rigidity EI in a transverse beam)
D_{xy} = the longitudinal torsional rigidity per unit width (corresponding to torsional rigidity GJ in a longitudinal beam)
D_{yx} = the transverse torsional rigidity per unit length (corresponding to torsional rigidity GJ in a transverse beam)
D_1 = the longitudinal coupling rigidity per unit width (which is the contribution of transverse flexural rigidity to longitudinal torsional rigidity through Poisson's ratio)
D_2 = the transverse coupling rigidity per unit length (which is the contribution of longitudinal flexural rigidity to transverse torsional rigidity through Poisson's ratio)
b = half width of the idealised plate
L = the span of the idealised plate

In slab-on-girder bridges, D_1 and D_2 are small and have little effect on load distribution. It is customary to ignore these rigidities in the calculation of α for slab-on-girder bridges. Figure 2.1 illustrates some of the notation.

The Morice and Little method, which is originally due to Guyon and Massonnet reported by Bares and Massonnet (1966), is based upon the harmonic analysis of

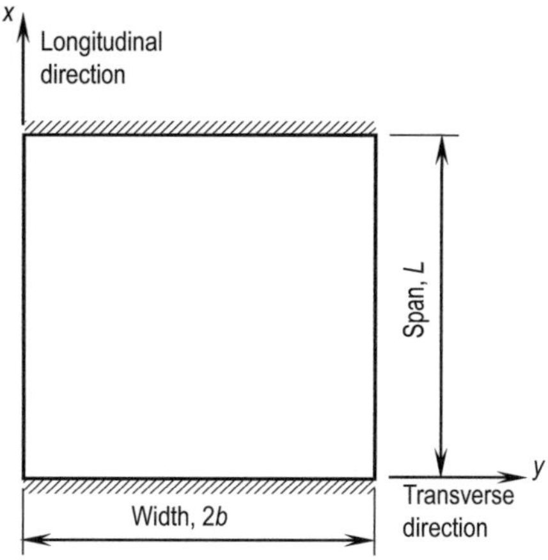

Fig. 2.1 Plan of a right bridge idealised as an orthotropic plate

orthotropic plates using only the first term of the harmonic series representing concentrated loads (discussed in Chap. 3). The basis of the method is the assumption that the deflected shape of a transverse section remains constant along the span irrespective of the longitudinal position of the load and the transverse section under consideration. The method uses charts of distribution coefficients K_α corresponding to nine different transverse reference stations and nine transverse positions of single concentrated loads. These coefficients are plotted in chart form against θ, and the charts are given for two values of α, namely 0.0 and 1.0.

Values of coefficients, K_α, for intermediate values of α are obtained by the following interpolation function:

$$K_\alpha = K_0 - (K_0 - K_1)(\alpha)^{0.5} \tag{2.3}$$

where K_0 and K_1 are the corresponding coefficients for α equal to 0.0 and 1.0, respectively. In using the method, the applied loads are converted into equivalent concentrated loads at the standard locations for which the charts are given. The distribution coefficients are then manually added for all the equivalent loads to give the final set of coefficients for the loading case under consideration. The exercise is, of course, repeated for each load case, and therefore requires extensive and tedious calculations.

To compensate for possible errors resulting from the representation of loads by only one harmonic, Morice and Little (1956) suggest that the computed longitudinal moments be increased by an arbitrary 10 %. Cusens and Pama (1975) improved the distribution coefficient method by taking seven terms of the harmonic series into account, and by extending the range of values of α up to 2.0. This method also uses an interpolation equation similar to Eq. (2.3).

It is interesting to examine the α-θ space with respect to various types of bridges. For practical bridges, α ranges between 0.0 and 2.0, and θ between 0.25 and 2.5. The ranges of α values for various types of bridges are shown in the α-θ space in

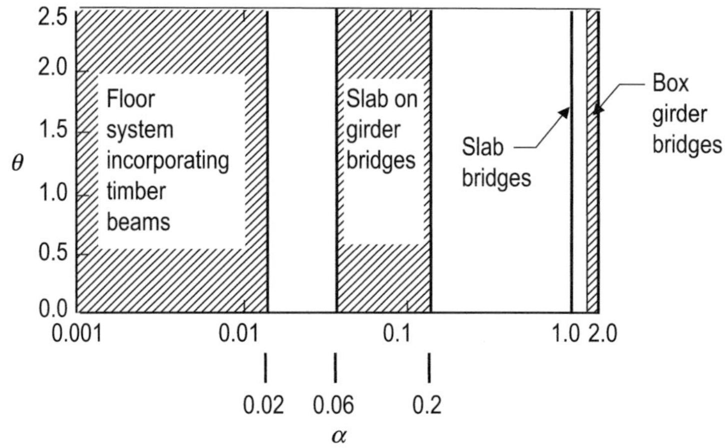

Fig. 2.2 The α-θ space for bridge superstructures idealized as orthotropic plates

Fig. 2.2. It can be seen that the various bridge types occupy distinctly separate zones. The space for bridges with timber beams occupies the space between $\alpha =$ nearly zero and 0.01; the space for slab-on-girder bridges is bracketed by values of α between 0.06 and 0.2; slab bridges have $\alpha = 1.0$; and box girder bridges have α close to 2.0.

In the Morice and Little method, since the values of α for slab-on-girder bridges are between 0.0 and 1.0, the distribution coefficients for these bridges have to be obtained by using Eq. (2.3). This equation is only an approximate design convenience and, irrespective of the accuracy of K_0 and K_1, can and does introduce significant errors, especially for bridges having smaller values of θ.

2.3 Simplified Methods of North America

Unlike the distribution coefficient methods, the simplified methods of bridge analysis used almost exclusively in North America provide only the maximum, i.e. the design, values of the various load effects at a given transverse section. Computation needed for these methods is only a fraction of that required for the distribution coefficient methods.

The North American simplified methods are permitted by the current and past design codes, being the AASHTO Specifications (1998, 2010), the CSA Code (1988), the Ontario Highway Bridge Design Code (1992) and the Canadian Highway Bridge Design Code (2000, 2006). These methods can be applied manually and can provide fairly reliable estimates of the design values of the various load effects in a very short period of time. The simplified methods of analysis are dependent upon the specification of the magnitude and placement of the design live loads, and accordingly are not always transportable between the various design codes.

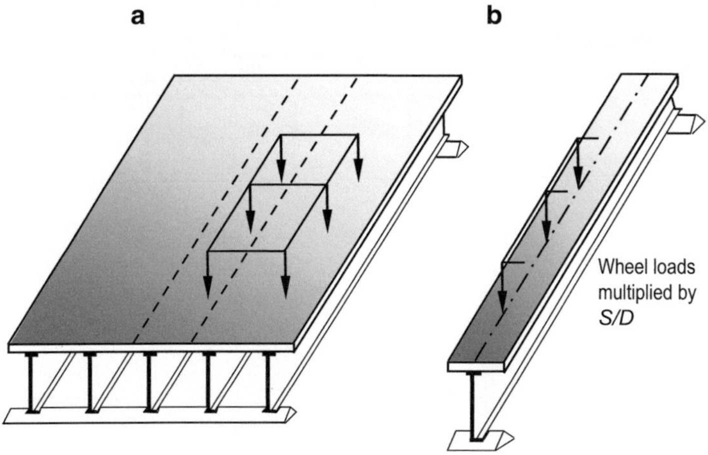

Fig. 2.3 Illustration of the simplified methods of North America: (**a**) actual bridge, (**b**) isolated 1-D model

Table 2.1 Some AASHTO (1989) *D* values

	D in m	
Bridge type	Bridge designed for one traffic lane	Bridge designed for two or more traffic lanes
Slab on steel or prestressed concrete girders	2.13	1.67
Slab on T-beams	1.98	1.83
Slab on timber girders	1.83	1.53

2.3.1 *Old AASHTO Method*

Most highway bridges in North America are designed by the AASHTO specifications. Design vehicles for these specifications consist of two- and three-axle vehicles having two lines of wheels the centres of which are 1.83 m apart. The old AASHTO specifications (1989) permitted a simplified method for obtaining live load longitudinal moments and shears, according to which a longitudinal girder, or a strip of unit width in the case of slabs, is isolated from the rest of the structure and treated as a one-dimensional beam. This beam, as shown in Fig. 2.3b, is subjected to loads comprising one line of wheels of the design vehicle multiplied by a load fraction (S/D), where S is the girder spacing and D, having the units of length, has an assigned value for a given bridge type. The resulting moments and shears are assumed to correspond to maximum girder moments and shears in the bridge. Values of D as specified in the AASHTO (1989) specifications for various cases of slab-on-girder bridges are given in Table 2.1.

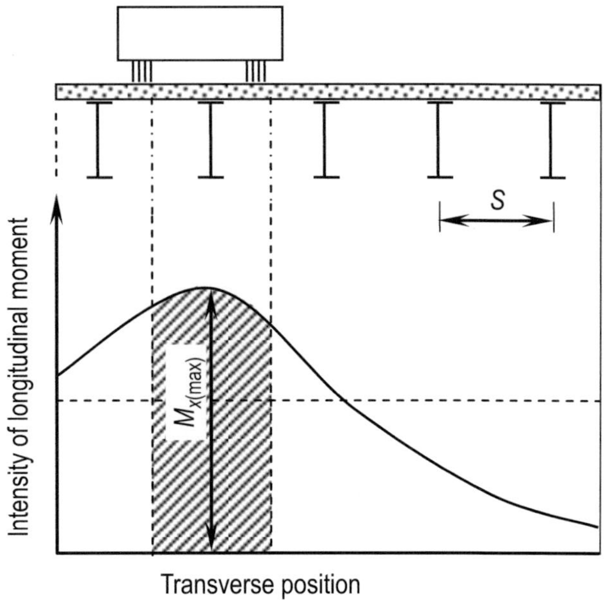

Fig. 2.4 Transverse distribution of longitudinal moment intensity

2.3.2 *Concept of D Method*

The concept of the factor D can be explained with reference to Fig. 2.4, which shows schematically the transverse distribution of live load longitudinal moment intensity in a slab-on-girder bridge at a cross-section due to one vehicle with two lines of wheels. The intensity of longitudinal moment, having the units of kN.m/m, is obtained by idealizing the bridge as an orthotropic plate.

It can be readily appreciated that the maximum girder moment, M_g, for the case under consideration occurs in the second girder from the left. The moment in this girder is equal to the area of the shaded portion under the moment intensity curve. If the intensity of maximum moment is $M_{x(max)}$ then this shaded area is approximately equal to $SM_{x(max)}$, so that:

$$M_g \simeq SM_{x(\max)} \tag{2.4}$$

It is assumed that the unknown quantity $M_{x(max)}$ is given by:

$$M_{x(\max)} = M/D \tag{2.5}$$

where M is equal to the total moment due to half a vehicle, i.e., due to one line of wheels. Substituting the value of $M_{x(max)}$ from Eq. (2.5) into Eq. (2.4):

$$M_g \simeq M(S/D) \tag{2.6}$$

Thus if the value of D is known, the whole process of obtaining longitudinal moments in a girder is reduced to the analysis of a 1-dimensional beam in which the loads of one line of wheels are multiplied by the load fraction (S/D).

2.3.3 New AASHTO Method

Earlier AASHTO D values for this extremely simple method were developed from results of extensive orthotropic plate analyses by Sanders and Elleby (1970). The simplicity of the method, however, does take its toll in accuracy. In the old AASHTO method, the value of D depends only on the bridge type; it is obvious, however, that the manner of load distribution in a long and narrow bridge is different from that in a short and wide bridge of the same type. The old AASHTO method is unable to cater for such factors as the aspect ratio of the bridge. It is noted that in its 1994 edition, the AASHTO specifications introduced another simplified method, which is similar in spirit to the D method, but is more accurate. In the new method, the ratio S/D is designated as g. For slab-on-girder bridges, the value of g is obtained as a function of (a) the girder spacing S in mm, (b) the span length L in mm, (c) the deck slab thickness t_S in mm, and (d) the longitudinal stiffness parameter K_g in mm^4, which is obtained as follows.

$$K_g = n\left(I_g + e_g^2 A\right) \tag{2.7}$$

where, n is the modular ratio E_{girder}/E_{deck}, I_g is the moment of inertia of the girder in mm^4, e_g is the girder eccentricity in mm, being the distance from the girder centroid to the middle of the deck slab, and A is the cross-sectional area of the girder in mm^2. For example, the value of g for moment in exterior girders of slab-on-girder bridges with steel or concrete girders and subjected to multiple lane loading is designated as mg_{moment} and is given by:

$$mg_{moment} = 0.075 + \left(\frac{S}{2900\,mm}\right)^{0.6}\left(\frac{S}{L}\right)^{0.2}\left(\frac{K_g}{Lt_S^3}\right)^{0.1} \tag{2.8}$$

2.3.4 Canadian Methods

2.3.4.1 Ontario Method I

When the development of the Ontario Highway Bridge Design Code (OHBDC) began in 1976, the committee which was entrusted with the task of writing the section on analysis was asked to develop and specify a method of analysis which was as simple as the old AASHTO method but far more accurate. The method, which was developed for the OHBDC and was specified in the editions published in 1979 and 1983, has come to be known as the α-θ method. As discussed by Bakht

Table 2.2 Values of D for longitudinal moments for the ultimate limit state specified by the CSA (1988) design code

	D in metres for bridge with no. of lanes =		
Bridge type	2	3	3 or more
Slab bridges and voided slab bridges	1.90	2.15	2.40
Concrete slab on girders	1.80	1.90	2.00
Timber flooring on girders	1.65	1.90	2.00
Multi cell box girders	1.80	2.05	2.40

and Jaeger (1985), in this method the contours of the values of D obtained from rigorous analyses are presented on charts which use the two characterizing parameters of orthotropic plates, being α and θ, as their axes. These parameters are the same as used in the distribution coefficient methods and are defined by Eqs. (2.1) and (2.2), respectively. Values of the plate rigidities used in these equations can be obtained by standard methods, e.g. Cusens and Pama (1975) and Bakht and Jaeger (1985).

The final value of D, which is used for analysing the bridge, denoted as D_d, is obtained from:

$$D_d = D\left\{\frac{1 + \mu C_f}{100}\right\} \tag{2.9}$$

where

$$\mu = \frac{W_e - 3.3}{0.6} \leq 1.0 \tag{2.10}$$

in which W_e is the design lane width in metres, and C_f is a factor, whose values are provided in chart-form on the α-θ space.

2.3.4.2 CSA Method

Despite the simplicity of the Ontario method, some designers were not happy with having to calculate the values of α and θ. When the 1988 edition of the Canadian Standards Association (CSA) bridge code was being developed by using the OHBDC as its model, it was decided to heed the above concern of the designers and present conservative estimates of the values of D depending upon the type and width of the bridge. A selection of the CSA (1988) values of D is presented in Table 2.2. It can be appreciated that in terms of accuracy, the CSA method lies between the AASHTO and Ontario methods.

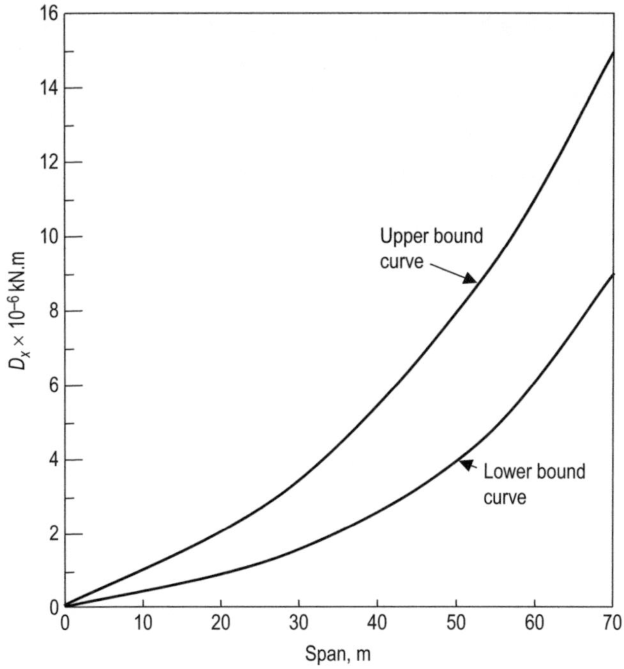

Fig. 2.5 D_x plotted against span length for slab-on-girder bridges in North America

2.3.4.3 Ontario Method II

Research done by Bakht and Moses (1988) has shown that the simplified method of Ontario incorporated in the 1979 and 1983 editions can be "simplified" further by recognizing mainly that the longitudinal flexural rigidity per unit width, D_x, of girder bridges in North America lies between the two bounds defined as follows:

$$D_x = 59,575L + 2,275L^2 \text{(upper bound)} \tag{2.11}$$

$$D_x = 9,250L + 1,790L^2 \text{(lower bound)} \tag{2.12}$$

where the span of the bridge L is in metres and D_x in kN.m. Figure 2.5 shows the upper and lower bounds of D_x for slab-on-girder bridges in North America.

The advantage of determining the upper and lower bound values of D_x for a bridge of a given span can be explained with the example of a specific four-lane slab-on-girder bridge, which has $L = 30$ m, and $2b = 14.4$ m; this bridge is analyzed by the semi-continuum method in Sect. 3.3.4. With reference to Fig. 2.2 it can be seen that α for a slab on girder bridge lies between 0.06 and 0.1. From Eqs. (2.11) and (2.12), the upper and lower bound values of D_x are found to be 3,834,750 and 1,888,500 kN.m, respectively. The value of D_y for this bridge with a 175 mm thick concrete deck slab is calculated to be 8930 kN.m, so that from Eq. (2.1) the upper

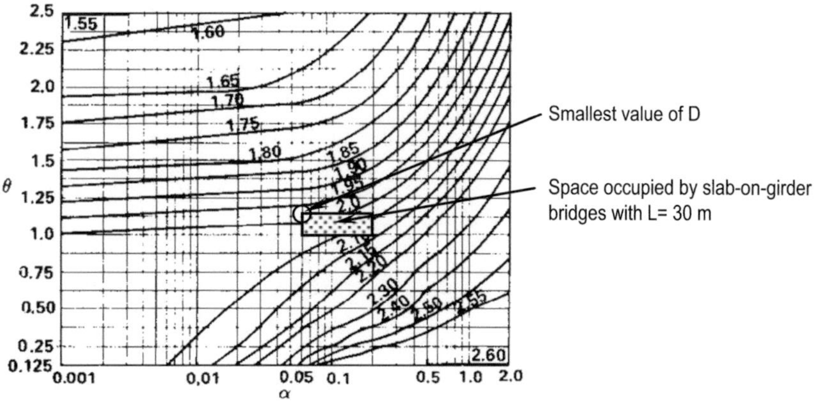

Fig. 2.6 *D* charts for four-lane bridges for ULS, OHBDC loading

and lower bound values of θ are found to be 1.17 and 0.98, respectively. The contours of the *D* values for four-lane bridges for internal portions specified in the OHBDC (1983) for the ULS are reproduced in Fig. 2.6, which also shows the very small rectangular area occupied by slab-on-girder bridges with a span of 30 m. From the trend of the contours of the *D* values plotted in this figure, it can be appreciated that within the small rectangular area, the smallest value of D, i.e. the most conservative value, is at the corner where the value α is the smallest and the value of θ the largest. If the values of *D* are obtained by rigorous analysis for α and θ representing this corner of the rectangle, then all other values of *D* corresponding to all other combinations of α-θ within the rectangle would be larger but by only a small amount. This principle was used to calculate the values of *D* and C_f as functions of span length *L*.

The expressions for *D* and C_f for longitudinal moments in two types of bridge on Class A highways are listed in Table 2.3 corresponding to the Ultimate Limit State (ULS).

Having obtained the values of *D* and C_f from the expressions given in Table 2.3, the design value of *D*, i.e. D_d, is obtained from Eq. (2.9).

Recognizing that the distribution of longitudinal moments is more benign than that of longitudinal shears, the OHBDC (1992) has specified that the values of D_d for longitudinal shears be obtained from a separate table which is reproduced herein as Table 2.4.

2.3.5 CHBDC Method

The successors to the OHBDC, the Canadian Highway Bridge Design Code (CHBDC 2000, 2006) have specified a slightly different simplified method of analysis. The purpose of this new method was to permit the designers to have a

Table 2.3 Expressions for D and C_f for longitudinal moments at the ULS in bridges on Class A highways

No. of design lanes	External/internal portion or girder	D(m) $3 < L \le 10$ m	$L > 10$ m	C_f (%)
(a) Slab bridges and voided slab bridges				
1	External	2.10	2.10	$16-(36/L)$
	Internal	$2.00+(3L/100)$	2.30	$16-(36/L)$
2	External	2.05	2.05	$20-(40/L)$
	Internal	$2.10-(1/L)$	$2.10-(1/L)$	$20-(40/L)$
3	External	$1.90+(L/20)$	$2.60-(2/L)$	$16-(30/L)$
	Internal	$1.45+(L/10)$	$2.65-(2/L)$	$16-(30/L)$
(b) Slab-on-girder bridges				
1	External	2.00	$2.10-(1/L)$	$5-(12/L)$
	Internal	$1.75+(L/40)$	$2.30-(3/L)$	$5-(12/L)$
2	External	1.90	$2.00-(1/L)$	$10-(25/L)$
	Internal	$1.40+(3L/100)$	$2.10-(4/L)$	$10-(25/L)$
3	External	1.90	$2.00-(1/L)$	$10-(25/L)$
	Internal	$1.60+(2L/100)$	$2.30-(5/L)$	$10-(25/L)$

Table 2.4 Values of D_d in metres for longitudinal shear for ultimate limit state for bridges on Class A highways

Bridge type	D_d in m for number of design lanes = 1	2	3	4 or more
Slab	2.05	1.95	1.95	2.15
Voided slab	2.05	1.95	1.95	2.15
Slab-on-girder	1.75	1.70	1.85	1.90
Stress-laminated wood decks	1.75	1.70	1.85	1.90

better feel of the load distribution characteristics of bridge than could be afforded by the somewhat abstract D method. The CHBDC method provides a multiplier to the average longitudinal moment or shear; this multiplier, which has to be always greater than 1.00 and which is denoted as F_m for moments and F_s for shears, gives an indication of the load distribution characteristics of a given bridge. If the multiplier is much larger than 1.00, the bridge has poor load distribution characteristics. On the other hand, a multiplier closer to 1.00 indicates that the bridge has good load distribution characteristics.

For girder-type bridges, the longitudinal moment M_g, or longitudinal shear V_g, in a girder due to design live loads is obtained by multiplying the average girder moment $M_{g\ avg}$, or the average girder shear $V_{g\ avg}$ by the multiplier F_m or F_v. Values of $M_{g\ avg}$, or $V_{g\ avg}$ for each girder are obtained by placing the design live loads in all

Table 2.5 Modification factors, R_L, for multi-lane loading specified by CHBDC (2006)

Number of loaded lanes	Highway class		
	A	B	C or D
1	1.00	1.00	1.00
2	0.90	0.90	0.85
3	0.80	0.80	0.70
4	0.70	0.70	–

design lanes and multiplying the loads by the appropriate modification factor for multi-lane loading, so that:

$$M_{g\,avg} = \frac{nM_tR_L}{N} \tag{2.13}$$

and

$$V_{g\,avg} = \frac{nV_tR_L}{N} \tag{2.14}$$

where

M_t = the maximum moment per design lane at the transverse section of the span under consideration
V_t = the maximum shear per design lane at the transverse section of the span under consideration
n = the number of design lanes in the bridge
R_L = the modification factor for multi-lane loading as shown in Table 2.5
N = the number of girders

The Highway classes referred to in Table 2.5 relate to the volume of average daily truck traffic (ADTT) per lane of the bridge. The ADTT for Class A, B, C and D Highways is more than 1000, between 250 and 1000, between 50 and 250, and less than 50, respectively.

The maximum girder moments M_g and shears V_g are obtained by multiplying the average moments or shears by the amplification factors F_m or F_v, respectively; these factors, which account for the transverse variation in maximum longitudinal moment or shear intensities, as compared to the average responses, are obtained by the following equations.

$$F_m = \frac{SN}{F\left(1 + \frac{\mu C_f}{100}\right)} \geq 1.05 \tag{2.15}$$

and

$$F_v = \frac{SN}{F} \geq 1.05 \tag{2.16}$$

Table 2.6 Expressions for F and C_f for longitudinal moments in slab-on-girder bridges for the ultimate limit state of CHBDC

Class of highway	No. of design lanes	External/internal girders	F, m For $L \leq 10$ m, but > 3 m	F, m For $L > 10$ m	C_f, %
A or B	1	External	3.30	$3.50 - (2/L)$	$5 - (12/L)$
		Internal	$3.30 + 0.05\,L$	$4.40 - (6/L)$	$5 - (12/L)$
	2	External	6.50	$6.80 - (3/L)$	$5 - (15/L)$
		Internal	$4.80 + 0.10\,L$	$7.20 - (14/L)$	$5 - (15/L)$
	3	External	8.30	$8.70 - (4/L)$	$10 - (25/L)$
		Internal	$6.70 + 0.08\,L$	$9.60 - (21/L)$	$10 - (25/L)$
	4	External	9.50	$10.00 - (5/L)$	$10 - (25/L)$
		Internal	$7.60 + 0.14\,L$	$11.20 - (22/L)$	$10 - (25/L)$
C or D	1	External	3.30	$3.50 - (2/L)$	$5 - (12/L)$
		Internal	$3.30 + 0.05\,L$	$4.40 - (6/L)$	$5 - (12/L)$
	2	External	6.10	$6.40 - (3/L)$	$5 - (15/L)$
		Internal	$4.80 + 0.10\,L$	$7.20 - (14/L)$	$5 - (15/L)$
	3	External	7.70	$8.10 - (4/L)$	$10 - (25/L)$
		Internal	$6.60 + 0.04\,L$	$8.80 - (18/L)$	$10 - (25/L)$

Table 2.7 Values for F for longitudinal shears in slab-on-girder bridges for the ultimate limit state of CHBDC

Class of highway	No. of design lanes	F, m
A or B	1	3.50
	2	6.10
	3	8.20
	4	9.50
C or D	1	3.50
	2	6.10
	3	7.60

where

S = centre-to-centre girder spacing in metres
$\mu = (W_e - 3.3)/0.6$
W_e = width of the design lane in metres

F is a width dimension that characterises load distribution for a bridge, and C_f is a correction factor in %. Both F and C_f for moments and shears are obtained from Tables 2.6 and 2.7, respectively, for slab-on-girder bridges for the ultimate limit state. It is noted that the values of F for external girders noted in Table 2.6 are applicable when the deck slab overhang is equal to or less than $0.5S$.

The Ontario II and CHBDC methods discussed above can also be used in conjunction with other design codes provided that the following conditions are met.

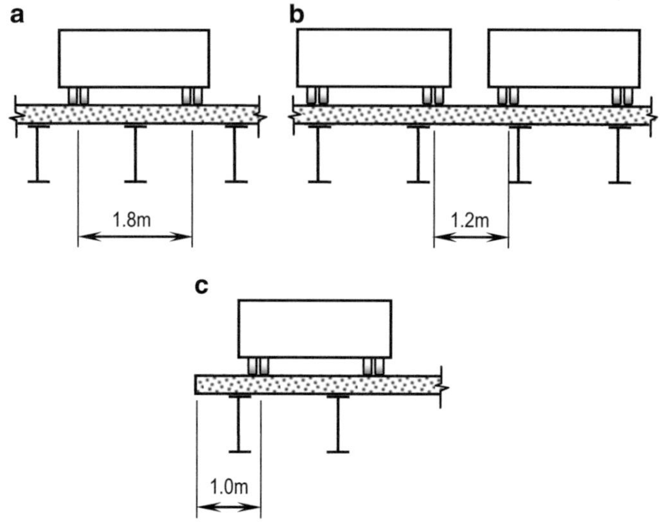

Fig. 2.7 Transverse spacing of longitudinal lines of wheels: (**a**) spacing between two lines of wheels of a vehicle, (**b**) spacing between two lines of wheels of adjacent vehicles, (**c**) minimum vehicle edge distance

(a) The design vehicle has two longitudinal lines of wheels the centres of which are transversely about 1.8 m apart, as illustrated in Fig. 2.7a.
(b) When two design vehicles are present on the bridge side-by-side, their adjacent longitudinal lines of wheels are about 1.2 m apart, centre to centre, as illustrated in Fig. 2.7b.
(c) The transverse distance between a longitudinal free edge of the bridge and the centre of the closest longitudinal line of wheels of the design vehicle, i.e. the vehicle edge distance, is not less than about 1.0 m.
(d) The reduction factors for multiple presences in more than one lane of the bridge are as shown in Table 2.5.

It is noted that the Ontario and CHBDC methods are applicable only to those bridges where the values of D_x lie below the upper-bound values defined by Eq. (2.11). In some jurisdictions, the bridges are considerably stiffer. Figure 2.8, for example, compares the values of D_x for slab-on-girder bridges in Hong Kong with those in North America (Chan et al. 1995). For such bridges, the Ontario and CHBDC methods should not be used as they will lead to unsafe results. When the conditions noted above are not met, a set of new simplified methods can be developed readily as explained in Sect. 2.4.

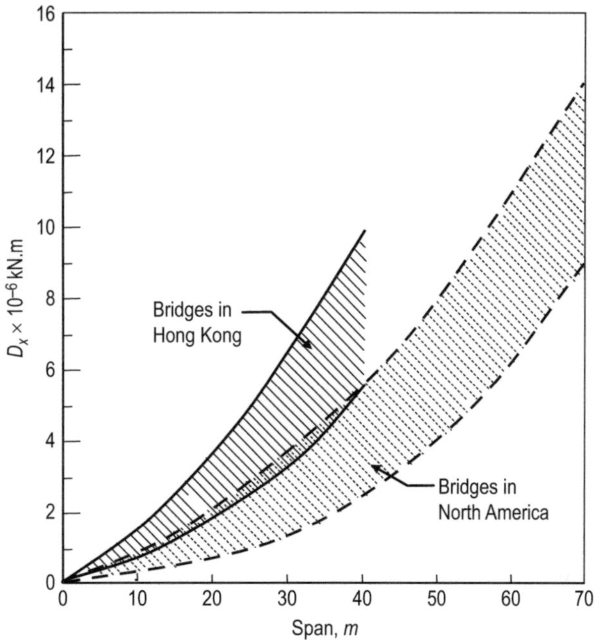

Fig. 2.8 D_x plotted against L for slab-on-girder bridges in Hong Kong and North America

2.4 Two Proposed Methods for Two-Lane Slab-On-Girder Bridges

The previous manual methods of bridge analysis, such as those by Morice and Little and by Cusens and Pama, were made simple by simplifying assumptions. On the other hand, the simplified methods presented in this chapter are based on the results of rigorous analyses, and accordingly are dependent upon the specification of design live loading. The design loadings used elsewhere in the world are significantly different from those of North America. Two simplified methods are presented in this section for the analysis of slab-on-girder bridges with two design lanes for two specific design loads. It is important to note that these methods give approximate values of maximum moments and shears in girders that should preferably be used only to verify the results of rigorous analyses. One of the presented methods was developed by analysing a selected number of bridges by the orthotropic plate method, and the other was developed by analysing the selected bridges by the semi-continuum method; both these rigorous methods of analyses are discussed in Chap. 3 along with their computer programs, which are can be downloaded from http://extras.springer.com. Other similar simplified methods could be developed by using these programs.

Fig. 2.9 70-*R* track loading

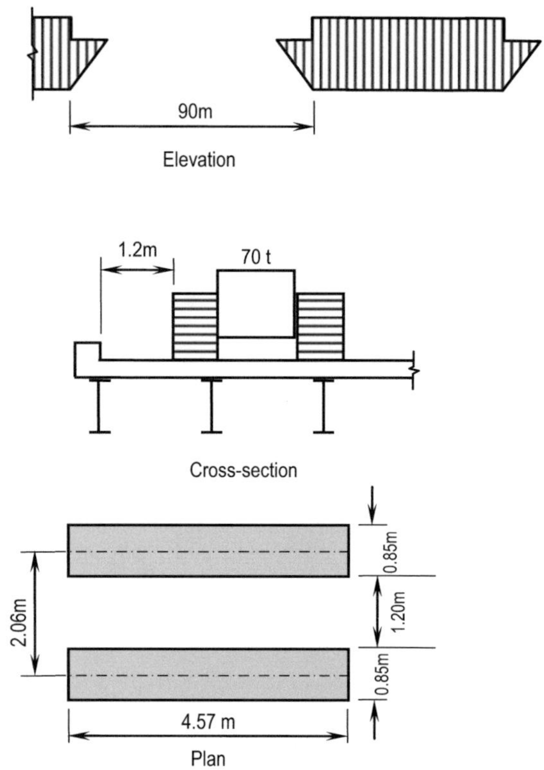

2.4.1 *Simplified Method for Indian Road Congress Bridge Design Loads*

The Indian Road Congress (IRC) specifies three design loads, being 70-R loading, Class A loading and Class B loading. Details of the 70-R loading are shown in Fig. 2.9, and those for Class A loading in Fig. 2.10. Class B loading is the same as Class A loading, except that its axle weights are 60 % of those of Class A loading, and its wheel contact areas are smaller than those for Class B loading. The transverse position of the 70R loading with respect to the nearest curb is shown in Fig. 2.9, and those for Class A and B loadings in Fig. 2.11. The latter figure also shows the transverse distance between the lines of wheels of two adjacent design loads.

A simplified method was developed for slab-on-girder bridges having a clear distance of 7.5 m between the curbs, so that each design lane has a width of 3.75 m.

The lane width was assumed to be 3.75 m, and the minimum curb width, *B*, was assumed to be 0.225 m, resulting in a total bridge width of 7.95 m. The upper limit of the curb width was assumed to be 0.5 m, which resulted in a total bridge width of 8.5 m. Fifteen bridges corresponding to each of these two widths, and having spans

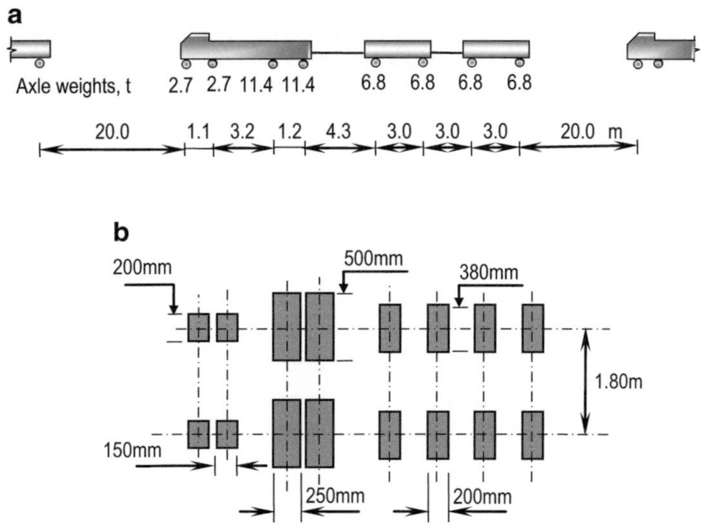

Fig. 2.10 Class A loading: (a) elevation, (b) plan, not to scale

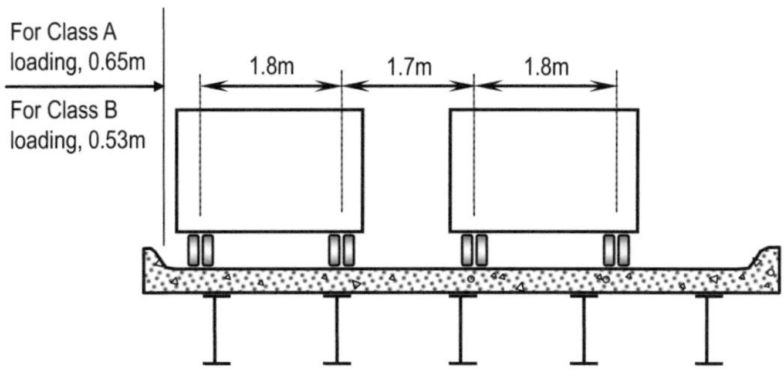

Fig. 2.11 Transverse vehicle positions for Class A and B loadings

of 20.0 m were analysed for *70-R*, *Class A* and *Class B* loadings by the orthotropic plate method of Cusens and Pama (1975), which is incorporated into a computer program PLATO, discussed in Chap. 3. The 15 bridges effectively covered the entire α-θ space for slab-on-girder bridges, as can be seen in Fig. 2.12.

For determining the values of D for longitudinal moments, the vehicles were placed in such longitudinal positions as would induce maximum total longitudinal moments. Similarly for shear D values, the vehicles were positioned to induce maximum longitudinal shears. The transverse vehicle positions were as shown in Figs. 2.9 and 2.11. The orthotropic plate method takes into account the finite size of a concentrated load, which has a significant effect on moment and shear intensities directly under a wheel load. To make the representation of loads as realistic as possible, a deck slab thickness of 200 mm was assumed. It was further assumed that the effective

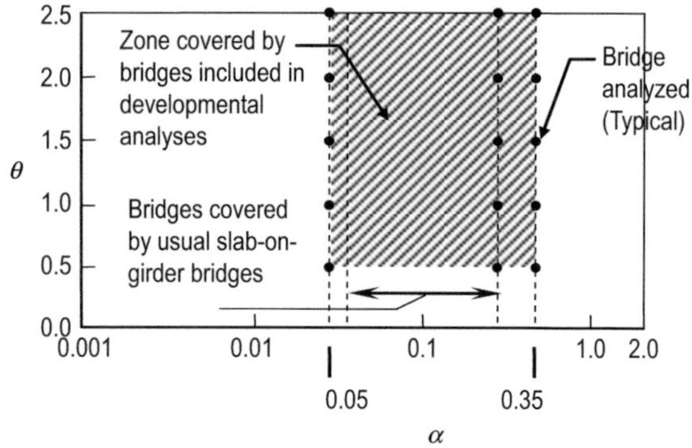

Fig. 2.12 Bridges analysed for the development of the proposed method for IRC loading

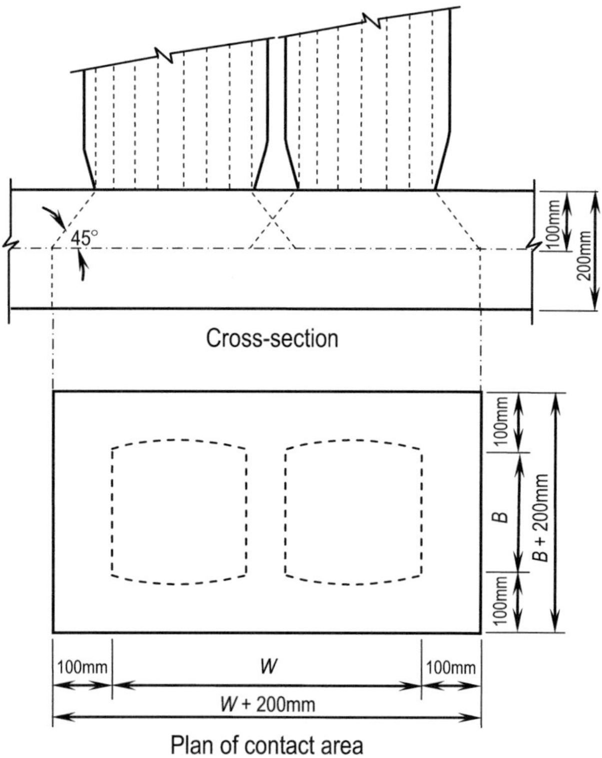

Fig. 2.13 Effective contact area of concentrated loads

size of a concentrated load is obtained by dispersing the surface contact area of the wheel load by 45° to the slab middle surface, as shown in Fig. 2.13.

From Fig. 2.4, it can be readily appreciated that the quantity $SM_{x\,(max)}$ is slightly more than the quantity represented by the shaded area under the curve, which is equal to the moment sustained by the girder. Such over-estimation of live load moments can be eliminated by taking an average of longitudinal moment intensity

Table 2.8 *D* values in m for longitudinal moments corresponding to *70-R Loading* and $B = 0.2$ m

	D in metres for $\theta =$				
α	0.5	1.0	1.5	2.0	2.5
0.05	2.72	2.37	2.19	2.14	2.13
0.20	2.90	2.46	2.22	2.18	2.18
0.35	3.03	2.53	2.29	2.22	2.20
Mean *D* values	2.88	2.45	2.23	2.18	2.17
Max. variation from mean	±5 %	±3 %	±3 %	±2 %	±2 %

over the width *S* and then multiplying this average moment intensity by *S* to obtain the girder moment. For all the analysed cases, the average maximum longitudinal moments and shears, $M'_{x(max)}$ and $V'_{x(max)}$ respectively, were obtained by averaging the corresponding quantities over a width of 2.0 m. It should be appreciated that this 2.0 m width is only to reduce the effect of "peakiness" of $M_{x(max)}$. A departure of actual girder spacing from this quantity would have negligible effect on $M'_{x(max)}$. From computer-calculated values of $M'_{x(max)}$, the governing value for longitudinal moments was obtained by the following equation, which is a rearranged form of Eq. (2.5), in which $M_{x(max)}$ is replaced by $M'_{x(max)}$:

$$D = M/M'_{x(\max)} \tag{2.17}$$

where *M* is the total moment due to one line of wheels or one track of the design loading. Similarly, *D* for longitudinal shear is given by:

$$D = V/V'_{x(\max)} \tag{2.18}$$

where *V* is the total shear due to one line of wheels or one track of the design loading. Each of the cases were analysed for 31 harmonics. Spot checks of results with 41 harmonics confirmed that the solutions were fully converged under 31 harmonics.

All *D* values for longitudinal moments were found to be relatively insensitive to small variations in α values, as may be seen for example in Table 2.8 which shows the *D* values corresponding to the *70-R* loading. Adopting a mean value of *D* for given θ values results in maximum errors or ±5 %. In the light of this observation, it was decided to eliminate α from consideration. Changes in curb widths did have a noticeable effect on *D* values for longitudinal moment, especially for *Class A* and *Class B* loadings, for which the code-specified minimum edge distances are unusually small.

The *D* values for moments are plotted in Fig. 2.14 for the three loadings, and for curb widths $B = 0.2$ and 0.5 m, respectively. A few spot checks indicated that a linear interpolation for intermediate curb widths provided results of reasonable accuracy.

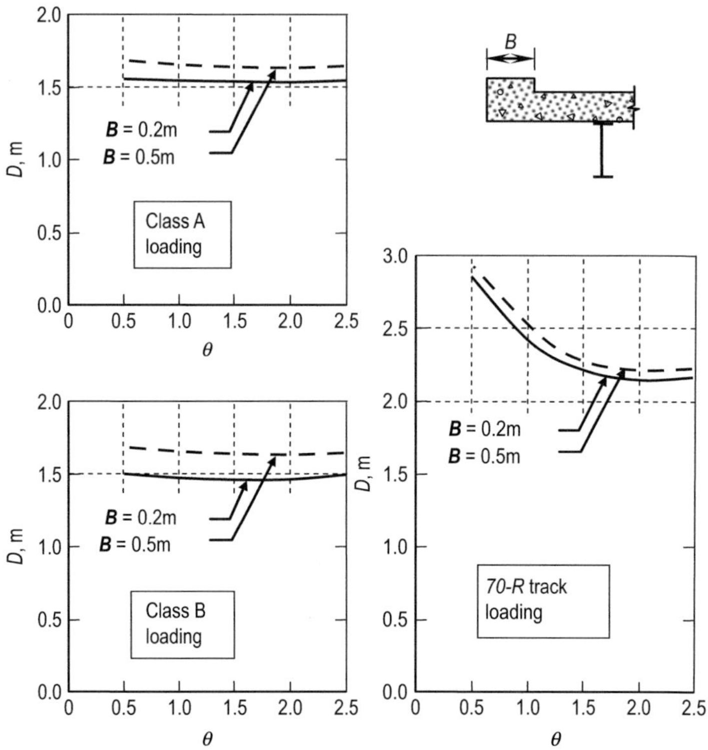

Fig. 2.14 *D* values for longitudinal moments for IRC design loads

The θ values of 0.5, 1.0, 1.5, 2.0, and 2.5 for which bridges were analysed, correspond to γ of 0.141, 0.035, 0.016, 0.009 and 0.006, respectively, where γ is given as follows:

$$\gamma = \left(D_y/D_x\right)^{0.5} \tag{2.19}$$

It was found that for the *70-R* loading, the *D* value for shear varied almost linearly with γ, resulting in the following simple relationship which gives the *D* value in metres:

$$D = 2.16 + 1.6\gamma \tag{2.20}$$

For *Class A* and *Class B* loadings, *D* values for shear were little affected by γ. It was found to be sufficiently accurate to adopt the single value of 1.6 m for *D* for all slab-on-girder bridges for both *Class A* and *Class B* loadings.

The method described above can be used for analysing slab-on-girder bridges with two design lanes, and subjected to *70-R*, *Class A* or *Class B* design loads, it being noted that the bridges concerned must satisfy the following conditions.

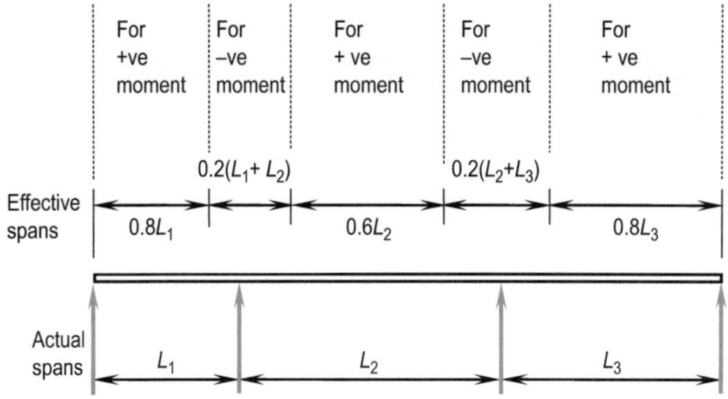

Fig. 2.15 Effective span lengths for calculating θ

(a) The width is constant or nearly constant and there are at least three girders in the bridge;
(b) the skew parameter $\varepsilon = (S \tan \psi)/L$ does not exceed 1/18 where S is girder spacing, L is span and ψ is the angle of skew;
(c) for bridges curved in plan, L^2/bR does not exceed 1.0, where R is the radius of curvature; L is span length; b is half width of the bridge;
(d) the total flexural rigidity of transverse cross-section remains substantially the same over at least the central 50 % length of each span;
(e) girders are of equal flexural rigidity and equally spaced, or with variations from the mean of not more than 10 % in each case; and
(f) the deck slab overhang does not exceed 60 % of S, and is not more than 1.8 m.

In cases where the above conditions are not fully met, engineering judgement should be exercised to ascertain if a bridge meets them sufficiently closely for the simplified method to be applicable.

The proposed method requires the following steps:

(a) Calculate values of D_x and D_y using Eqs. (2.21) and (2.22), respectively.
(b) Obtain the value of θ from Eq. (2.2). The effective spans for continuous bridges can be obtained from Fig. 2.15 for the purpose of calculating θ.
(c) Corresponding to the type of loading and values of θ and curb width B, read the value of D for moments from Fig. 2.14. For curb widths larger than 0.5 m, use the D value corresponding to $B = 0.5$ m.
(d) Calculate live load longitudinal moment at any section by multiplying the total moment at that section due to one line of wheels, or one track, of the relevant loading by the load fraction S/D.
(e) For longitudinal shear, use a D value of 1.60 m for *Class A* and *Class B* loadings. For class *70-R* loading, obtain the value of D from Eq. (2.20) corresponding to the γ value given by Eq. (2.19). The same value of D for longitudinal shear is applicable for both single and continuous spans.
(f) Similarly to longitudinal moments, obtain longitudinal shear per girder by multiplying the total shear for half the design loading by the load fraction (S/D).

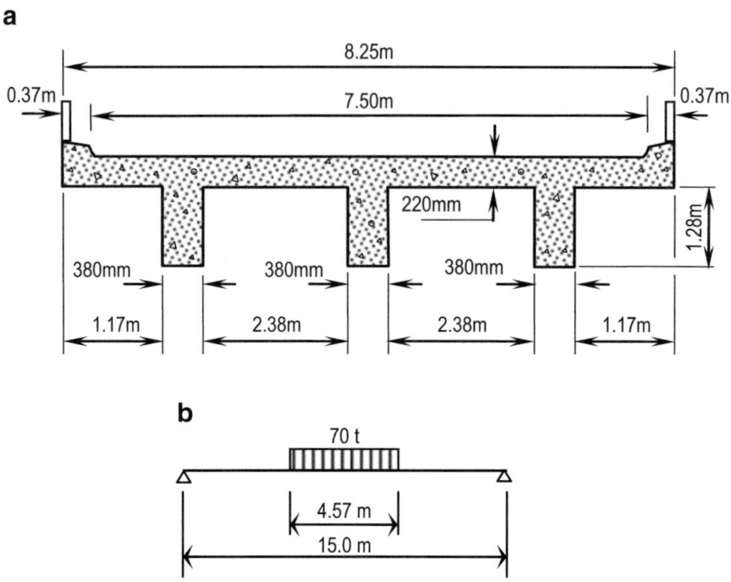

Fig. 2.16 Details of example: (**a**) cross-section, (**b**) simply supported beam under 70-R track loading

The longitudinal flexural rigidity D_x, of a bridge is the product of E and i, where E is modulus of elasticity of deck slab concrete, and i is the longitudinal moment of inertia per unit width in the units of deck slab concrete. For obtaining a value for i, the total moment of inertia of the cross-section of the bridge, I, should be calculated in terms of deck slab concrete. The parameter i is then obtained by dividing I by the bridge width. Thus:

$$D_x = \frac{EI}{2b} \tag{2.21}$$

For bridges having fewer than five intermediate diaphragms per span, the transverse flexural rigidity is obtained by ignoring contributions from diaphragms, so that for slab thickness t:

$$D_y = Et^3/12 \tag{2.22}$$

The contribution of diaphragms to transverse flexural rigidity should be taken into consideration only when engineering judgement shows that their contribution can be realistically assumed to be uniformly distributed along the span. Neglecting contributions of diaphragms is a safe-side assumption for nearly all practical bridges.

To illustrate the use of the method, the example of a single-span T-beam bridge is presented; the cross-section of the bridge is shown in Fig. 2.16a, and the 70-R loading on the 15 m simply supported span in Fig. 2.16b.

Table 2.9 Steps in calculation of maximum moments due to 70-R loading

S, m	D, m (from Fig. 2.14)	S/D	Mid-span moment due to one line of tracks, t.m	Max. girder moment at mid-span, t.m
2.76	2.57	2.76/2.57 = 1.07	111.25	1.07 × 111.25 = 119.0

The total moment of inertia, I, of the bridge cross-section is calculated to be 0.66 m^4. Hence $D_x = E \times 0.66/8.25 = 0.08E$. The transverse rigidity, D_y, is calculated by ignoring contributions of any diaphragms, so that: $D_y = E \times 0.22^3/12 = 0.00087E$. The half width b and span L for the bridge are 4.125 and 15.0 m, respectively. Therefore, θ is given by:

$$\theta = \frac{4.125}{15}\left[\frac{E \times 0.08}{E \times 8.87 \times 10^{-4}}\right]^{0.25} = 0.85$$

The 70-R track loading, shown in Fig. 2.16b leads to the mid-span moment = 70 (7.5 − 1.14) = 222.5 t.m, so that the mid-span moment due to one line of tracks = 0.5 × 222.5 = 111.25 t.m. The steps in calculating the maximum mid-span moment due to the 70-R loading are listed in Table 2.9, which shows that the maximum live load longitudinal moment per girder is equal to 119.0 t·m.

The above-cited bridge is similar to the one used by Krishna and Jain (1977) to illustrate the use of the Morice and Little method, according to which the maximum longitudinal moment per girder due to the 70-R loading is found to be 110 t·m.

2.4.2 *Simplified Method for HB Design Loads*

A simplified method for two-lane slab-on-girder bridges in Hong Kong was presented by Chan et al. (1995); details of this method, which was developed for the British HB loading, are presented in the following with the hope that it would prove useful for preliminary design of bridges designed by the British document BS5400.

As shown in Fig. 2.17, the HB loading comprises four axles, with four wheels in each axle. The weights of the wheels are governed by the units of the HB loading. In Great Britain, most highway bridges are designed for 45 units of HB loading, so that each wheel has a load of 112.5 kN, giving a load of 45 kN per axle. Each unit of HB loading is equal to 10 kN. It should, however, be noted that the units of the design loading does not affect the simplified method presented in the following.

Chan et al. (1995) have shown that D_x, the longitudinal flexural rigidity per unit width, of slab-on-girder bridges in Hong Kong lies between two bounds defined by the following equations.

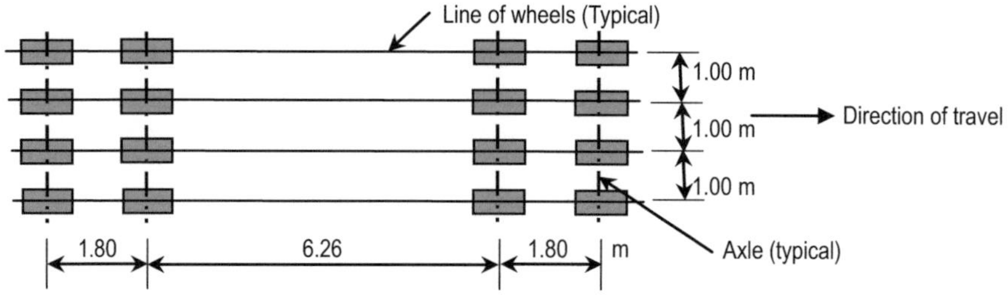

Fig. 2.17 Plan of HB loading

$$D_x = 48,000L + 5,100L^2(\text{upper bound}) \quad (2.23)$$

$$D_x = 2,000L + 3,650L^2(\text{lower bound}) \quad (2.24)$$

where the span of the bridge L is in metres and D_x in kN.m. The two bounds defined by the above equations are illustrated in Fig. 2.8. From a study of a large number of slab-on-girder highway bridges in Hong Kong, it was determined that the ranges of the various parameters which influence the transverse load distribution characteristics of a bridge are as follows.

(a) The deck slab thickness, t, varies between 150 and 230 mm, with the usual value being 200 mm.
(b) The centre-to-centre spacing of girders, S, varies between 0.2 and 2.0 m, with the usual value being 1.0 m.
(c) The lane width, W_e, varies between 3.2 and 3.8 m, with the usual value being 3.5 m.
(d) The vehicle edge distance, VED, being the transverse distance between the centre of the outermost line of wheels of the HB loading and the nearest longitudinal free edge of the bridge, varies between 0.75 and 5.00 m, with the usual value being 1.00 m.

From the above observations, the following values of the various parameters were adopted for the developmental analyses conducted for developing the simplified method: $t = 200$ mm; $S = 1.0$ m; $W_e = 3.5$ m; and VED = 1.00 m. In addition, it was assumed that the deck slab overhang beyond the outer girders was 0.55 m.

Bridges with spans of 10, 20, 30 and 40 m were selected for the developmental analyses. The flexural rigidities of their girders were calculated from the mean values of D_x given by Eqs. (2.22) and (2.23). The selected idealised bridges under the HB loading were analysed by the semi-continuum method, which is incorporated in computer program SECAN, discussed in Chap. 3. For each bridge, the HB vehicle was placed so as to induce maximum moments at the mid-span. The maximum moment, designated as $M_{g(max)}$ was calculated for each case for each of the external and internal girders.

From Eq. (2.6), it can be shown that D, which has the units of length and provides a measure of the transverse load distribution characteristics of bridge as discussed earlier, is given by:

$$D = MS/M_{g(max)} \tag{2.25}$$

Chan et al. (1995), having plotted the values of D from the above analyses against the span length L, found that these values of D are related to L according to the following equations with a reasonable degree of accuracy, provided that the design value of D, i.e. D_d, is corrected by Eq. (2.30).

For internal girders having $L < 25$ m:

$$D = 1.20 - 3.50/L \tag{2.26}$$

For internal girders having $L \geq 25$ m:

$$D = 1.06 \tag{2.27}$$

For external girders having $L < 25$ m:

$$D = 0.95 + 2.10/L \tag{2.28}$$

For external girders having $L \geq 25$ m:

$$D = 1.03 \tag{2.29}$$

The correcting equation for obtaining D_d is as follows.

$$D_d = D\left(1 + \frac{\mu C_w}{100}\right) \tag{2.30}$$

in which μ is given by:

$$\mu = \frac{3.5 - W_e}{0.25} \tag{2.31}$$

where W_e is the width of the design lane in meters.

Chan et al. (1995) have provided charts in which C_w is plotted as functions of L for internal and external girders. However, C_w can also be obtained fairly accurately by the following equations.

For internal girders:

$$C_w = \frac{6L}{40} \tag{2.32}$$

For external girders:

$$C_w = \frac{17.5 - 8.5(L - 10)}{10} \tag{2.33}$$

It is noted that in the above equations, both D and L are in metres.

The following conditions must be met for applying the above simplified method.

(a) The value of D_x, calculated from Eq. (2.21), lies between the upper and lower bound values given by Eqs. (2.23) and (2.24), respectively.
(b) The bridge has two design lanes, with the lane width, W_e, being ≥ 3.5 m.
(c) The width is constant or nearly constant, and there are at least three girders in the bridge.
(d) The skew parameter $\varepsilon = (S \tan \psi)/L$ does not exceed 1/18 where S is girder spacing, L is span and ψ is the angle of skew.
(e) For bridges curved in plan, L^2/bR does not exceed 1.0, where R is the radius of curvature; L is span length; $2b$ is the width of the bridge.
(f) The total flexural rigidity of transverse cross-section remains substantially the same over at least the central 50 % length of each span.
(g) Girders are of equal flexural rigidity and equally spaced, or with variations from the mean of not more than 10 % in each case.
(h) The deck slab overhang does not exceed $0.6\,S$, and is not more than 1.8 m.

In cases where the above conditions are not fully met, engineering judgement should be exercised to ascertain if the bridge meets them sufficiently closely for the simplified method to be applicable.

The following steps of calculation are required in calculating maximum moments in internal an external girders due to HB loading.

(a) Calculate the value of D from the relevant of Eqs. (2.26, 2.27, 2.28, and 2.29); for simply supported spans, L is the actual span length, and for continuous span bridges, the effective L for different spans is obtained from Fig. 2.15.
(b) For the design lane width, W_e, obtain μ from Eq. (2.31) and C_w from the relevant of Eqs. (2.32) and (2.33), and thereafter obtain D_d from Eq. (2.30).
(c) Isolate one girder and the associated portion of the deck slab, as illustrated in Fig. 2.3b, and analyse it by treating it as a one-dimensional beam under one line of wheels of the HB loading, shown in Fig. 2.17. The moment thus obtained at any transverse section of the beam is designated as M.
(d) For of the internal and internal girders, obtain the maximum moment at the transverse section under consideration by multiplying M with (S/D_d), where D_d is as obtained in Step (b) for the relevant of the internal and external girders.

The use of the above method is illustrated by analysing an actual two-span continuous steel girder bridge in Hong Kong, the Canton Road Duplication Bridge (Chan et al. 1995). Since this bridge was included in determining the upper and lower bound values of D_x for bridges in Hong Kong (Fig. 2.8), there is no need to check if the value of its D_x falls within the two bounds. In the following example, maximum positive, i.e. sagging, moments in the internal girders of the bridge under 45 units of HB loading are determined by the simplified method. The various parameters of the bridge are as noted in the following.

Actual span lengths, $L = 22.0$ and 22.0 m
Bridge width, $2b = 8.7$ m
Design lane width $W_e = 3.35$ m
Deck slab thickness = 0.18 m
Girder spacing $S = 1.48$ m
Deck slab overhang = 0.65 m
Vehicle edge distance VED = 1.25 m

The parameters for the simplified analysis are calculated as follows.

From Fig. 2.15, effective span $L = 0.8 \times 22.0 = 17.6$ m
From Eq. (2.26), $D = 1.20 - 3.5/17.6 = 1.001$ m
From Eq. (2.31), $\mu = (3.5 - 3.35)/0.25 = 0.6$
From Eq. (2.32), $C_w = 2.6$ %
From Eq. (2.30), $D_d = 1.001 \times (1 + 0.6 \times 2.5/100) = 1.017$ m
Load fraction $S/D_d = 1.48/1.017 = 1.455$

From beam analysis, the maximum positive moment in the two-span beam due to one line of 45 units of HB loading, M, is found to be 1160 kN.m. By multiplying this moment with load fraction 1.445, the maximum positive moment in any of the internal girders of the bridge was found to be 1688 kN.m.

A rigorous analysis of the above problem by the semi-continuum and grillage methods gave the maximum moments of 1503 and 1636 kN.m, respectively (Chan et al. 1995). This observation confirms that the moments given by the simplified method are 12 % larger than those given by the semi-continuum method and 3 % larger than those given by the grillage method. It is discussed in Chap. 3 that while the two rigorous methods of analysis are similar in idealisation, the semi-continuum method is believed to be more accurate than the grillage method.

2.5 Analysis of Two-Girder Bridges

In two-girder bridges and those bridges which comprise two main longitudinal members, such as trusses, the transverse distribution analysis is usually done by simple static apportioning of the loads to the two main longitudinal members. For example, considering the cross-section of the two-girder bridge shown in Fig. 2.18a, with a girder spacing S and a concentrated load P located on the deck slab at a distance eS from the left girder, it is usual to assume, as shown in Fig. 2.18b, that the left and right girders receive loads $(1 - e)\,P$ and eP, respectively. It is recalled that these loads are the same as the reactions of a beam simply supported by the two girders.

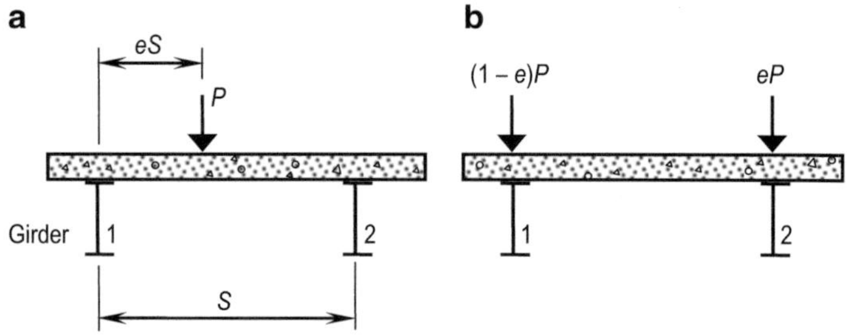

Fig. 2.18 The usual method of apportioning loads to girders in a two-girder bridge (**a**) single load between two girders, (**b**) static apportioning of load without considering torsional rigidities

Transverse load distribution analysis by static apportioning as described above is based upon the implicit assumption that the bridge has negligible torsional rigidities in both the longitudinal and transverse directions of the bridge. In practice, even though the torsional rigidities of girders themselves may be negligible, the torsional rigidity of the deck slab, in both the longitudinal and transverse directions, can be substantial. The neglect of these torsional rigidities can make the analysis by static apportioning somewhat conservative. The objective of this section is to present a very simple, yet accurate, method of apportioning loads to the girders of two-girder, right bridges, based upon the semi-continuum method, the general treatment of which is described in Chap. 3. It is emphasized that knowledge of the general semi-continuum method is not needed in order to be able to use the proposed method.

2.5.1 Two-Girder Bridges

The case of a right, simply-supported bridge with two main girders supporting a concrete deck slab, is considered first. Consistent with usual practice, transverse deflections of the solid concrete deck slab due to shear are assumed to be negligible.

Figure 2.18a shows a two-girder bridge carrying a longitudinal line load at a distance *eS* from girder 1. By using the general semi-continuum method described by Jaeger and Bakht (1989), it can be shown that the distribution coefficients for longitudinal bending moment and torsion are given by:

$$\rho_1 = \frac{\frac{\eta}{2}(1 - e + \lambda + 2\mu) + \mu(1 + \lambda - 3e^2 + 2e^3)}{\frac{\eta}{2}(1 + 2\lambda + 4\mu) + \mu(1 + 2\lambda)} \tag{2.34}$$

$$\rho_2 = \frac{\frac{\eta}{2}(e + \lambda + 2\mu) + \mu(\lambda + 3e^2 - 2e^3)}{\frac{\eta}{2}(1 + 2\lambda + 4\mu) + \mu(1 + 2\lambda)} \tag{2.35}$$

$$\rho_1^* = \frac{e(1 - e)}{2(\frac{\eta}{6} + \mu)} + \frac{(\frac{1}{2} - e)\{(1 + 2\lambda)e(1 - e) - \eta\}}{\frac{\eta}{2}(1 + 2\lambda + 4\mu) + \mu(1 + 2\lambda)} \tag{2.36}$$

$$\rho_2^* = \frac{-e(1-e)}{2\left(\frac{\eta}{6}+\mu\right)} + \frac{\left(\frac{1}{2}-e\right)\{(1+2\lambda)e(1-e)-\eta\}}{\frac{\eta}{2}(1+2\lambda+4\mu)+\mu(1+2\lambda)} \tag{2.37}$$

where ρ_1 and ρ_2 are the distribution coefficients for longitudinal bending moments in girders 1 and 2, respectively; and ρ_1^* and ρ_2^* are the distribution coefficients for longitudinal torsional moments. It is noted that the left girder is referred to as girder 1, the right girder as girder 2, and that the quantity e is positive when the load is to the right hand side of girder 1.

In the above equations, η, λ and μ are the dimensionless characterizing parameters of the bridge defined by:

$$\eta = \frac{12}{\pi^4}\left\{\frac{L}{S}\right\}^3 \frac{LD_y}{EI} \tag{2.38}$$

$$\lambda = \frac{1}{\pi^2}\left\{\frac{L}{S}\right\}^2 \frac{SD_{yx}}{EI} \tag{2.39}$$

$$\mu = \frac{1}{\pi^2}\left\{\frac{L}{S}\right\}^2 \frac{GJ}{EI} \tag{2.40}$$

in which L = bridge span; S = girder spacing; EI = the combined flexural rigidity of one girder and the associated portion of the deck slab; GJ = the combined torsional rigidity of one girder and the associated portion of the deck slab; D_y = the transverse flexural rigidity per unit length of the deck slab; and D_{yx} = the transverse torsional rigidity per unit length of the deck slab. For a deck slab of thickness t, modulus of elasticity E_c and shear modulus G_c, the values of D_y and D_{yx} are given by:

$$D_y = \frac{E_c t^3}{12} \tag{2.41}$$

$$D_{yx} = \frac{G_c t^3}{6} \tag{2.42}$$

It is recalled that for analysis by the semi-continuum method, the applied loading is represented by harmonic series, and that Eqs. (2.38), (2.39), and (2.40) are applicable for only the first term of the loading series. The characterizing parameters for higher terms, which are not utilized for the development of the proposed method, are obtained by replacing π in these equations by $m\pi$, where m is the harmonic number.

A slab-on-girder bridge carrying a single concentrated load is considered. For such a case, the ratio of the moment induced in a girder to the total moment at the transverse section under consideration varies from section to section; furthermore, this ratio of girder moments is not the same as the ratio for girder shears (Jaeger and Bakht 1989). For these reasons, the girder moments and shears cannot be directly derived simply by multiplying the total moments and shears by ρ_1 and ρ_2.

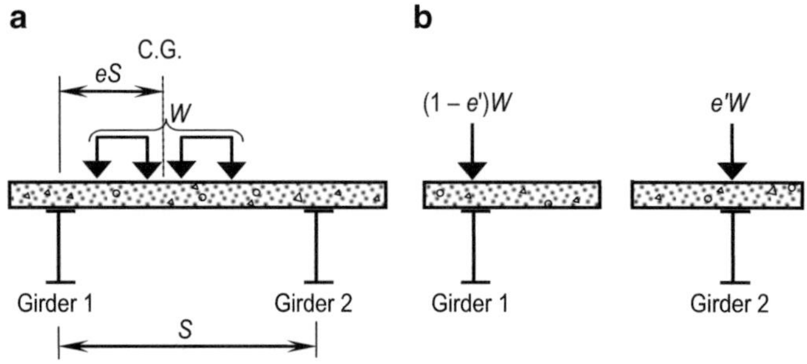

Fig. 2.19 Notation used in conjunction with the proposed method (**a**) four loads between two girders, (**b**) static apportioning of load by considering torsional rigidities

Variation of the transverse distribution patterns of longitudinal responses along the span is caused mainly because some girders are directly loaded while others are not. In a two-girder bridge, most of the applied loads are located transversely between the two girders, as a result of which the situation of only some girders carrying the load directly is eliminated. It can be shown that in a two-girder bridge which carries several concentrated loads, the coefficients ρ_1 and ρ_2, which are strictly true only for the first harmonic, nevertheless represent with very good accuracy the fractions of loads transformed to the girders.

A longitudinal line load P at a distance eS from girder 1 is considered as shown in Fig. 2.18a. In light of the above discussion, it can be appreciated that this load is effectively transferred as $\rho_1 P$ and $\rho_2 P$ on girders 1 and 2 respectively. By denoting $e^* = \rho_2$, it follows that these loads can be written $(1 - e')\,P$ and $e'P$, which are the loads that would be obtained if the external load were located at a distance $e'S$ from girder 1, and were then apportioned statically in the usual manner. Using Eq. (2.34), the equation for e' can be written as:

$$e' = \frac{\frac{\eta}{2}(e + \lambda + 2\mu) + \mu\left(\lambda + 3_e{}^2 - 2_e{}^3\right)}{\frac{\eta}{2}(1 + 2\lambda + 4\mu) + \mu(1 + 2\lambda)} \tag{2.43}$$

In the case of a single concentrated load, eS is the distance of the load measured in the transverse direction from the left girder. When there are two or more concentrated loads on a transverse line, eS becomes the distance of the centre of gravity of the loads from the left girder as illustrated in Fig. 2.19a. The approach of apportioning loads to girders by using e' as defined by Eq. (2.43) can also be used in the case of multiple loads on a transverse line. It is noted that both e and e' are measured positive on the right hand side of the left hand girder, i.e. girder number 1.

The use of the proposed method can be illustrated with the help of the example shown in Fig. 2.19a. As shown in this figure, there are four concentrated loads on a transverse line with a total weight W. The centre of gravity of these loads is a distance eS from girder 1. Using the values of η, λ and μ obtained from Eqs. (2.38), (2.39), and (2.40), respectively, and e, the value of e' is obtained from Eq. (2.43). As shown in Fig. 2.18b, the four loads can be transformed as single concentrated loads

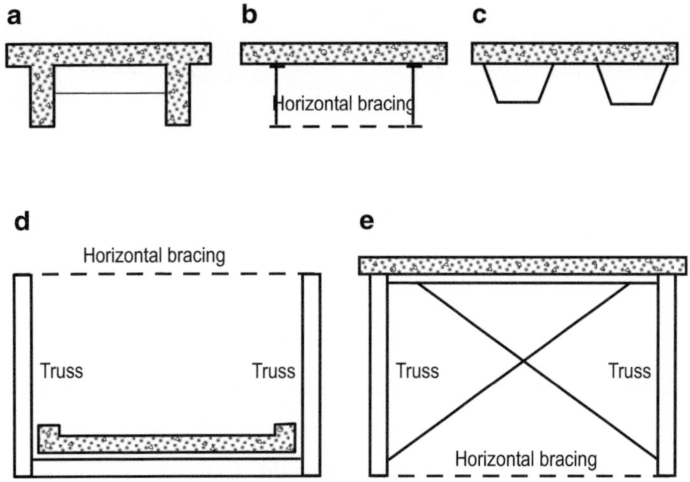

Fig. 2.20 Bridges with two main longitudinal members: (**a**) girder bridge without *horizontal* bracing, (**b**) girder bridge with *horizontal* bracing, (**c**) box girder bridge with two spines, (**d**) through truss bridge, (**e**) deck truss bridge

of weights $(1 - e')W$ and $e'W$ on girders 1 and 2, respectively, both acting on the same section which contains the four applied loads. Thereafter, each girder can be analysed in isolation under the action of the transformed loads.

For the case shown in Fig. 2.18b, equilibrium can be maintained only if the two girders have torsional couples T_1T_2. While it is usual to ignore the effect of these couples in design, they can be derived from the distribution coefficients ρ_1^* and ρ_2^*.

2.5.2 Calculation of Stiffnesses

It will be appreciated that the simplified method proposed above is applicable to a variety of bridges including (a) two-girder bridges without horizontal bracing; (b) two-girder bridges with horizontal bracing; (c) box girder bridges with two spines; (d) through truss bridges; and (e) deck truss bridges. The cross-sections of these bridges are shown in Fig. 2.20a–e, respectively. It is noted that the semi-continuum method can also be applied for the analysis of single cell box girders, as discussed later.

While the proposed method is simple enough to be applied without any difficulty, the calculation of the stiffnesses needed for obtaining the characterizing parameters needs care, and is not always self-evident. The procedures for calculating these stiffnesses for certain kinds of bridges are given below, mainly because not all of them are available in readily-available references.

The longitudinal flexural rigidity *EI* of girder bridges having uniform section along the span can be obtained in the usual manner and needs no explanation. When

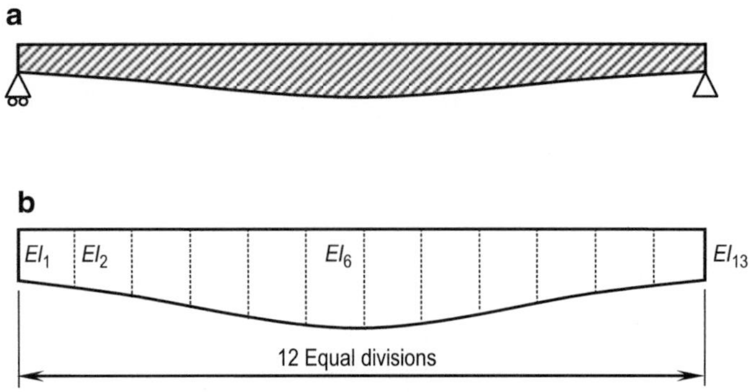

Fig. 2.21 A girder with variable flexural rigidity: (**a**) elevation, (**b**) flexural rigidity

the flexural rigidity of a girder varies along the span, the following expression proposed by Jaeger and Bakht (1989) can be used to obtain the equivalent uniform flexural rigidity EI_e:

$$EI_e = \left(\frac{\pi}{72}\right)\left\{\begin{array}{l} 0.0000EI_1 + 1.0352(EI_2 + EI_{12}) \\ +1.0000(EI_3 + EI_{11}) + 2.8284(EI_4 + EI_{10}) \\ +1.7320(EI_5 + EI_9) + 3.8636(EI_6 + EI_8) \\ +2.0000EI_7 \end{array}\right\} \quad (2.44)$$

where EI_1, $EI_2, \ldots$, EI_{12} are the flexural rigidities of the girder at longitudinal locations identified in Fig. 2.21.

For truss bridges, the equivalent EI can be obtained by seeking equivalence between the maximum truss and equivalent beam deflections under uniformly distributed loads.

In the absence of transverse diaphragms, or transverse floor beams, the transverse flexural rigidity per unit length, D_y is obtained from Eq. (2.41). However, when these transverse members are present, their contribution to D_y may be significant, especially when they are closely spaced. When the deck slab is supported on both longitudinal main members and transverse beams, and the spacing of transverse beams is less than about $0.75S$, D_y can be calculated from:

$$D_y = \frac{EI_t}{S_t} \quad (2.45)$$

where EI_t is the flexural rigidity of a transverse beam and the associated portion of the deck slab, and S_t is the spacing of these beams.

When the deck slab is supported only by transverse beams, as it usually is in the floor systems of truss bridges, Eq. (2.41) can be used for calculating D_y even when S_t is greater than S.

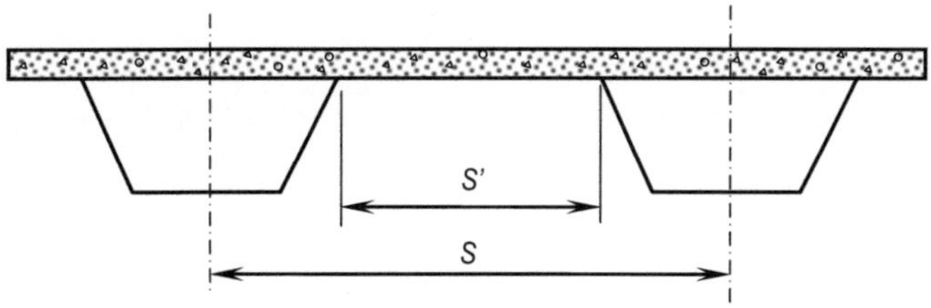

Fig. 2.22 Cross-section of a bridge with two box girders

For twin-cell box girder bridges, the simplified analysis can be performed by idealizing each box girder by a one-dimensional longitudinal beam having the same *EI* and *GJ* as the box girder and the associated portion of the deck slab (taken from centre to centre of the individual cells). However, this will underestimate its ability to transfer loads laterally; to correct for this inconsistency, the value of D_y can be enhanced as follows:

$$D_y = \frac{E_c t^3}{12}\left(\frac{S}{S'}\right)^3 \tag{2.46}$$

where, as shown in Fig. 2.22, *S* is the centre to centre spacing of the two box girders and S' is the clear transverse span of the deck slab.

For a two-girder bridge without horizontal bracing at the bottom flange level, the longitudinal torsional rigidity of a girder and the associated portion of the deck slab is estimated simply as the sum of the torsional rigidities of the girder and the deck slab, so that:

$$GJ = G_g J_g + G_c b \frac{t^3}{6} \tag{2.47}$$

where G_g and J_g are the shear modulus of the girder material and the torsional inertia of the girder, respectively; G_c is the shear modulus of the deck slab material; *b* is half the width of the bridge; and *t* is the slab thickness.

The longitudinal torsional rigidity of a thin walled box girder having a closed section can be obtained from the following equation:

$$GJ = G_c \left\{ \frac{4A^2}{\oint \frac{ds}{n_s t'}} \right\} \tag{2.48}$$

where *A* is the area enclosed by the median line passing through the walls of the closed section; and *t'* is the thickness of the steel girder; $\oint \frac{ds}{t'}$ refers to the contour

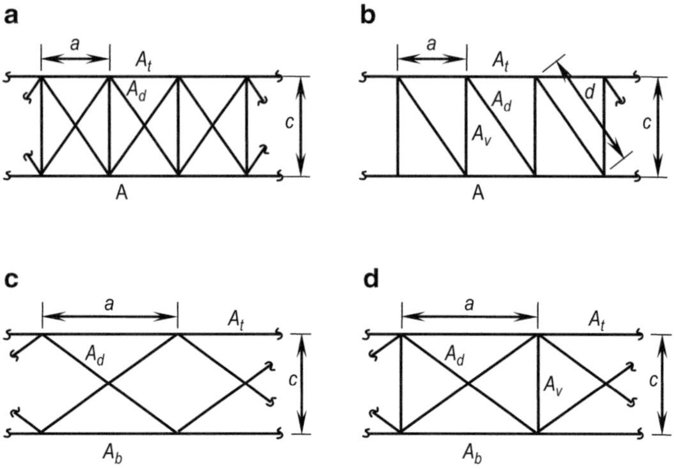

Fig. 2.23 Various frame configurations (**a**) K-type, (**b**) N-type, (**c**) X-type without transverse members, and (**d**) X-type with transverse members

integral along the median line of the reciprocal of the wall thickness; and n_s is the ratio of the shear modulus of the deck slab material and the material of the wall under consideration.

The bottom flanges of steel girder bridges are usually connected by horizontal bracing, which is also referred to as wind bracing. This bracing has the effect of closing the cross-section and thus enhancing considerably the longitudinal torsional rigidity of the bridge. This enhancement of the longitudinal torsional rigidity of the bridge by closing the section also takes place in through-truss bridges and deck-truss bridges. As shown in Fig. 2.20d, the horizontal bracing at the top closes the section in through-truss bridges, whilst in deck truss bridges the section is closed by the horizontal bracing at the bottom, as shown in Fig. 2.20e.

The case is now considered of a closed-section box member in which one or more walls of the member are composed of a framework such as a truss or a system of horizontal bracing. In such a case, as noted by Kollbrunner and Basler (1969), the framework can be idealized as a plate whose thickness t depends upon the configuration of the framework and the cross-sectional areas of the various chord members. For the K-type of framework shown in Fig. 2.23a, the equivalent thickness t is given by:

$$t = \frac{E_s}{G_s} \frac{ac}{\frac{d^3}{A_d} + \frac{a^3}{3}\left\{\frac{1}{A_t} + \frac{1}{A_b}\right\}} \tag{2.49}$$

For the N-type of framework shown in Fig. 2.23b, t is given by:

$$t = \frac{E_s}{G_s} \frac{ac}{\frac{d^3}{A_d} + \frac{c^3}{A_v} + \frac{a^3}{12}\left\{\frac{1}{A_t} + \frac{1}{A_b}\right\}} \tag{2.50}$$

For the X-type of framework shown in Fig. 2.23c, t is given by:

$$t = \frac{E}{G_s} \frac{ac}{\frac{d^3}{2A_d} + \frac{a^3}{12}\left\{\frac{1}{A_t} + \frac{1}{A_b}\right\}} \tag{2.51}$$

When the X-type of framework incorporates members perpendicular to the main members as shown in Fig. 2.23d, the idealized thickness t can be found either by ignoring these transverse members and thus using Eq. (2.50), or by ignoring the diagonal members in compression in which case the system becomes identical to the one shown in Fig. 2.23b. For this latter case t can be obtained from Eq. (2.49).

In the above equations, E_s and G_s are respectively the elastic and shear modulus of the material of the framework; a, c and d are the chord lengths identified in Fig. 2.23; and A_t, A_b A_d and A_v are the cross-sectional areas of the chord members, also identified in Fig. 2.23. For horizontal bracing in girder bridges, A_t and A_b can be taken as the areas of cross-section of the bottom flanges of the relevant girders.

For all the bridges discussed so far, it is conservative to assume that only the deck slab provides the transverse torsional rigidity, so that D_{yx} can always be obtained from Eq. (2.41).

2.5.3 Numerical Example

As an illustration, the proposed simplified method is used to analyse a two-girder bridge under two eccentrically-placed vehicles. Various details of the bridge are given in Fig. 2.24 including the areas of cross-sections of the various members of the horizontal bracing system; other relevant details are given below.

Span, L	= 18.0 m
Girder spacing, S	= 8.75 m
Flexural rigidity, EI, of a girder	= 5.886×10^6 kN•m^2
Modular ratio, m_s	= 10
Floor beam spacing, S_t	= 4.5 m
Flexural rigidity, EI_t, of floor beam	= 1.320×10^6 kN•m^2

The bracing system is the same type as is shown in Fig. 2.23d. The equivalent thickness for this system can be obtained either from Eqs. (2.50) or (2.51), which give $t = 0.42$ mm and 0.36 mm, respectively. Both of these values are conservative estimates of the effective thickness. The smaller thickness would give an even safer side estimate of the load distribution characteristics of the structure. Accordingly, this value was chosen.

Using $t = 0.36$ mm and other relevant properties of the bridge cross-section, the total longitudinal torsional stiffness of the bridge is found to be 3.480×10^6 kN·m^2. By assigning half of this torsional rigidity to each girder, the effective girder GJ becomes 1.74×10^6 kN·m^2.

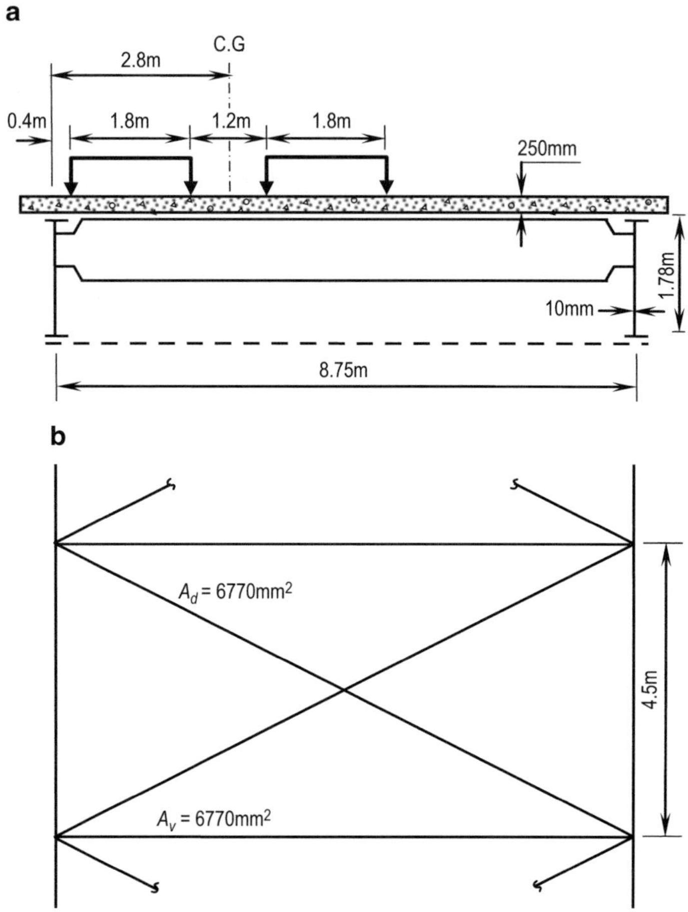

Fig. 2.24 Details of a two-girder bridge: (**a**) cross-section, (**b**) plan

Using Eq. (2.45) $D_y = 386{,}667$ kN·m, and from Eq. (2.42) $D_{yx} = 22{,}100$ kN·m. By using the various stiffnesses and other properties of the bridge, the values of the non-dimensional parameters η, λ and μ are found to be 0.962, 0.014 and 0.127, respectively.

As shown in Fig. 2.24a, the centre of gravity of the four axle loads is 2.8 m from the left girder. Hence $e = 2.8/8.75 = 0.32$. Substituting this value of e and the characterizing parameters calculated above into Eq. (2.43), one obtains $e' = 0.372$.

If the loads were apportioned to the two girders in the usual static manner, girder 1 would have been called upon to sustain 0.68 times the total load. By contrast, load distribution analysis by the proposed method shows that this fraction is 0.628. This reduction represents a 7.6 % reduction in maximum live load effect. Such a reduction in maximum live load effects may not be of great consequence for the design of new bridges, but it can prove to be very useful in the evaluation of the load carrying capacity of existing ones.

References

AASHTO (1989) Standard specifications for highway bridges. American Association of State Highway and Transportation Officials, Washington, DC

AASHTO (1994) Standard specifications for highway bridges. American Association of State Highway and Transportation Officials, Washington, DC

AASHTO (1998) Standard specifications for highway bridges. American Association of State Highway and Transportation Officials, Washington, DC

AASHTO (2010) Standard specifications for highway bridges. American Association of State Highway and Transportation Officials, Washington, DC

Bakht B, Jaeger LG (1985) Bridge analysis simplified. McGraw-Hill, New York

Bakht B, Moses F (1988) Lateral distribution factors for highways bridges. J Struct Eng, ASCE 114(8)

Bares R, Massonnet C (1966) Analysis of beam grids and orthotropic plates by the Guyon-Massonnet-Bares method. Cross Lockbywood and Sons, London

Chan THT, Bakht B, Wong MY (1995) An introduction to simplified methods of bridge analysis for Hong Kong. HKIE Trans 2(1):1–8

CHBDC (2000) Canadian highway bridge design code CAN/CSA-S6-00, 06. CSA International, Toronto

CHBDC (2006) Canadian highway bridge design code CAN/CSA-S6-00, 06. CSA International, Toronto

CSA (1988) Design of highway bridges, CAN/CSA-S6-88. Canadian Standards Association, Toronto

Cusens AR, Pama RP (1975) Bridge deck analysis. Wiley, London

Jaeger LG, Bakht B (1989) Bridge analysis by microcomputer. McGraw-Hill, New York

Kollbrunner CF, Basler K (1969) Torsion in structures. Springer, New York

Krishna J, Jain OP (1977) Plain and reinforced concrete, vol 11. Nem Chand and Bros, Roorkee

Morice PB, Little G (1956) The analysis of right bridge decks subjected to abnormal loading. Cement and Concrete Association, Wexham Springs

Ontario Highway Bridge Design Code (1979) Ministry of transportation of Ontario. Downsview

Ontario Highway Bridge Design Code (1983) Ministry of transportation of Ontario. Downsview

Ontario Highway Bridge Design Code (1992) Ministry of transportation of Ontario. Downsview

Sanders WW Jr, Elleby HA (1970) Distribution of wheel loads on highway bridges. National co-operative highway research program, report no 83, Washington, DC

Chapter 3
Analysis by Computer

3.1 Introduction

As discussed in the introduction to Chap. 2, the term *analysis* is being used in this book for the determination of load effects in the various components of a bridge superstructure subjected to vehicle loads. The methods employed for bridge analysis range in complexity from the overly simplified D-type method of AASHTO (1989) to highly complex finite element methods. The earlier AASHTO simplified method of analysis, because of being too simple, is often excessively conservative. The finite element methods, which require fairly complex computer programs, on the other hand are prone to common errors of idealization and interpretation of results; the former difficulty is discussed with the help of specific examples by Mufti et al. (1994). The large quantity of numerical output associated with finite element analyses also tends to deprive the designer of the physical feel of the behaviour of the structure.

Details of two rigorous methods of analysis, the semi-continuum and orthotropic plate methods, are presented in this chapter. The former method, while retaining the accuracy and versatility of other refined methods, is extremely efficient. It is an added advantage that it enables the designer to retain the physical feel for the structural behaviour. The latter method, which takes account of the size of the applied loads, is very useful for the analysis of slab and voided slab type of bridges.

3.2 The Semi-Continuum Method

Two mathematical idealizations, namely grillage and orthotropic plate, are frequently used for live load analysis of bridge superstructures. The former idealization is a discrete one in which the flexural and torsional stiffnesses of the actual structure in both the longitudinal and transverse directions are concentrated in

B. Bakht, A. Mufti, *Bridges*, DOI 10.1007/978-3-319-17843-1_3

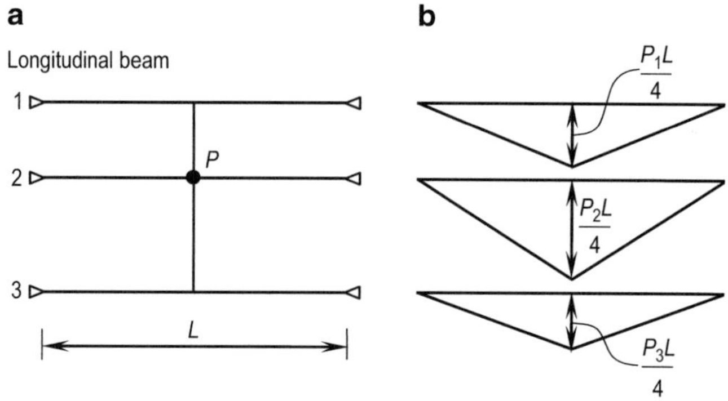

Fig. 3.1 An assembly of three longitudinal and one transverse beams: (**a**) plan, (**b**) bending moment diagram

discrete beams of the grillage (see e.g., Jaeger and Bakht 1982). The latter idealization, on the other hand, is a continuous one in which the various stiffnesses are uniformly distributed in both the longitudinal and transverse directions. Such an idealisation is used in the orthotropic plate method of Cusens and Pama (1975), which is discussed in this chapter along with a computer program.

In a third type of idealization, a slab-on-girder bridge is represented by discrete longitudinal members and a continuous transverse medium. This type of idealization, which is referred to here as the semi-continuum idealization, is a closer representation of a slab-on-girder type of bridge than the other two. Hendry and Jaeger (1955) and Jaeger (1957) used the semi-continuum idealization to analyse bridges with negligible torsional stiffnesses.

A generalized form of the Hendry and Jaeger method, which can now take account of the torsional effects in both the longitudinal and transverse directions, is now available. The basics of the semi-continuum method of analysis are explained in this section by using very simple beam type mathematical models. Engineers who have a liking for the mathematical kind of structural analysis are referred to the text book on the subject by Jaeger and Bakht (1989).

3.2.1 2-D Assembly of Beams

A two-dimensional (2-D) assembly of torsionless beams shown in Fig. 3.1 is considered. As can be seen in this figure, there are three simply-supported parallel longitudinal beams of span *L*, which are interconnected at their mid-spans by a transverse beam. The middle longitudinal beam, i.e. beam 2, carries a concentrated load *P* at its mid-span.

In the absence of the transverse beam, all of the applied load would have been carried by beam 2, whose bending moment diagram will be the familiar triangular one with the maximum value of *PL*/4. This bending moment diagram will be

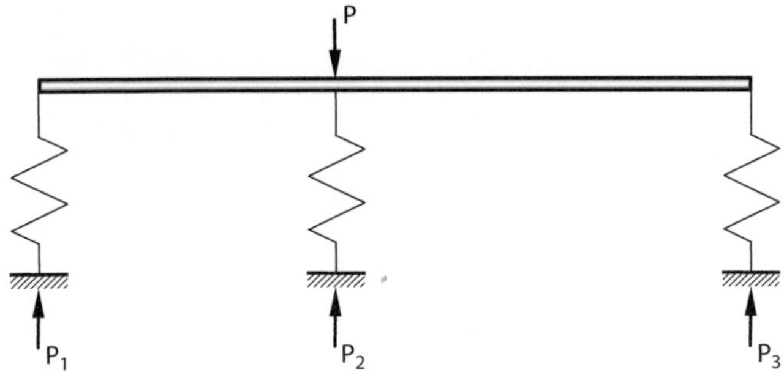

Fig. 3.2 A beam on three springs

referred to as the free bending moment diagram. The presence of the transverse beam will ensure that a portion of the applied load will be shared by the two beams which do not carry the external load directly. How much of the externally applied load is distributed to the outer beams will depend upon the beam rigidities and their spans and spacing. If the flexural rigidity of the transverse beam is a very small fraction of the rigidities of longitudinal beams and the spacing of longitudinal beams are relatively large, then hardly any load will be transferred to the beams not carrying the direct load.

If beams 1 and 3 respectively accept loads P_1 and P_3, then the bending moment diagrams for the three beams will be as shown in Fig. 3.1. It is obvious that the load retained by directly loaded beam is $P_2 = P - P_1 - P_3$. As shown in Fig. 3.1, the bending moment diagrams of all the beams have the same shape, i.e. triangular, as the shape of the *free* bending moment diagram. Similarly, the shear force diaphragms of all the longitudinal beams have the same shape as the *free* shear force diagram.

It can be shown that the 2-D assembly of beams of Fig. 3.1 can be idealized as a one-dimensional (1-D) beam resting on three springs.

The analysis of a beam on springs, shown in Fig. 3.2 can readily give the values of P_1, P_2 and P_3, which are indeed the vertical reactions in the three springs. The ratios P_1/P, P_2/P and P_3/P are called the *distribution coefficients* and are designated as f_1, f_2 and f_3, respectively. It can be appreciated that deflections, bending moments and shear forces along any beam can be determined by multiplying its *free* counterparts by its distribution coefficient.

The analysis of a 2-D assembly of beams by a 1-D beam has been made possible by the fact that the various load effects in all the beams have exactly the same shape as their respective *free* counterparts. Because of this convenience, a distribution coefficient remains applicable to all the responses along the length of a beam.

The 2-D assembly of beams of Fig. 3.1 is considered again with the addition of transverse beams at the quarter-span and three-quarter span sections. This new assembly of beams is shown in Fig. 3.3. Once again, all the beams are assumed to have zero torsional rigidity.

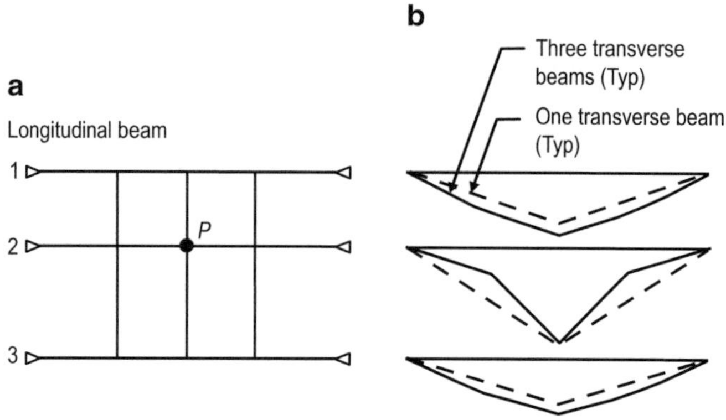

Fig. 3.3 An assembly of three longitudinal and three transverse beams: (**a**) plan, (**b**) bending moment diagrams

Before the introduction of the additional transverse beams, the deflections of the middle beam at the quarter-span and three-quarter span points are larger than the corresponding deflections in the two outer beams. The introduction of transverse beams at these points has the effect of reducing the differential deflection between the middle and outer beams, and in doing so an upward force ΔP_2 is introduced at each of the two points in the middle beam. The outer beams each receive a downward force of ΔP_1 and ΔP_3 at the two points. It is readily verified, as will be demonstrated later, that the introduction of the additional transverse beams also has the effect of slightly changing the previous values of the interactive forces P_1, P_2 and P_3.

As can be seen in Fig. 3.3, the introduction of the two additional transverse beams and the consequent appearance of reactive forces at the quarter-span and three-quarter-span points changes the shapes of the bending-moment diagrams. Instead of being triangular as they were for the case shown in Fig. 3.1, the bending-moment diagrams now become polygonal; for the outer girders the polygon is convex downward, while for the middle girder the polygon is concave upward in each half of the span. Thus, even in the case of only three longitudinal girders and only three transverse beams it is obvious that the fraction of *free* bending moment which is accepted by any one girder varies appreciably along the span.

By losing the convenience of the various beam responses having the same shapes as the respective *free* responses, the concept of the distribution coefficient can no longer be applied in a strict sense. Consequently, the 2-D assembly of the beams cannot be analysed exactly with the help of a 1-D beam idealization. As noted later in Sect. 3.2.3, elimination of one dimension from the idealization has a dramatic effect on the reduction of the computing time.

In a real-life bridge, the deck slab can be represented by an infinite number of conceptual transverse beams. For such a case, it can be shown that, for a single-point load at the mid-span, the bending-moment diagram for the loaded girder has the shape shown in Fig. 3.4a and for a girder not carrying a load directly the shape is as shown in Fig. 3.4b.

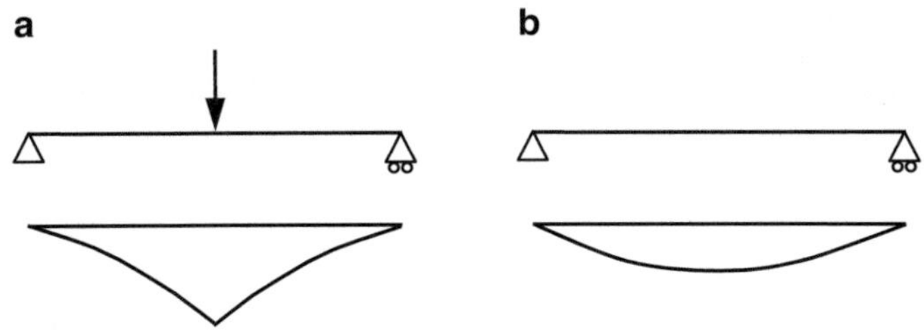

Fig. 3.4 Bending moment diagrams: (**a**) in the directly-loaded beam, and (**b**) in a beam not carrying a load directly

3.2.2 *Harmonic Analysis of Beams*

A point load P on a simply supported beam of span L can be represented as a continuous load of intensity, P_x given by the following expression.

$$P_x = \frac{2P}{L}\left(\sin\frac{\pi c}{L}\sin\frac{\pi x}{L} + \sin\frac{2\pi c}{L}\sin\frac{2\pi x}{L} + \ldots\right) \tag{3.1}$$

where c is the distance of the point load from the left-hand support and x is the distance along the span, also measured from the left-hand support.

Thus according to Eq. (3.1), a point load is equivalent to the sum of an infinite number of distributed loads, each of which corresponds to a term of the series and is a continuous function of x. For example, if $c = L/4$ in Eq. (3.1), the representation of a point load at the quarter-span position is obtained. This series is shown diagrammatically in Fig. 3.5. As shown in the figure, the load corresponding to the first term (or harmonic) has the shape of a half sine wave, and that corresponding to the second has the shape of two half sine waves, and so on. For the particular case shown in Fig. 3.5, the contribution to the P_x series is zero for harmonic numbers that are divisible by 4.

From elementary small deflection beam theory, it is well known that the load intensity P_x, bending moment M_x shear force V_x and slope θ_x of a beam of uniform flexural rigidity EI are related to its deflection w by the following equations.

$$\left.\begin{aligned} P_x &= EI\frac{d^4w}{dx^4} \\ V_x &= -EI\frac{d^3w}{dx^3} \\ M_x &= -EI\frac{d^2w}{dx^2} \\ \theta_x &= \frac{dw}{dx} \end{aligned}\right\} \tag{3.2}$$

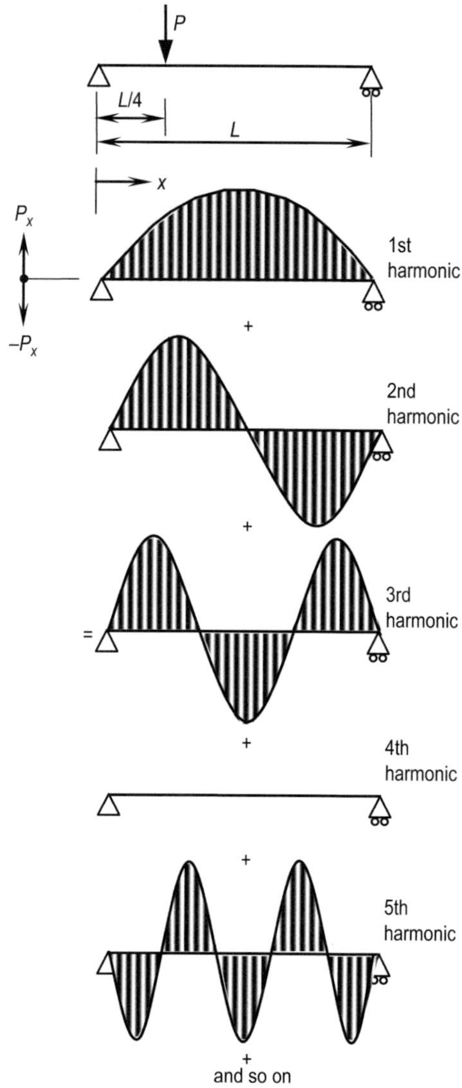

Fig. 3.5 Representation of a point load on a simply supported beam by a harmonic series

Thus values of V_x, M_x, and θ_x, and w can be obtained by successively integrating the right-hand side of Eq. (3.2) with respect to x. The end conditions of a simply supported beam are such that all constants of integration are equal to zero, and the following equations are obtained for a beam of flexural rigidity EI.

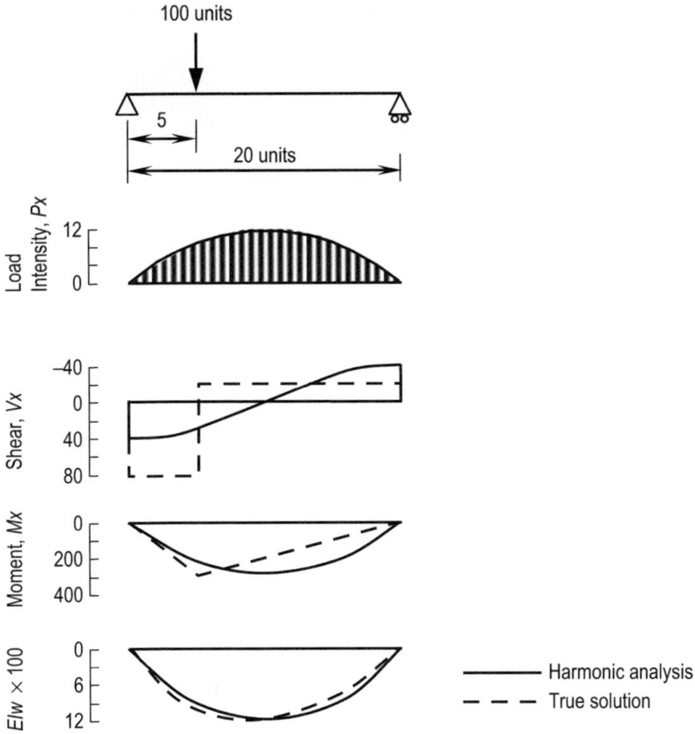

Fig. 3.6 Various responses in a simply supported beam due to first harmonic load

$$
\left.
\begin{aligned}
V_x &= \frac{2P}{\pi} \sum_{n=1}^{n=\infty} \frac{1}{n} \sin \frac{n\pi c}{L} \cos \frac{n\pi x}{L} \\
M_x &= \frac{2PL}{\pi^2} \sum_{n=1}^{n=\infty} \frac{1}{n^2} \sin \frac{n\pi c}{L} \sin \frac{n\pi x}{L} \\
\theta_x &= \frac{2PL^2}{\pi^3 EI} \sum_{n=1}^{n=\infty} \frac{1}{n^3} \sin \frac{n\pi c}{L} \cos \frac{n\pi x}{L} \\
w &= \frac{2PL^3}{\pi^4 EI} \sum_{n=1}^{n=\infty} \frac{1}{n^4} \sin \frac{n\pi c}{L} \sin \frac{n\pi x}{L}
\end{aligned}
\right\} \tag{3.3}
$$

Figure 3.6 shows the values of P_x, V_x, M_x, and EIw for the first harmonic in a beam of 20-unit length and subjected to a load of 100 units at the quarter span. The figure also compares these first harmonic values with the respective true solutions. These true solutions are shown in dashed lines and are, of course, statically determinate. They are referred to herein as *free* solutions; for example, the *free* bending-moment diagram is the familiar triangular one. The *free* bending-moment is designated as M_L, the *free* shear force as V_L, etc. It can be seen in Fig. 3.6 that the first harmonic component of P_x does not resemble the single concentrated load at all, yet the deflections due to this first harmonic are fairly close to the actual (*free*) ones.

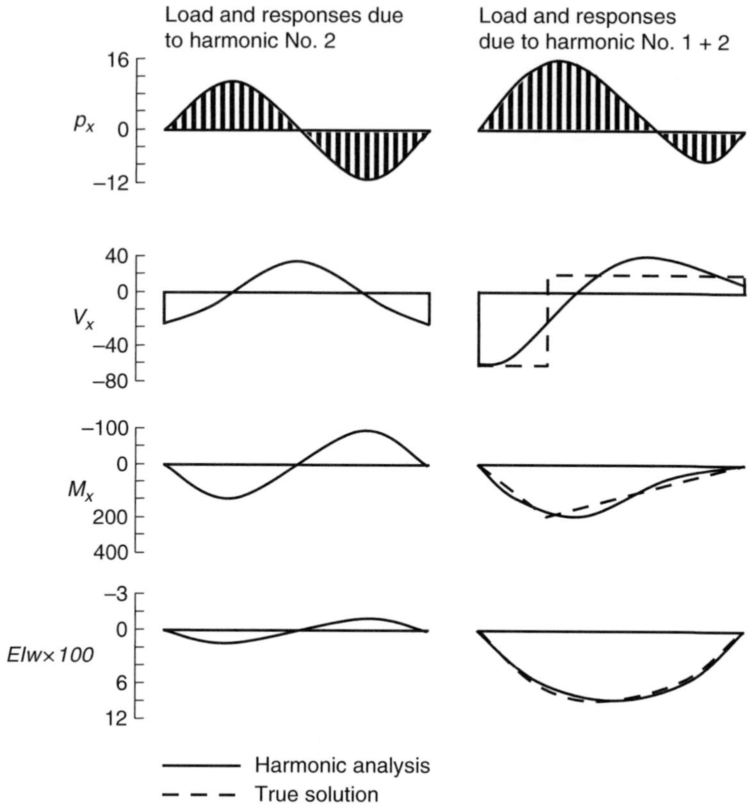

Fig. 3.7 Beam responses due to the second harmonic load and due to the first two harmonic loads added together

By adding the effects of higher harmonics, it can be readily demonstrated that responses corresponding to higher derivatives of deflections converge more slowly to the true, i.e. *free*, solutions. For example, Fig. 3.7 shows P_x, V_x, M_x, and EIw due to the second harmonic loading and also those due to the first two harmonics added together. The deflection due to the latter loading has now converged to almost exactly the free shape, but the bending moment has some way to go and the shear force even further.

It can be shown (Jaeger and Bakht 1989) that a load uniformly distributed over a length $2u$, which is smaller than the span L of a simply-supported beam, as shown in Fig. 3.8 can be represented by the following series expression:

$$P_x = \frac{4P}{\pi}\sum_{n=1}^{n=\infty}\frac{1}{n}\sin\frac{n\pi c}{L}\sin\frac{n\pi u}{L}\sin\frac{n\pi x}{L} \tag{3.4}$$

where P is the total load and c is the distance of the centre of gravity of the load from the same support from which x is measured.

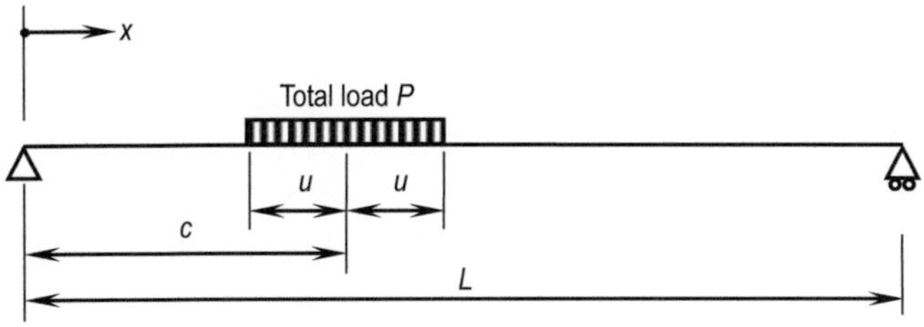

Fig. 3.8 Uniformly distributed load over a part of a simply supported beam

Fig. 3.9 First harmonic load on one girder in a three-girder bridge: (**a**) first harmonic loading on Girder 2; (**b**) loads received by the three girders

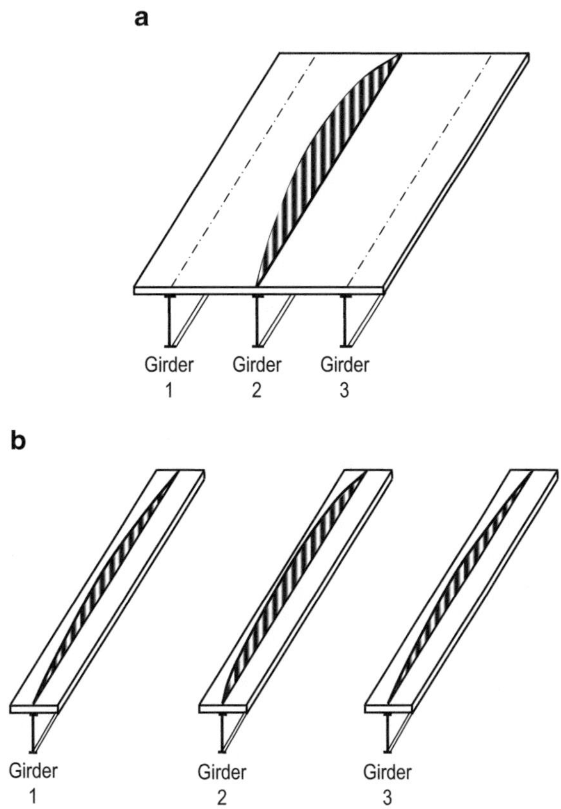

3.2.3 Basis of the Method

The significance of the representation of loads by harmonic series to the semi-continuum method is now discussed. Consider a set of parallel longitudinal beams interconnected by a transverse medium, as shown in Fig. 3.9a. One of these beams is subjected to an externally applied load which corresponds to one term of the kind of harmonic series represented by Eq. (3.1) or (3.4). For the case shown in Fig. 3.9a,

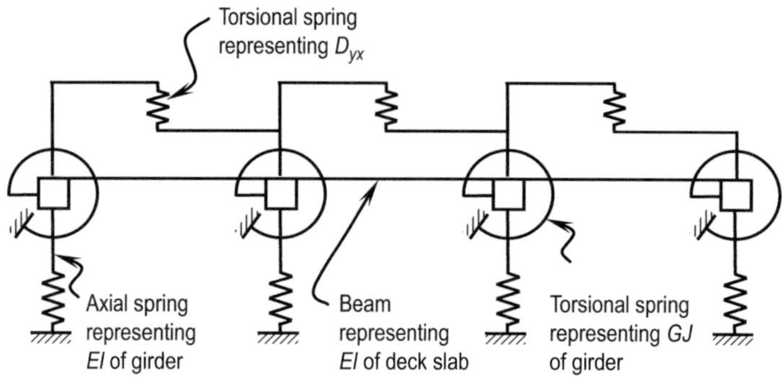

Fig. 3.10 Idealisation of a slab-on-girder bridge by an assembly of springs

the first harmonic loading is considered, which is in the shape of a half sine wave. It can be shown that for the loading represented by any term of a harmonic series, the loads accepted by all the longitudinal beams will have exactly the same shape as that of the applied load, as illustrated in Fig. 3.9b. Consequently, the deflection and rotation profiles of all the longitudinal beams will have the same shapes and will be related to each other by scalar multipliers which are similar to the distribution coefficients f_1, f_2 etc. discussed in Sect. 3.2.1.

The implication of similar deflection patterns of longitudinal beams is that the assembly of beams can be analysed *exactly* by considering only a transverse slice of the structure. In this way, the 2-D semi-continuum idealization of a slab-on-girder bridge can be analysed with the help of a 1-D assembly of a beam and springs; this process is similar in concept to the analysis of the assembly of three longitudinal and one transverse beams discussed in Sect. 3.2.1 with the help of Figs. 3.1 and 3.2.

The simple case of three longitudinal beams with one transverse beam of Fig. 3.1 has been represented by a beam resting on three springs shown in Fig. 3.2. Similarly, by representing the loads through a harmonic series, a transverse slice of a unit width of a number of longitudinal beams inter-connected by a continuum can be represented by the kind of assembly shown in Fig. 3.10. It is noted that these springs account for torsional stiffnesses in both the longitudinal and transverse directions as well as the longitudinal and transverse flexural stiffness. The behaviour of the assembly shown in Fig. 3.10 is characterized by three dimensionless parameters which are defined in the following for structures with equally-spaced girders of equal stiffness.

$$\eta = \frac{12}{(m\pi)^4}\left(\frac{L}{S}\right)^3 \frac{LD_y}{EI} \tag{3.5}$$

$$\lambda = \frac{1}{(m\pi)^2}\left(\frac{L}{S}\right)^2 \frac{SD_{yx}}{EI} \tag{3.6}$$

$$\mu = \frac{1}{(m\pi)^2}\left(\frac{L}{S}\right)^2 \frac{GJ}{EI} \tag{3.7}$$

in which:

$L =$ the span of the longitudinal beams
$S =$ the spacing of longitudinal beams
$EI =$ flexural rigidity of each longitudinal beam
$GJ =$ torsional rigidity of each longitudinal beam
$D_y =$ transverse flexural rigidity of the transverse medium per unit length
$D_{yx} =$ transverse torsional rigidity of the transverse medium per unit length
$m =$ harmonic number under consideration

It can be seen that the values of the characterizing parameters change with every harmonic.

Analysis of only a transverse slice of the idealization, by eliminating one dimension from the analysis of the two-dimensional assembly, reduces the number of unknowns dramatically. Jaeger and Bakht (1989) have shown that for each harmonic loading, the semi-continuum idealization with N longitudinal beams has only $2N$ unknowns, thus requiring the solution of $2N$ simultaneous equations, it being noted that the torsional parameters λ and μ are added together to a new parameter β according to Eq. (3.8). A grillage idealization, on the other hand, requires the solution of an extremely large number of equations.

3.2.3.1 Distribution Coefficients

By solving the bank of $2N$ simultaneous equations discussed above, one can obtain the values of the distribution coefficient $\rho_{q,n}$, where the subscript q refers to the longitudinal beam, or girder, to which the coefficient applies and n to the loaded girder. Expressions for the various distribution coefficients can be obtained readily for bridges having equally-spaced girders with equal stiffness in terms of two characterizing parameters being η, which is defined by Eq. (3.5), and β which is the sum of λ and μ, so that:

$$\beta = \lambda + \mu \tag{3.8}$$

in which λ and μ are defined by Eqs. (3.6) and (3.7), respectively.

Expressions for distribution coefficients in a three-girder bridge, the cross-section of which is shown in Fig 3.11a, are as noted in the following equations in which $\rho_{i,j}$ is the distribution coefficient for girder, i, when the load is on girder j as shown in Eqs. (3.9 and 3.10):

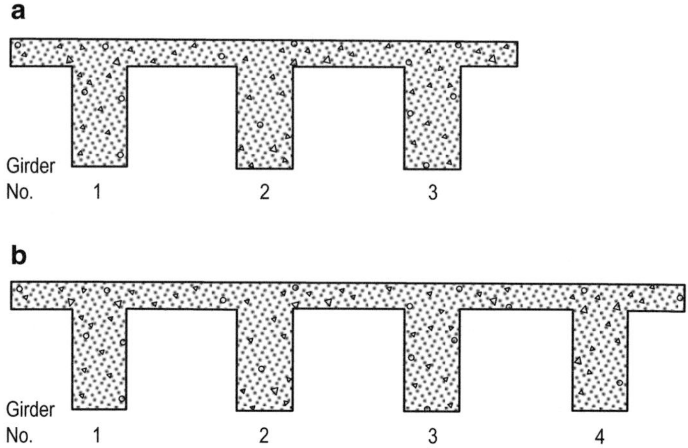

Fig. 3.11 Identification numbers for girders (**a**) in three-girder bridges, (**b**) in four-girder bridges

1. For load on girder 1

$$\left.\begin{aligned} \rho_{1,1} &= \frac{2+(\eta+4\beta)}{4+3(\eta+4\beta)} + \frac{1}{2(1+\beta)} \\ \rho_{2,1} &= \frac{(\eta+4\beta)}{4+3(\eta+4\beta)} \\ \rho_{3,1} &= \frac{2+(\eta+4\beta)}{4+3(\eta+4\beta)} - \frac{1}{2(1+\beta)} \end{aligned}\right\} \quad (3.9)$$

2. For load on girder 2

$$\left.\begin{aligned} \rho_{1,2} &= \frac{(\eta+4\beta)}{4+3(\eta+4\beta)} \\ \rho_{2,2} &= \frac{4+(\eta+4\beta)}{4+3(\eta+4\beta)} \\ \rho_{3,2} &= \rho_{1,2} \end{aligned}\right\} \quad (3.10)$$

Expressions for distribution coefficients in a four-girder bridge, whose cross-section is shown in Fig. 3.11b, are noted in the following:

1. For load on girder 1

$$\begin{aligned} \rho_{1,1} &= \frac{0.5(5+5\beta+0.5\eta)}{\Delta_1} + \frac{1.5(1+3\beta+1.5\eta)}{\Delta_2} \\ \rho_{2,1} &= \frac{0.5(5\beta+0.5\eta)}{\Delta_1} + \frac{1.5(\beta+0.5\eta)}{\Delta_2} \\ \rho_{3,1} &= \frac{0.5(5\beta+0.5\eta)}{\Delta_1} - \frac{1.5(\beta+0.5\eta)}{\Delta_2} \\ \rho_{4,1} &= \frac{0.5(5+5\beta+0.5\eta)}{\Delta_1} - \frac{1.5(1+3\beta+1.5\eta)}{\Delta_2} \end{aligned} \quad (3.11)$$

where

$$\left.\begin{aligned} \Delta_1 &= 5 + 10\beta + \eta \\ \Delta_2 &= 3 + 12\beta + 5\eta + 3\eta\beta + 6\beta^2 \end{aligned}\right\} \tag{3.12}$$

2. For load on girder 2

$$\begin{aligned} \rho_{1,2} &= \frac{0.5(5\beta + 0.5\eta)}{\Delta_1} + \frac{1.5(\beta + 0.5\eta)}{\Delta_2} \\ \rho_{2,2} &= \frac{0.5(5 + 5\beta + 0.5\eta)}{\Delta_1} + \frac{1.5\left(1 + \beta + \frac{\eta}{6}\right)}{\Delta_2} \\ \rho_{3,2} &= \frac{0.5(5 + 5\beta + 0.5\eta)}{\Delta_1} - \frac{1.5\left(1 + \beta + \frac{\eta}{6}\right)}{\Delta_2} \\ \rho_{4,2} &= \frac{0.5(5\beta + 0.5\eta)}{\Delta_1} - \frac{1.5(\beta + 0.5\eta)}{\Delta_2} \end{aligned} \tag{3.13}$$

where Δ_1 and Δ_2 are as given by Eq. (3.12).

As shown by Jaeger and Bakht (1989), the expressions for the coefficients given above can be used to analyse bridges rigorously even by manual calculations.

3.2.3.2 Convergence of Results

It is well known that deflections in beams due to loads represented by harmonic series converge faster than moments, which in turn converge faster than shears. As shown by Jaeger and Bakht (1985), as many as 20 terms of the harmonic series may have to be considered to obtain shears with a reasonable degree of accuracy in a beam subjected to several point loads.

From the slow convergence of results in harmonic analysis of beams, it has often been wrongly concluded by some engineers that the results obtained by the semi-continuum method are also slow to converge. By the technique shown below, the convergence of results by the semi-continuum method can be hastened to such an extent that up to 99 % convergence of even of shears can be obtained by considering only the first five harmonics.

The case of the three-girder bridge with negligible torsional rigidities in both the longitudinal and transverse directions is used conveniently to demonstrate this technique. For a torsionless bridge $\beta = 0.0$, and a typical value of 3.0 is assumed for η for the first harmonic. From Eq. (3.5), the values for η for the second and third harmonics are found to be 0.187 and 0.037 respectively. For the first harmonic, Eq. (3.10) gives $\rho_{1,2}$ and $\rho_{2,2}$ to be 0.23 and 0.54 respectively, which shows that 54 % of the loading represented by the first harmonic is retained by the loaded girder and the rest is distributed to the other two girders. For the second harmonic, i.e. for η of 0.187, $\rho_{1,2}$ and $\rho_{2,2}$ are found to be 0.04 and 0.92 respectively. This

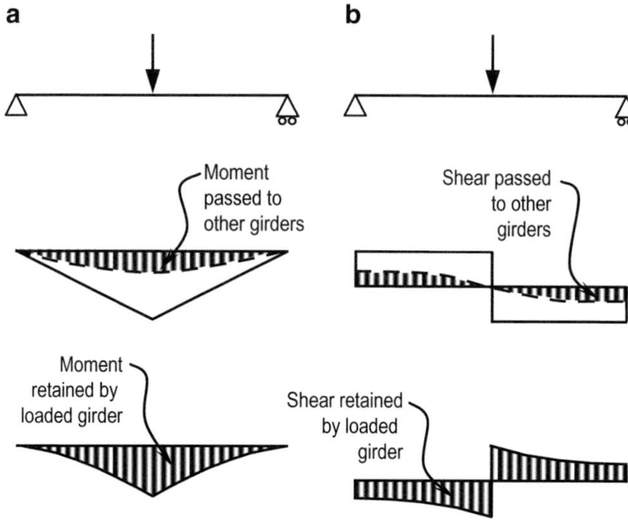

Fig. 3.12 Illustration of the technique of achieving quick convergence: (**a**) bending moment diagram, (**b**) shear force diagram

shows that 92 % of the loading represented by the second harmonic is retained by the loaded girder. Similarly, it can be shown that 98 % of the load represented by the third harmonic is retained by the loaded girder. Extensive analyses have confirmed that in most practical cases, virtually all effects of loading due to fourth and higher harmonics are retained by the externally-loaded girder.

By taking advantage of this property of harmonic loads, the convergence of results can be hastened considerably by subtracting from *free* response diagrams those load effects, which are distributed to other girders. Figure 3.12a illustrates the use of this technique for obtaining moments in a loaded girder of a multi-girder bridge; it shows the free moment diagram of the girder carrying a point load at its mid-span, and the diagram of first harmonic moments distributed to the other girders. The net bending moments retained by the loaded girder are then obtained as the difference between the two diagrams. The diagram of the net bending moments, as shown in Fig. 3.12a, is instructive; its shape is no longer triangular as it is for the free moment diagram. From the differences in the shape of the two bending moment diagrams, it can be readily seen that the ratio of moment retained by a girder to the free moment, is not constant along the span. It is noted that the simplified methods of analysis discussed in Chap. 2, are based on the assumption that this ratio, which is usually referred to as the distribution factor, has the same value along the span.

The process of obtaining the shear force diagram in the loaded girder is similarly illustrated in Fig. 3.12b. Considering only the first harmonic, the shape of the shears distributed to the other girders is that of a cosine curve. Shears retained by the loaded girder are obtained by subtracting from the free shear diagram those shears which are distributed to other girders. It is interesting to note in the diagram of the retained shears, shown in Fig. 3.12b, that in the immediate vicinity of the load, virtually no shear is distributed to the other girders; however, the portion of shears distributed to other girders increases as the reference point moves away from the load. Unlike shears, moments are distributed well in the vicinity of the load, indicating that different responses are distributed in differently in different girders or beams.

3.2.4 Structures with Intermediate Supports

Structures with randomly spaced intermediate supports cannot be analysed directly by the basic semi-continuum method. The problem can, however, be tackled by using the well-known force method which requires the following steps of calculation (Bakht and Jaeger 1986).

(a) Remove all intermediate supports, and by treating the structure as simply supported at the two ends, find deflections at the intermediate support locations due to the applied loading by the basic semi-continuum method.
(b) Again, treating the structure as simply supported at its two ends, find the forces at each of the intermediate support locations which would bring the structure at these locations back to their original positions.
(c) The structure with intermediate supports can now be analysed by the basic semi-continuum method, as a simply supported structure which is subjected to downward applied loading and, usually, upward reactions of the intermediate supports calculated in step (b).

The procedure discussed above is applicable to only those bridges in which the longitudinal flexural and torsional rigidities are constant along the length of the bridge. Many continuous span slab-on-girder bridges, however, have variable girder depths and hence variable longitudinal flexural rigidities. For such cases, the following procedure is proposed.

(a) Find the equivalent flexural rigidity EI_e of the girders by using Eq. (2.44), the notation for which is illustrated in Fig. 2.21 in Chap. 2. It is noted that the entire length of a girder between the two simple supports is used to calculate EI_e.
(b) Analyse the continuous bridge with equivalent girders by the semi-continuum method for bridges with intermediate supports.
(c) Analyse the continuous bridge as a beam of varying moment of inertia representing the actual bridge and also as a beam of constant equivalent moment of inertia. Denote the ratio of the responses in the beams of the former and the latter respectively as F. For a given position along the beam, the value of F for a certain response, namely, moment, shear, or deflection, is given by:

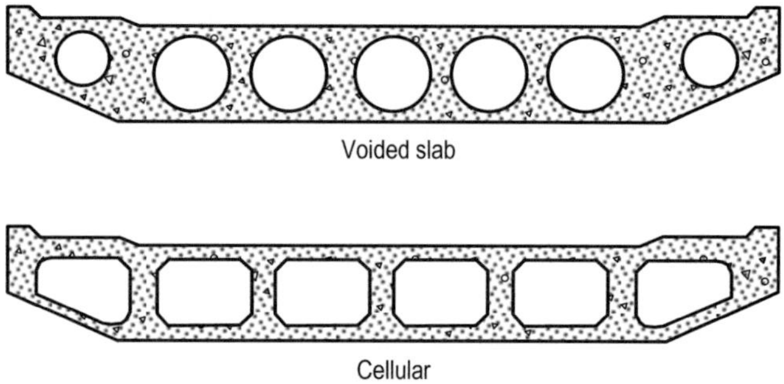

Fig. 3.13 Cross-sections of bridges which are weak in transverse shear

$$F = \frac{R_v}{R_e} \tag{3.14}$$

where R_v and R_e are respectively the responses in beams of varying and equivalent constant flexural rigidities.

(d) Multiply the responses obtained by the semi-continuum method in (b) above by the appropriate value of F to obtain the final values of responses.

3.2.5 *Shear-weak Grillages*

In the load distribution analysis of bridges, it is usual to assume that the deflections of the various bridge components arise from bending effects only, and that the deflections due to shear deformations are negligible. Accordingly, and quite appropriately, most methods of bridge analysis do not take account of shear deformations. There is one category of bridge, however, for which the shear deformations of certain components can be significant in affecting the bridge behaviour under vehicle loads; this category includes concrete voided slab bridges and cellular bridges. Typical cross-sections of these bridges are shown in Fig. 3.13.

The semi-continuum method described in the text book by Jaeger and Bakht (1989) is based on the assumption that the shear deflections are negligible. In a subsequent work, Jaeger and Bakht (1990) have extended the scope of the semi-continuum method to include shear deformations in the transverse medium. The inclusion of shear-weak transverse medium is based on the assumption that the shear deflections of the transverse continuum, whilst influencing girder deflections, do not change the rotations of the girders. These shear deflections are thus of the nature shown in Fig. 3.14. At a cursory glance this assumption may seen unrealistic, but it is justifiable by close logical reasoning and is further borne out by the fact that the method gives results that are virtually identical with those of the grillage method, whose accuracy is already well established.

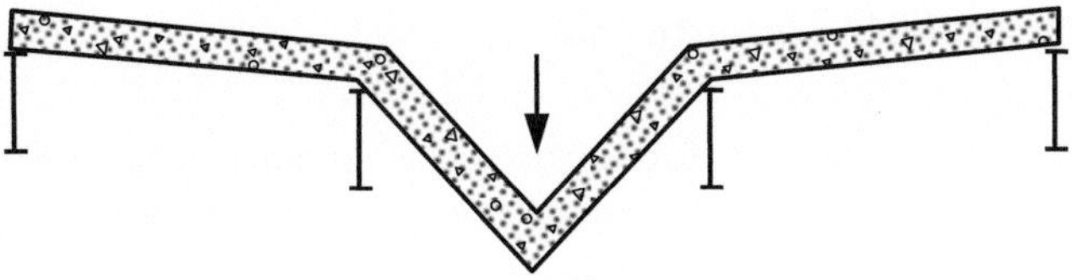

Fig. 3.14 Assumed shear deflections of the deck slab

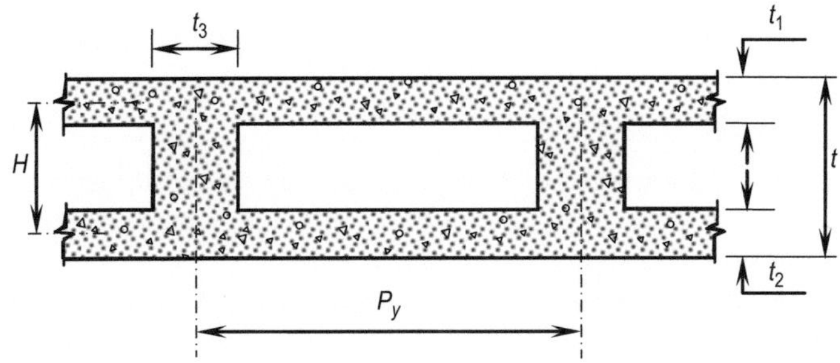

Fig. 3.15 Notation for use in conjunction with Eq. (3.15)

The inclusion of the finite shear rigidity, S_y, of the transverse medium in the semi-continuum method results in changes to the $2N$ simultaneous equations discussed in Sect. 3.2.3; these changes are relatively minor and do not affect the efficiency of the method. For cellular structures, S_y can be calculated from the following expression:

$$S_y = \frac{E_c\left(t_1^3 + t_2^3\right)}{P_y^2}\left[\frac{t_3^3 P_y}{t_3^3 P_y + H\left(t_1^3 + t_2^3\right)}\right] \tag{3.15}$$

where E_c is the modulus of elasticity of concrete and the remainder of the notation is as illustrated in Fig. 3.15.

Jaeger and Bakht (1990) have also shown that the shear deformations of longitudinal beams can be accounted for by replacing their actual flexural rigidity EI by an equivalent flexural rigidity EI_E which is given by:

$$EI_E = \frac{EI}{\left\{1 + \frac{(m\pi)^2 EI}{L^2 GA}\right\}} \tag{3.16}$$

in which m is the harmonic number, L is the span, G is the shear modulus and A is the shear area of the cross-section of the longitudinal beam.

3.2.6 *Intermediate Diaphragms*

Jaeger and Bakht (1993) have formulated a technique which enables the semi-continuum method to take account of the transverse diaphragms. This technique makes use of the familiar force method, and is based upon the determination, of the unknown reactive forces between the girders and the diaphragms through a bank of simultaneous equations. The main, and new, consideration in the formulation of the equations is that of compatibility of deflections between the basic semi-continuum structure and the diaphragms.

It has been shown by Jaeger and Bakht (1993) that a bridge with N longitudinal beams and M transverse diaphragm can be handled by the semi-continuum method through the following steps of calculations.

(a) Remove all the transverse diaphragms and analyse the bridge under applied loading by the basic semi-continuum method to obtain deflections at the intersections of all the girders and diaphragms.
(b) Analyse the bridge without diaphragms again, this time to obtain the deflection coefficients at each intersection due to loads applied successively at each intersection of girders and diaphragms.
(c) Using the flexural stiffness of the diaphragm, construct and solve the flexibility matrix defined by Jaeger and Bakht (1993) to obtain the interactive forces at the $M \times N$ intersections.
(d) Analyse the bridge without the diaphragms under the interactive forces obtained in Step (c).
(e) Superimpose the results of analysis in Step (a) with those obtained in Step (d); the results thus obtained will correspond to the bridge with diaphragms.

As shown by Jaeger et al. (1998), and discussed in Sect. 3.3, the computer program incorporating the semi-continuum method can handle transverse diaphragms, idealised as torsion-less beams of uniform flexural stiffness. Mufti et al. (1998) have given methods for the calculation of the flexural rigidity of various kinds of diaphragms.

3.3 Computer Program Secan

The basic semi-continuum methods and its modifications are applied most effectively through a computer program called SECAN. This section provides some details of the program, which can be downloaded from http://extras.springer.com.

3.3.1 Installation

SECAN, which can be downloaded from http://extras.springer.com, is written in FORTRAN90 to run on personal computers having a Vista operating system and math co-processor. The root directory of the disk contains Secanin.exe and Secan.exe. To install SECAN on a computer, first make a directory for SECAN. Then copy the files from the website into the SECAN directory.

3.3.2 Input Data

The program SECANIN is used for the preparation of the input data file. It can be executed in Windows. The user will be prompted to enter the values of different parameters required by SECAN. The SECANIN program creates an ASCII file called "secan.dat", which can be used directly by SECAN. This file can also be created, or modified, by using a text editor.

The notation for data input to SECAN is illustrated with the help of an example presented in Fig. 3.16, which shows a bridge with four girders having four intermediate supports and four transverse diaphragms, both placed randomly. The responses of the example bridge are investigated at one reference section under a truck loading which comprises two longitudinal lines of loads with four wheels in each longitudinal line. Figure 3.16 also shows the scheme for girder Nos. The interactive program SECANIN, which prompts the user for information, produces the data shown in Table 3.1. All data must be in a consistent set of units.

3.3.3 Example of Use

The Lord's bridge, located in a small municipality in Ontario, Canada, is a single-span, single-lane bridge, comprising a laminated timber decking supported by eight rolled steel girders. The elevation and cross-section of the bridge are shown in Fig. 3.17 along with the relevant details. A load test on the Lord's Bridge is described by Bakht and Mufti (1991). It was found in this test that uncertainties relating to the determination of the effective span length and the degree of composite action of the girders with the timber decking, prevented exact analytical prediction of the absolute values of girder deflections. As shown in Fig. 3.18, the measured values of mid-span girder deflections are bound by the results from SECAN obtained for the extreme conditions of span length and composite action.

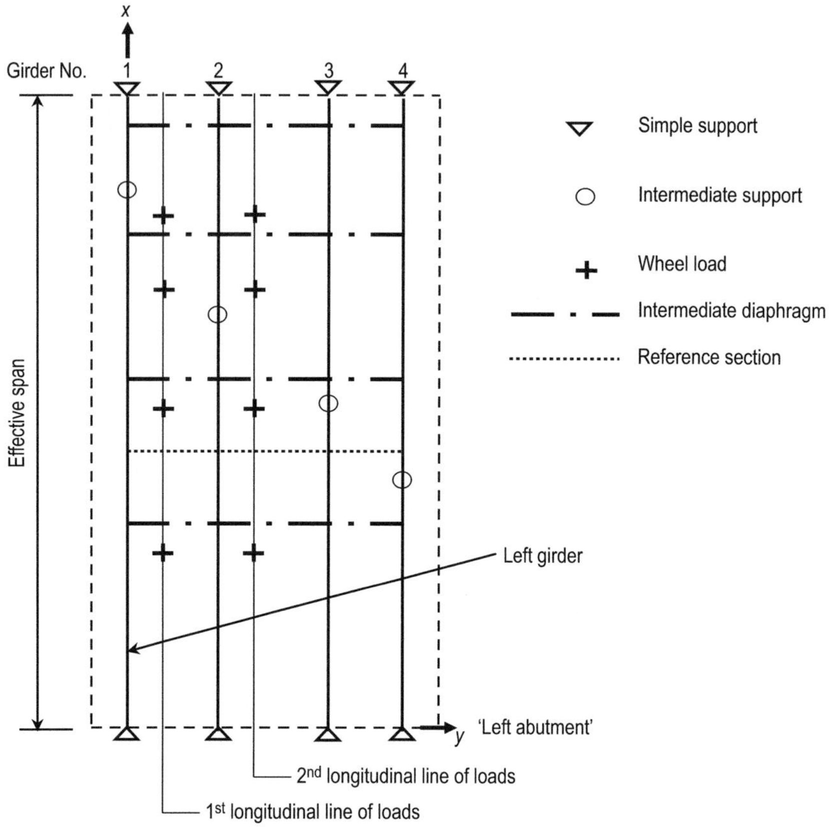

Fig. 3.16 Notation for data input to SECAN

It is noted that the discrete girder deflections in Fig. 3.18 and distribution factors in Fig. 3.19, are joined by continuous curves merely to facilitate visual interpretation.

The comparisons given in Fig. 3.18 do not provide a convenient means of ascertaining the validity of SECAN. It can be provided more conveniently by the comparison of distribution factors (DF) which are the non-dimensional ratios of the actual and average girder responses. The transverse distributions of DF for measured girder deflections are compared in Fig. 3.19 for both eccentric and central truck loads with those obtained by SECAN.

It can be seen in Fig. 3.19 that the distribution factors corresponding to SECAN compare very well with those obtained from the test results. These comparisons appear even more convincing in light of the fact that the corresponding experimental values of DF for the two eccentric load cases, whose transverse positions on the bridge are mirror images of each other, are not exactly the same.

Table 3.1 Details of data input to SECAN

Input ref. no.	Input parameter	Notes
1	No. of bridges	The program can handle a maximum of 15 bridges at a time
2	Title	
3	No. of harmonics	The current version of the program can deal with a maximum of 40 harmonics
4	No. of girders (N_g)	A maximum of 30 girders can be handled by the program
5	Span length	In a bridge with intermediate supports, the span length is the distance between the two outer supports (see Fig. 3.16)
6	*E* value of girder material	
7	*G* value of girder material	
8	No. of diaphragms (N_d)	A maximum of ten diaphragms can be handled by the program
9	No. of intermediate supports (N_s)	
10	Girder spacing	($N_g - 1$) values are required. If girder spacings are equal, SECANIN requires only one value to calculate the rest
11	Moments of inertia of girders	N_g values are required. If the moments of inertia are equal, SECANIN requires only one value to calculate the rest
12	Torsional inertias of girders	N_g values are required. If the torsional inertias are equal, SECANIN requires only one value to calculate the rest
13	Slab thickness	
14	*E* value for deck slab material	
15	*G* value for deck slab material	
16	Equivalent shear area	The equivalent shear are is that of the transverse medium. Zero indicates that transverse shear deformations are to be ignored
17	*E* value of diaphragm material	Not required if $N_d = 0$
18	Spacing of intermediate diaphragm/s from left support	Distances of N_d diaphragm from left support, if $N_d \geq 1$
19	Moments of inertia of diaphragm/s	Moments of inertia of N_d diaphragm from left support, if $N_d \geq 1$
20	Prescribed settlements at N_s intermediate supports	N_s values, if $N_s \geq 1$. Intermediate support should be assigned sequential numbers, and the numbering scheme should be followed for all data for intermediate supports
21	Flexibilities of N_s intermediate supports	N_s values, if $N_s \geq 1$
22	Girder Nos. under which each successive intermediate support is located	N_s girder Nos. are required. Each intermediate support must lie under a girder.

(continued)

Table 3.1 (continued)

Input ref. no.	Input parameter	Notes
		There can be more than one intermediate support under a girder
23	Distances of N_s intermediate supports from the left abutment	The 'left abutment' is at $x = 0$ This input is not required if $N_s = 0$
24	No. of load cases (N_{lc})	The program can handle a maximum of five load cases
25	No. of loads in one longitudinal line (N_l)	The longitudinal line containing concentrated wheel loads is defined in Fig. 3.16. The program can handle a maximum of 20 loads in one longitudinal line
26	Weights of N_l loads	The weights are numbered in ascending order of their values of x
27	Distances of N_l loads in one longitudinal line from the left abutment	These are x values of the centres of wheels
28	Number of lines of loads (N_{ll})	All lines of loads must be similar. The program can handle a maximum of 20 lines of loads
29	Number of transverse reference sections (N_{ref})	The program can handle a maximum of 20 reference sections
30	N_{ref} values of the distances of the reference sections from the left abutment	These are the x values of the reference sections
Data for (25) through (30) has to be repeated for additional ($N_{lc} - 1$) load cases.		

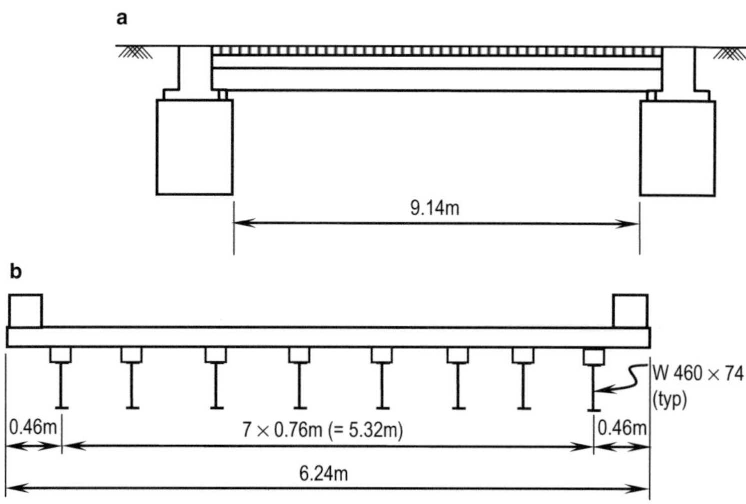

Fig. 3.17 Details of the Lord's Bridge: (**a**) elevation, (**b**) cross-section

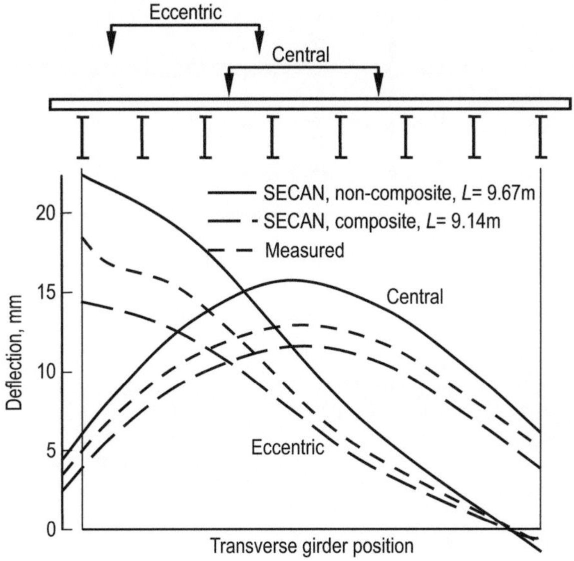

Fig. 3.18 Mid-span deflections

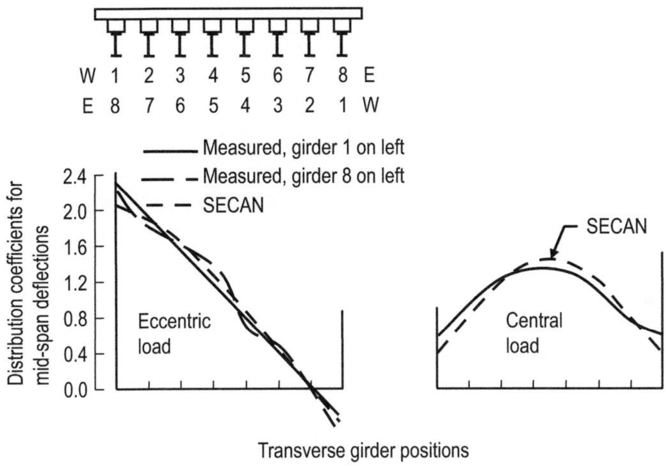

Fig. 3.19 Distribution coefficients for mid-span deflections

3.3.4 Comparison with Grillage Analysis

The versatile grillage method of analysis has been used for more than four decades for the analysis of bridge superstructures (e.g., Sawko 1968; Jaeger and Bakht 1982; Hambley 1994), and is generally regarded as a highly reliable method with respect to the accuracy of its results. The highly-efficient semi-continuum method is not as well-known as the grillage method. To use the semi-continuum method with confidence, engineers often compare its results with those obtained by the grillage

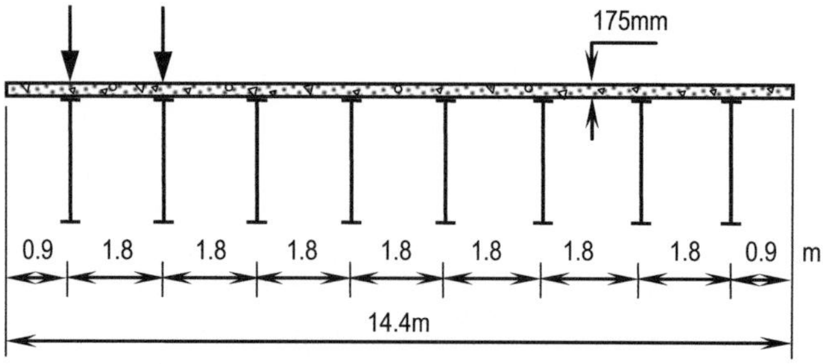

Fig. 3.20 Cross-section of a four-lane bridge

Fig. 3.21 Longitudinal position of loads in one line of wheels

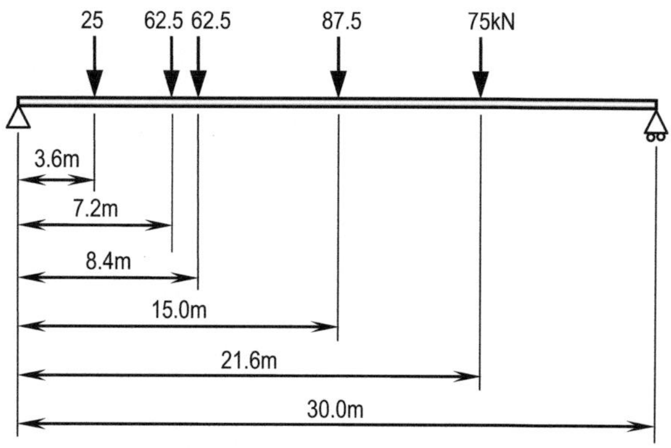

method. In the case of discrepancy between the two set of results, it is usually assumed that the semi-continuum method is in error. With the help of a specific example, it is demonstrated in the following that the semi-continuum method is more accurate than the grillage method and is less prone to errors of idealization.

The example is that of a composite four-lane slab-on-girder bridge with a span of 30 m. As shown in Fig. 3.20, the bridge has a 175 mm thick concrete deck slab supported by 8 steel girders which are spaced at 1.8 m. The various relevant properties of the bridge are noted in the following:

Modulus of elasticity of girder material = 200×10^6 kN/m^2
Shear modulus of girder material = 100×10^6 kN/m^2
Moment of inertia of composite girder = 0.04 m^4
Torsional inertia of composite girder = 0.0002 m^4
Modulus of elasticity of deck slab material = 20×10^6 kN/m^2
Shear modulus of deck slab material = 10×10^6 kN/m^2

The bridge described above is subjected to a five-axle truck having two lines of wheels at a centre-to-centre transverse spacing of 1.8 m. The load is placed on the bridge in such a way that the two lines of wheels are exactly above the two left-hand outer girders, as also shown in Fig. 3.20. The longitudinal positions of the loads, and their magnitudes, on one line of wheels are shown in Fig. 3.21.

Table 3.2 Girder moments at mid-span

Details of analyses	Moments in kN · m in girder no.							
	1	2	3	4	5	6	7	8
SECAN 1 harmonic	1509	1001	480	139	−14	−51	−40	−17
SECAN 3 harmonics	1508	1004	477	140	−14	−51	−40	−17
SECAN 5 harmonics	1508	1005	476	140	−14	−51	−40	−17
Grillage, 7 transverse beams	1526	1005	478	134	−21	−57	−42	−16
Grillage, 15 transverse beams	1519	1007	476	136	−21	−54	−42	−16
Grillage, 31 transverse beams	1518	1007	475	137	−17	−53	−41	−16

The bridge was analysed by SECAN as well as by a standard grillage program. For each analysis, the idealised structure comprised eight longitudinal beams. In the semi-continuum idealization, the deck slab is represented by a continuous medium, which is equivalent to an infinite number of transverse beams. The grillage analysis was, however, conducted by representing the deck slab with only 7, 15, and 31 transverse beams, respectively. The SECAN analysis was done by considering 1, 3, and 5 harmonics, respectively. Girder moments at the mid-span of the bridge obtained by the various analyses are noted in Table 3.2. From the results presented in this table, it can be seen that the SECAN results are hardly affected by the number of harmonics considered in the analysis; this observations attests to the rapid convergence of the results. Table 3.2 also shows that the maximum grillage moment, which occurs in an outer girder, corresponding to only 7 transverse beams is only 1.12 % larger than the corresponding SECAN moment; the gap between moments by the grillage and semi-continuum methods narrows down to 0.66 and 0.59 % when the number of transverse of transverse beams in the grillage idealization is increased to 15 and 31, respectively.

It is obvious that a grillage idealization with an extremely large number of transverse beams should lead to virtually the same results as those given by the semi-continuum method. The example of the four-lane bridge subjected to an eccentric vehicle was chosen for the above comparison because the transverse load distribution characteristics of this bridge are expected to be highly non-uniform, so that any errors in analysis would be intensified. The fact that the comparison between the two methods of analysis is excellent for even such a difficult case validates yet again the accuracy of the semi-continuum method of analysis.

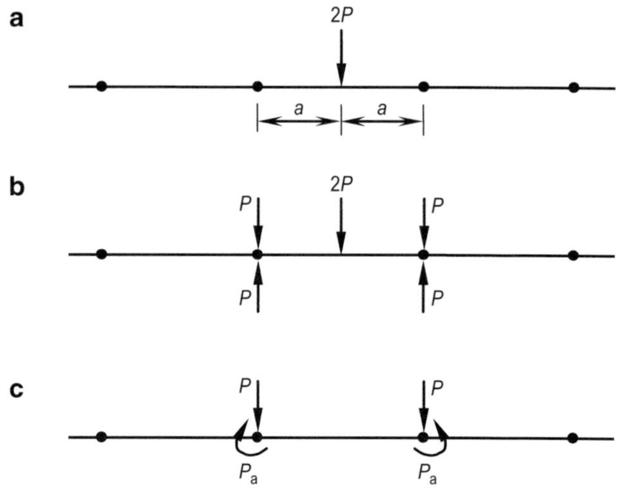

Fig. 3.22 Apportioning of loads to nodes: (**a**) actual load position; (**b**) additional self equilibrating vertical loads on nodes; (**c**) nodal loads and moments

3.3.5 Idealization of Loads

The good comparison between the results of grillage and semi-continuum methods in the example noted above was made possible by placing the loads directly on the longitudinal beams. As explained in the following, the idealization of loads usually employed in grillage analysis is inaccurate because of which its results can be in error.

When the position of a load does not coincide with a grillage node as shown in Fig. 3.22a, the equivalent nodal loads should ideally be obtained by a scheme which ensures that the statics of the transformation are satisfied; such a scheme is illustrated with the help of Fig. 3.22b, c. It can be appreciated that the static apportioning of the loads incorporates not only nodal loads but also nodal moments. It is common, however, to ignore the latter. Jaeger and Bakht (1982) have observed that the neglect of the nodal moments in the transverse direction can result in significant errors of load idealization.

A slab-on-girder bridge, having the cross-section as shown in Fig. 3.23a and a simply supported span of 15 m, was analysed by both the grillage and the semi-continuum methods for live loads, the transverse positions of which are also shown in Fig. 3.23a. For the grillage analyses, the loads were apportioned to the longitudinal beams without taking account of the nodal transverse moments discussed above. The semi-continuum analysis incorporated in SECAN rigorously takes account of loads applied between the longitudinal beams, thus making it unnecessary to apportion the loads to locations of longitudinal beams. It was hardly surprising that for the example under discussion, there was a significant difference between the results of the grillage and semi-continuum methods. The transformed loads employed in the grillage analysis are illustrated schematically in Fig. 3.23b. It is interesting to note that an analysis by SECAN for the loads shown in this figure gave practically the same girder moments as those obtained by the grillage analysis; the girder moments for the loads shown in Fig. 3.23a were, however, significantly

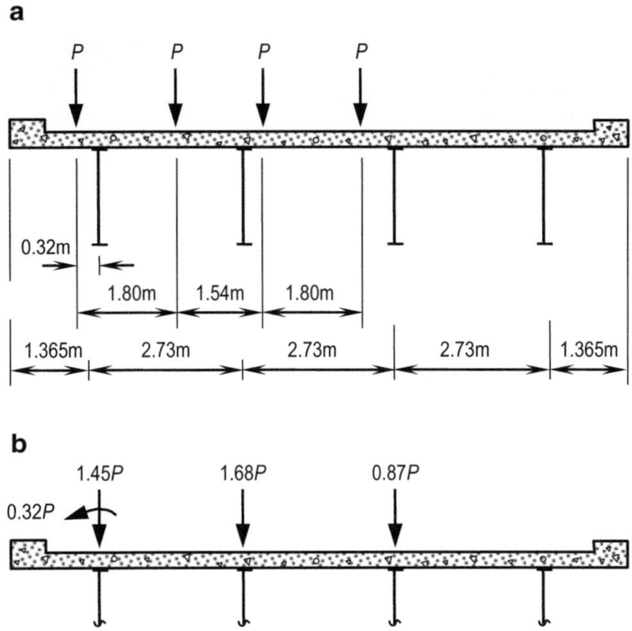

Fig. 3.23 Transverse position of loads: (**a**) actual position of loads; (**b**) 'equivalent' loads on longitudinal beams employed in grillage analysis

different than those for the loads shown in Fig. 3.23b. This example illustrates that the likely errors in the idealization of loads to grillage nodes are avoided in analysis by SECAN.

The comparisons given in Fig. 3.19 and in Table 3.2, confirm that SECAN, and hence, the semi-continuum method of analysis, can predict reliably the load distribution characteristics of bridges.

3.3.6 Example of Data Output by SECAN

The data file could be prepared directly according to the scheme of Table 3.3. Alternatively, the interactive program SECANIN could be used to create the data file. An example of the input data file for the bridge of Figs. 3.20 and 3.21 is reproduced in Fig. 3.24. After the data file either has been created by SECANIN, or a revised existing SECAN.DAT file has been saved, the icon for the SECAN application file is clicked. The execution ends almost instantly, and the results are stored in a filled called SECAN.RES. The results files first echo-prints the input data, a part of which is reproduced in Fig. 3.25 for the bridge under consideration.

SECAN calculates moments, shears and deflections due to the applied loading at the specified reference sections. For the bridge under consideration, only one reference section was specified at 15.0 m from the 'left-hand' support, i.e. at the mid-span of the 30 m long bridge. The moment output is reproduced in Fig. 3.26.

```
            1
1 Test example of Figs. 3.20 and 3.21
5           8      30.0000         2.000000E+08     1.000000E+08   0         0
      1.80000          1.80000          1.80000          1.80000
1.80000         1.80000          1.80000
     4.000000E-02     4.000000E-02     4.000000E-02     4.000000E-02
4.000000E-02    4.000000E-02     4.000000E-02
     4.000000E-02
     2.000000E-04     2.000000E-04     2.000000E-04     2.000000E-04
2.000000E-04    2.000000E-04     2.000000E-04
     2.000000E-04
     0.175000         2.000000E+07     1.000000E+07     0.00000
            1         5
25.0000         62.5000          62.5000          87.5000          75.0000
3.60000         7.20000          8.40000          15.0000          21.6000
            2
      0.00000          1.80000
            1
      15.0000
```

Fig. 3.24 Input created by SECANIN for the illustrative example

```
BRIDGE NAME       :      Test example of Figs. 3.20 and 3.21
NUMBER OF HARMONICS   :       5
**********************
                         NUMBER OF GIRDERS          :       8
                         SPAN OF THE BRIDGE         :    30.00
GIRDERS INFORMATION   :
**********************   MODULUS OF ELASTICITY :0.2000E+09
                         SHEAR MODULUS              :0.1000E+09
                         NO. OF COLUMNS             :     0
                         GIRDER       MOMENT OF    TORSIONAL
                           NO.          INERTIA      INERTIA
                           1          0.4000E-01   0.2000E-03
                           2          0.4000E-01   0.2000E-03
                           3          0.4000E-01   0.2000E-03
                           4          0.4000E-01   0.2000E-03
                           5          0.4000E-01   0.2000E-03
                           6          0.4000E-01   0.2000E-03
                           7          0.4000E-01   0.2000E-03
                           8          0.4000E-01   0.2000E-03
                          GIRDER SPACINGS
                           PANEL        GIRDER     DISTANCE FROM
                           NO.         SPACING     L.H. GIRDER
                           1            1.80          1.80
                           2            1.80          3.60
                           3            1.80          5.40
                           4            1.80          7.20
                           5            1.80          9.00
                           6            1.80         10.80
                           7            1.80         12.60
```

Fig. 3.25 Echo-printing of bridge data by SECAN

```
REF. POINT NO.                          1
 DIST. FROM L.H.SUPPORT:               15.0
    GIRDER NO.                            MOMENTS
----------------------------------------------------------------------
        1                   0.1508E+04
        2                   0.1005E+04
        3                   0.4763E+03
        4                   0.1398E+03
        5                   -.1388E+02
        6                   -.5100E+02
        7                   -.3986E+02
        8                   -.1650E+02
```

Fig. 3.26 Girder responses by SECAN at mid-span of the example bridge

It can be seen that the SECAN output for girder moments is consistent with the moments given in Table 3.2 for 5 harmonics.

3.4 The Orthotropic Plate Method

Cusens and Pama (1975) have presented a rigorous method for analyzing a rectangular orthotropic plate simply supported on two opposite edges and subjected to rectangular patch loads. It is recalled that an orthotropic plate has two different set of flexural and torsional rigidities in the two perpendicular directions, being x and y. As shown in Fig. 3.27, the rectangular plate has span L, width $2b$, and the x and y directions are referred to as the longitudinal and transverse directions, respectively.

3.4.1 Basis of the Orthotropic Method

The various rigidities of the orthotropic plate are noted in the following.

D_x = Longitudinal flexural rigidity per unit width of the plate.
D_y = Transverse flexural rigidity per unit length of the plate.
D_{xy} = Longitudinal torsional rigidity per unit width of the plate.
D_{yx} = Transverse torsional rigidity per unit length of the plate.
D_1 = Longitudinal coupling rigidity per unit width of the plate.
D_2 = Transverse coupling rigidity per unit length of the plate.

An element of the plate under the action of bending moments, M_x and M_y, twisting moments, M_{xy} and M_{yx}, and shearing forces, Q_x and Q_y, is shown in Fig. 3.28; this figure also defines the axis system used in the analysis.

Consideration of equilibrium of the moments and forces on the plate element shown in Fig. 3.28 leads to the following classical 4th order differential equation for an elastic orthotropic plate with small deflections (e.g. Timoshenko and Woinowsky-Kreiger 1959).

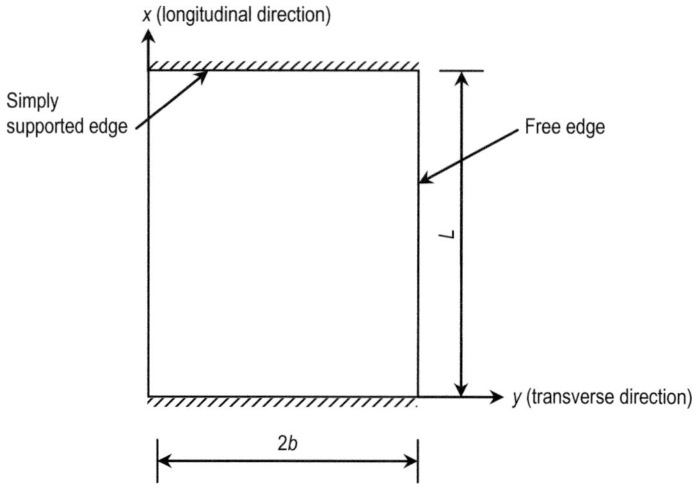

Fig. 3.27 A rectangular orthotropic plate, simply supported on two opposite edges

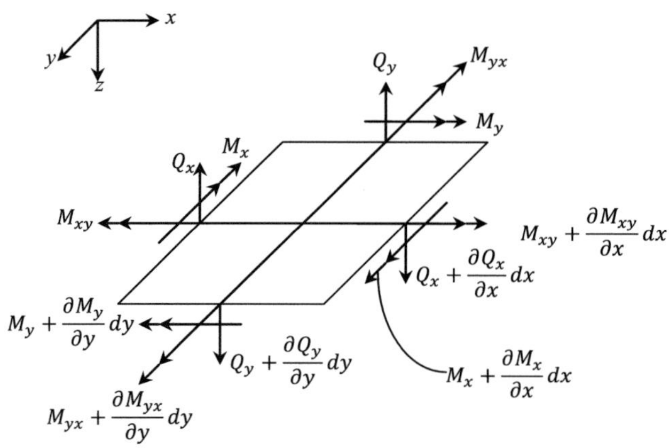

Fig. 3.28 An element of a plate subjected to moments and shear forces

$$D_x \frac{\partial^4 w}{\partial x^4} + 2H \frac{\partial^4 w}{\partial x^2 \partial y^2} + D_y \frac{\partial^4 w}{\partial y^4} = P(x, y) \tag{3.17}$$

In the above equation:

$$H = 0.5\left(D_{xy} + D_{yx} + D_1 + D_2\right) \tag{3.18}$$

The load $P(x,y)$ is expressed as a function of x and y. For example, $P(x,y)$ for a point load P can be expressed by Eq. (3.1), which was discussed in connection with the semi-continuum method, and which is reproduced in the following.

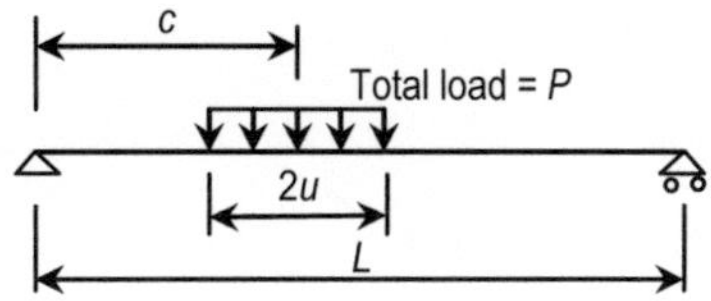

Fig. 3.29 A simply supported beam under a load uniformly distributed over a length $2u$

$$P_x = \frac{2P}{L}\sum_{n=1}^{\infty} \sin\frac{n\pi c}{L}\sin\frac{n\pi x}{L} \tag{3.19}$$

Similarly, as shown in Fig. 3.29a load P distributed uniformly over a length $2u$ and with its centre at $x = c$ is expressed as:

$$p_x = \sum_{n=1}^{\infty} \frac{4P}{n\pi}\sin\frac{n\pi c}{L}\sin\frac{n\pi u}{L}\sin\frac{n\pi x}{L} \tag{3.20}$$

The solution of Equation (3.17) for w is obtained by adding the particular part w_1 with the homogeneous part w_0, so that:

$$w = w_1 + w_0 \tag{3.21}$$

The particular solution is obtained for a plate of infinite width, shown in Fig. 3.30. The solution to Eq. (3.17) taken in the following form satisfies the condition of simple supports.

$$w_1 = \sum_{n=1}^{\infty} Ae^{\varphi_n y}\sin\frac{n\pi}{L}x \tag{3.22}$$

Substituting the above solution in Eq. (3.17), the following characteristic equation is obtained for the unloaded portion of the plate.

$$D_y\varphi_n^4 - 2H\left(\frac{n\pi}{L}\right)^2\varphi_n^2 + D_x\left(\frac{n\pi}{L}\right)^4 = 0 \tag{3.23}$$

The four roots of the characteristic equation are:

$$\varphi_{n\ 1,\ 2,3,4} = \pm\frac{n\pi}{L}\sqrt{\frac{H}{D_y} \pm \sqrt{\left(\frac{H}{D_y}\right)^2 - \left(\frac{D_x}{D_y}\right)}} \tag{3.24}$$

By examining these roots, Cusens and Pama (1975) show that four cases are possible: (a) $H^2 > D_xD_y$; (b) $H^2 = D_xD_y$, (c) $H^2 < D_xD_y$ and (d) $H = 0.0$. Using these expressions, the characterising parameter α, discussed in Chap. 1, can be calculated by Eq. (2.1). Referring to Fig. 2.2, it can be seen that Case (a) includes slab-on-girder bridges and bridges with timber beams, Case (b) is for slab and voided slab bridges, and Case (c) includes box-girder type of bridges. Case (d) is for

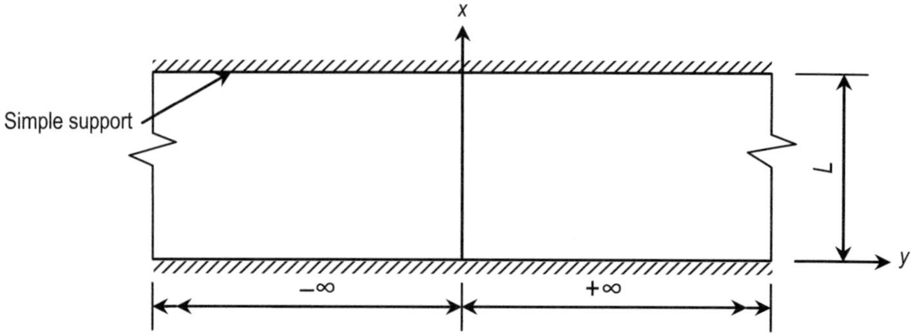

Fig. 3.30 An orthotropic plate with infinite width

articulated plates for which the transverse flexural rigidity D_y is zero, and consequently $\alpha = \infty$.

Equations for w_1 for the positive roots are provided by Cusens and Pama by assuming that both the slope and transverse shear at $y = 0$ are zero.

To generalize the derivation for the particular solution, it is assumed that the orthotropic plate has a finite width $2b$, and has two symmetrical edge beams at its free longitudinal edges. As shown in Fig. 3.31, each edge beam has flexural rigidity EI and torsional rigidity GJ. Also as shown in this figure, the y_o axis is at the centre of the plate, and the y_1 axis passes through the load, whose transverse distances from the left and right free edges of the plate are $\eta_2 b$ and $\eta_1 b$, respectively. Cusens and Pama (1975) provide the particular solution for w_0, and show that the complete solution for w is obtained by adding w_1 and w_0. The equation for w for a uniformly distributed load over a length $2u$ (Fig. 3.29) is as follows.

$$w = \sum_{n=1}^{\infty} \frac{\frac{4P}{L} \sin \frac{(n\pi c)}{L} \sin \frac{n\pi u}{L} \sin \frac{n\pi x}{L}}{D_x \left(\frac{n\pi}{L}\right)^4 2b} K_1 \tag{3.25}$$

where the equation for K_1, which includes four constants of integration, is provided by Cusens and Pama (1975). The constants of integration are obtained by assuming that the moments at $y_1 = -\eta_2 b$ and $\eta_1 b$ are respectively equal to $+GJ\left(\frac{\partial^3 w}{\partial x^2 \partial y}\right)$ and $-GJ\left(\frac{\partial^3 w}{\partial x^2 \partial y}\right)$; and the shears at $y_1 = -\eta_2 b$ and $\eta_1 b$ are respectively equal to $-EI\left(\frac{\partial^4 w}{\partial x^4}\right)$ and + $EI\left(\frac{\partial^4 w}{\partial x^4}\right)$.

Cusens and Pama (1975) have further considered the case of the load distributed uniformly over an area $2u \times 2v$, shown in Fig. 3.32. It is assumed the total load P is distributed uniformly over this area, so that by integrating the deflection function over the area $2u \times 2v$, Eq. (3.25) becomes:

$$w = \frac{PL^4}{uv\pi^5 D_x} \sum_{n=1}^{\infty} \frac{1}{n^5} \sin \frac{n\pi c}{L} \sin \frac{n\pi x}{L} \sin \frac{n\pi u}{L} K_1^* \tag{3.26}$$

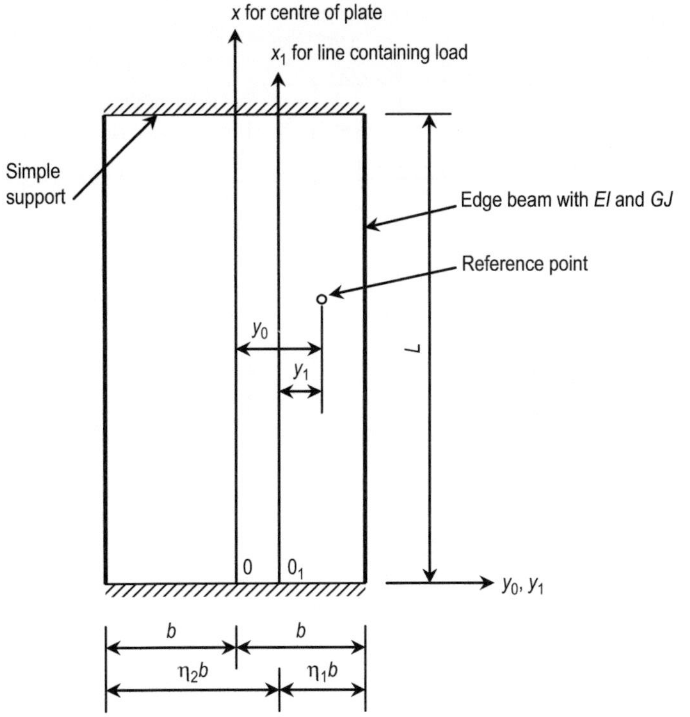

Fig. 3.31 An orthotropic plate with finite width and edge beams

where

$$K_1^* = \frac{1}{2b} \int_{y_1 - v}^{y_1 + v} K_1 d_y \tag{3.27}$$

The equation for the patch load leads to the following equations for bending moments M_x and M_y, twisting moments M_{xy} and M_{yx} and supplemented shear forces V_x and V_y. It is noted that the supplemented shear forces V_x and V_y are nearly the same as the shear forces Q_x and V_y, with the only difference that V_x also includes the effect of D_{xy} and V_y the effect of D_{yx}.

$$M_x = \frac{PL^2}{uv\pi^3} \sum_{n=1}^{\infty} \frac{1}{n^3} \sin\frac{n\pi c}{L} \sin\frac{n\pi x}{L} \sin\frac{n\pi u}{L} \left(K_1^* - \frac{D_1}{D_x} K_2^* \right) \tag{3.28}$$

$$M_y = -\frac{PL^2}{uv\pi^3} \sum_{n=1}^{\infty} \frac{1}{n^3} \sin\frac{n\pi c}{L} \sin\frac{n\pi x}{L} \sin\frac{n\pi u}{L} \left(\frac{D_y}{D_x} K_2^* - \frac{D_1}{D_x} K_1^* \right) \tag{3.29}$$

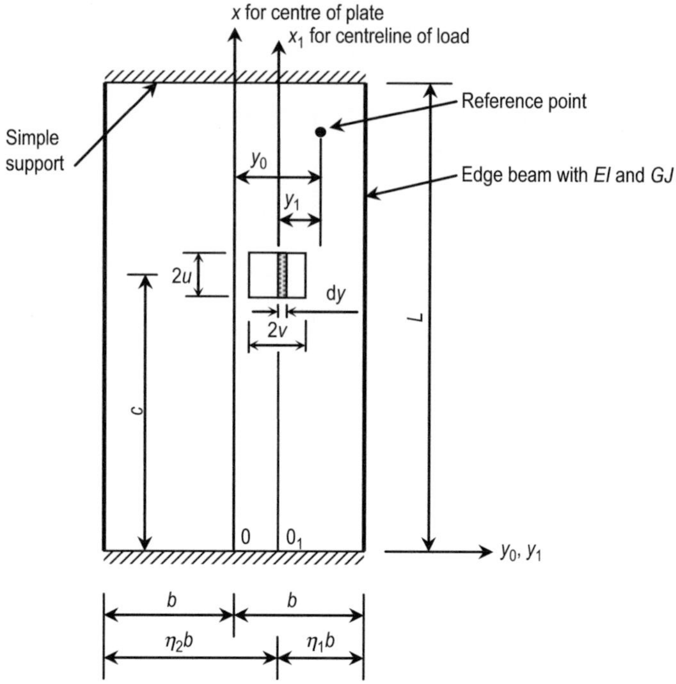

Fig. 3.32 An orthotropic plate with finite width and edge beams under a rectangular patch load

$$M_{xy} = \frac{PL^2}{uv\pi^3}\sum_{n=1}^{\infty}\frac{1}{n^3}\sin\frac{n\pi c}{L}\cos\frac{n\pi x}{L}\sin\frac{n\pi u}{L}\left(\frac{D_{xy}}{D_y}K_3^*\right) \quad (3.30)$$

$$M_{yx} = -\frac{PL^2}{uv\pi^3}\sum_{n=1}^{\infty}\frac{1}{n^3}\sin\frac{n\pi c}{L}\cos\frac{n\pi x}{L}\sin\frac{n\pi u}{L}\left(\frac{D_{yx}}{D_y}K_3^*\right) \quad (3.31)$$

$$V_x = \frac{PL}{uv\pi^2}\sum_{n=1}^{\infty}\frac{1}{n^2}\sin\frac{n\pi c}{L}\cos\frac{n\pi x}{L}\sin\frac{n\pi u}{L}\left(K_1^* - \frac{D_2 + D_{xy} + D_{yx}}{D_y}K_2^*\right) \quad (3.32)$$

$$V_y = -\frac{PL}{uv\pi^2}\sum_{n=1}^{\infty}\frac{1}{n^2}\sin\frac{n\pi c}{L}\sin\frac{n\pi x}{L}\sin\frac{n\pi u}{L}\left(K_4^* - \frac{D_2 + D_{xy} + D_{yx}}{D_y}K_3^*\right) \quad (3.33)$$

Expressions for coefficients K_1^*, K_2^*, K_3^* and K_4^*, which are governed by the position of the reference point with respect to that of the load, are provided by Cusens and Pama (1975).

For handling intermediate supports, which are also assumed to have rectangular contact areas with the orthotropic plate, are handled by the force method that has been discussed in Sect. 3.2.4 in conjunction with the semi-continuum method.

Convergence of Results

Unlike the semi-continuum method, the orthotropic plate method does not have a technique for quick convergence of results. For a plate without intermediate supports and with several applied patch loads, it is usually sufficient to consider only 15 harmonic for nearly complete convergence of results. However, the demand for the number of harmonic for nearly complete convergence increases with the number of intermediate supports. In some cases it might be necessary to consider up to 41 harmonics. The surest way of deciding whether the results have converged fully is to re-analyse the plate under a higher number of harmonics and see if the results under the previous number of harmonics are nearly the same as under the higher harmonic.

Calculation of Plate Rigidities

The CHBDC (2006) and some text books, e.g. Bakht and Jaeger (1985), provide detailed expressions for calculating the various plate rigidities for different types of superstructures. The expressions for calculating these rigidities are reproduced in the following for slab and slab-on-girder type of bridges; in these expressions, E_c, G_c and ν_c are respectively the modulus of elasticity, shear modulus and Poisson's ratio of concrete; E_g and G_g are respectively the modulus of elasticity and shear modulus of the girder material; t is the thickness of the concrete slab or deck slab; and S is the spacing of girders.

For concrete slab bridges:

$$D_x = D_y = \frac{E_c t^3}{12(1 - v_c^2)} \tag{3.34}$$

$$D_{xy} = D_{yx} = \frac{G_c t^3}{6} \tag{3.35}$$

$$D_1 = D_2 = \frac{\upsilon_c E_c t^3}{12(1 - v_c^2)} \tag{3.36}$$

For composite slab-on-girder bridges:

$$D_x = \frac{E_g I_g}{S} \tag{3.37}$$

$$D_y = \frac{E_c t^3}{12} \tag{3.38}$$

$$D_{xy} = \frac{G_g J}{S} + \frac{G_c t^3}{6} \tag{3.39}$$

$$D_{yx} = \frac{G_c t^3}{6} \tag{3.40}$$

$$D_1 = D_2 = \nu_c D_y \tag{3.41}$$

In the above equations, I_g is the composite moment of inertia of the girder and the associated portion of the deck slab, and J is the torsional inertia of the girder, which is usually ignored for steel girders.

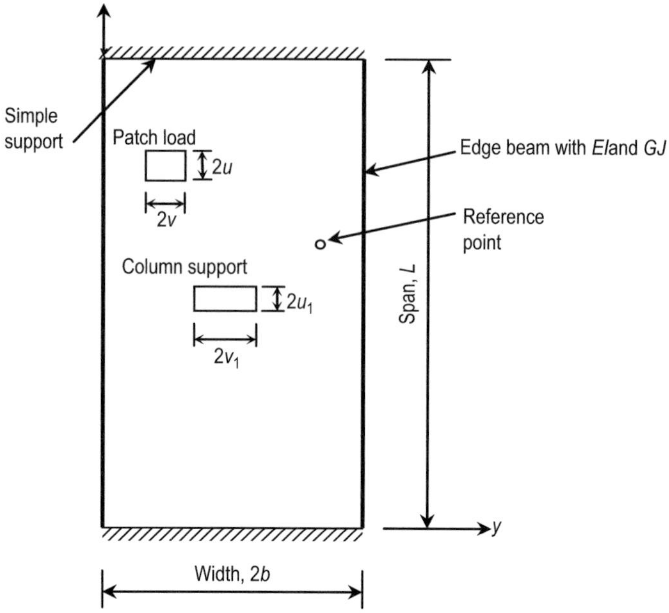

Fig. 3.33 An orthotropic plate with finite width, edge beams and intermediate supports subjected to rectangular patch loads

3.4.2 Computer Program PLATO

The orthotropic plate theory discussed in Sect. 3.4.1 has been incorporated in a computer program PLATO. As shown in Fig. 3.33, PLATO can analyse the effect of rectangular patch loads on a rectangular plate simply supported on two opposite edges. The plate can have symmetrical edge beams at its longitudinal free edges, and intermediate supports with rectangular contact area. The details of the program are provided in the following; a copy of the program can be downloaded from http://extras.springer.com.

PLATO is written in FORTRAN90 to run on personal computers having a Windows operating system and math co-processor. The root directory of the file contains Plato.exe and Platoin.exe. To install PLATO on a computer, first make a directory for PLATO. Then copy the files downloaded from http://extras.springer.com into the PLATO directory. The Example directory in the PLATO folder contains the input files and the output files for some examples.

3.4.3 Data Input for PLATO

The data input to PLATO can be prepared according to the scheme of Table 3.3. Alternatively, the data can be prepared by PLATOIN, an interactive program. The

Table 3.3 Details of data input to PLATO

Input ref. no.	Input parameter	Notes
1	Title	
2	Span of plate	As shown in Fig. 3.32, the span of the plate is the distance between its two simple supports
3	Width of the plate	If there are edge beams, the width of the plate is the distance between the centrelines of edge beams
4	No. of harmonics	Recommended minimum No. of harmonics = 15. For further advice on the No. of harmonics see Sect. 3.4.1
5	Longitudinal flexural rigidity, D_x	See Sect. 3.4.1 for calculation of these plate rigidities
6	Transverse flexural rigidity, D_y	
7	Longitudinal torsional rigidity, D_{xy}	
8	Transverse torsional rigidity, D_{yx}	
9	Longitudinal coupling rigidity D_1	
10	Transverse coupling rigidity, D_2	
10	Control letter for edge beams: = n for no edge beams; = y if edge beams are present.	The edge beams at the two free longitudinal edges of the plate must have the same properties
11	*E* of edge beam material	Skip these input parameters if there are no edge beams
12	*I* of edge beam	
13	*G* of edge beam material	
14	*J* of edge beam material	
15	Control letter for intermediate supports: = n for no intermediate supports; = y if intermediate supports are present	
16	No. of intermediate supports (N_s)	Skip this input parameter if there are no intermediate supports
17	Length of column in *x* direction	(a) Skip these input parameters if there are no intermediate supports (b) these input parameters have to be repeated N_s No. of times (c) if these parameters are the same for all intermediate supports, PLATOIN requires the data for only the first intermediate support
18	Width of column in *y* direction	
19	Flexibility of column	
20	Prescribed settlement of support at the centre of column	
21	*x*-coordinate of the centre of the intermediate support	(a) Skip these input parameters if there are no intermediate supports (b) these input parameters have to be repeated N_s No. of times
22	*y*-coordinate of the centre of the intermediate support	
23	Control letter for type of loading: = u if the whole plate is covered by a uniformly distributed load; = p if there are one more rectangular patch loads	
24	Intensity of load per unit area if control letter is u	Skip this input parameter if the control letter is p

(continued)

Table 3.3 (continued)

Input ref. no.	Input parameter	Notes
25	No. of patch loads (N_{pl}) in one longitudinal line	Skip this input parameter if the control letter is u
26	Magnitude of patch load	These parameters have to be repeated N_{pl} No. of times. If PLATOIN is used, these data have to be entered only once
27	Dimension of patch load in x direction	
28	Dimension of patch load in y direction	
29	x dimension of centre of patch load	
30	No. of longitudinal lines of loads (N_{ll})	
31	y dimensions for N_{ll} lines of loads	
32	No. of longitudinal reference sections (N_{lref})	
33	No. of equidistant stations in one longitudinal reference sections	These input parameters are not required if $N_{lref}=0$
34	y coordinates of N_{lref} sections	
35	No. of transverse reference sections (N_{tref})	
36	No. of equidistant stations in one transverse reference sections	These input parameters are not required if $N_{tref}=0$
37	x coordinates of N_{tref} sections	
38	No. of discrete reference sections (N_{dref})	
39	x coordinate of discrete reference section	These parameters have to be repeated N_{dref} No. of times, but are not required if $N_{dref}=0$
40	y coordinate of discrete reference section	

data prepared by PLATOIN is stored in a file called Plato.dat. The user can either input data directly in this file or modify the data created by PLATOIN.

3.4.4 Example of Use

The example of the slab-on-girder bridge of Figs. 3.20 and 3.21 is presented to illustrate the use of PLATO. The plan of the orthotropic plate idealizing the example bridge is shown in Fig. 3.34.

From the data presented in Sect. 3.3.4, the various plate rigidities of the example are calculated as follows by using Eqns. (3.37, 3.38, 3.39, 3.40, 3.41).

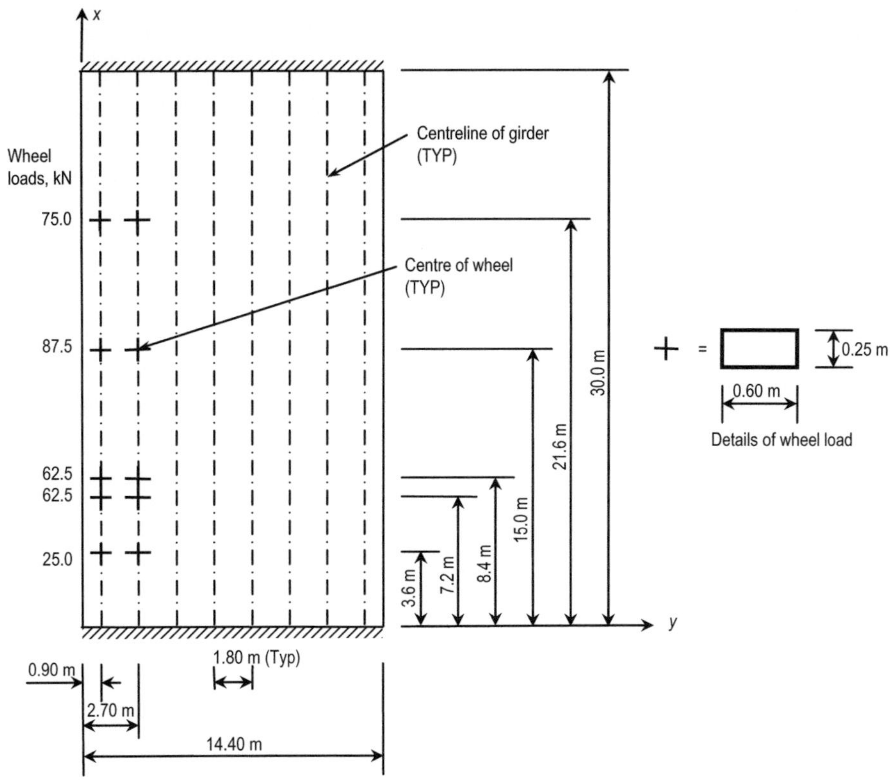

Fig. 3.34 Plan of orthotropic plate analysed by PLATO

$$
\begin{aligned}
D_x &= 200 \times 10^6 \times 0.04/1.8 &&= 4,444,444 \text{ kN.m}\\
D_y &= 20 \times 10^6 \times 0.175^3/12 &&= 8932 \text{ kN.m}\\
D_{xy} &= 100 \times 10^6 \times 0.0002/1.8 + 10 \times 10^6 \times 0.175^3/6 &&= 20,043 \text{ kN.m}\\
D_{yx} &= 10 \times 10^6 \times 0.175^3/6 &&= 8932 \text{ kN.m}\\
D_1 = D_2 &= 0.15 \times 8932 &&= 1340 \text{ kN.m}
\end{aligned}
$$

In the above calculations, the Poisson's ratio ν for concrete is assumed to be 0.15. PLATOIN is used to create the input data for the orthotropic plate with a span of 30.0 m and a width of 14.4 m. The individual wheel loads, which were assumed to be point loads in the semi-continuum analysis, are assumed in the orthotropic plate analysis to be rectangular 0.25×0.60 m patch loads, with the former direction being the x direction. As for analysis by SECAN, the load effects are investigated at $x = 15.0$ m, i.e. at the mid-span, at 17 equidistant reference sections, so that there is a reference section at each girder location and at 0.9 m on each side of it. The data input created by PLATOIN is reproduced in Fig. 3.35.

PLATO echo-prints the input data, which is reproduced in Fig. 3.36 for the example under consideration.

Fig. 3.35 Data input created by PLATOIN

```
Example of Fig. 3.34
           30.000000
           14.400000
              15
       4444444.000000
         8932.000000
        20043.000000
        8932.000000
        1340
        1340
n
n
p
                 5
n
           25.000000
        2.500000E-01
        6.000000E-01
            3.600000
           62.500000
        2.500000E-01
        6.000000E-01
            7.200000
           62.500000
        2.500000E-01
        6.000000E-01
            8.400000
           87.500000
        2.500000E-01
        6.000000E-01
           15.000000
           75.000000
        2.500000E-01
        6.000000E-01
           21.600000
                 2
        9.000000E-01
            2.700000
                 0
                 1
                17
           15.000000
                 0
```

The bending and twisting moments calculated by PLATO and reproduced in Fig. 3.37, are used to calculate the girder moments by integrating the area under the M_x curves over the 1.80 m width represented by each girder. The integration was done by assuming that the variation of the intensity of longitudinal moment over the 0.90 m width is linear. The girder moments thus calculated are compared in Table 3.4 with the moments obtained by SECAN.

At a cursory glance, it might seem that the difference between the girder moments predicted by SECAN and PLATO are not small. A graphical representation of the two sets of girder moments presented in Fig. 3.38 will, however, show that the two predictions are indeed very close to each other.

It can be seen in Fig. 3.38 that the maximum girder moment predicted by PLATO is slightly smaller than that by SECAN; this outcome is not surprising

```
Bridge name               = Example of Fig. 3.34
 Span                     =       30.00000
 Width                    =       14.40000
 No. of harmonics         =    15
 MATERIAL PROPERTIES OF ORTHOTROPIC PLATE
 Dx                       =        .44444E+07
 Dy                       =        .89320E+04
 Dxy                      =        .20043E+05
 Dyx                      =        .89320E+04
 D1                       =        .13400E+04
 D2                       =        .13400E+04
 LOAD DATA
 Number of patch loads =   10
Load No.  Magnitude         x             y         x-dimension    y-dimension
1        .25000E+02      3.60000        .90000         .25000         .60000
2        .62500E+02      7.20000        .90000         .25000         .60000
3        .62500E+02      8.40000        .90000         .25000         .60000
4        .87500E+02     15.00000        .90000         .25000         .60000
5        .75000E+02     21.60000        .90000         .25000         .60000
6        .25000E+02      3.60000       2.70000         .25000         .60000
7        .62500E+02      7.20000       2.70000         .25000         .60000
8        .62500E+02      8.40000       2.70000         .25000         .60000
9        .87500E+02     15.00000       2.70000         .25000         .60000
10       .75000E+02     21.60000       2.70000         .25000         .60000
```

Fig. 3.36 Echo printing of data by PLATO

```
 BENDING AND TWISTING MOMENTS AT REFERENCE POINTS
 Stn    x       y          Mx            My             Mxy            Myx
 1    15.00    .00    .89477E+03    .95348E-06     .41176E-01    -.18350E-01
 2    15.00    .90    .80136E+03    .21497E+01     .48522E-01    -.21624E-01
 3    15.00   1.80    .65910E+03    .14166E+01     .15323E+00    -.68284E-01
 4    15.00   2.70    .54967E+03    .21998E+01     .29768E+00    -.13266E+00
 5    15.00   3.60    .39758E+03   -.21237E+01     .36015E+00    -.16050E+00
 6    15.00   4.50    .27242E+03   -.51021E+01     .29266E+00    -.13042E+00
 7    15.00   5.40    .16817E+03   -.61366E+01     .14816E+00    -.66027E-01
 8    15.00   6.30    .88410E+02   -.57802E+01     .42577E-01    -.18974E-01
 9    15.00   7.20    .34004E+02   -.47503E+01    -.41548E-02     .18515E-02
 10   15.00   8.10    .75187E+00   -.35297E+01    -.14877E-01     .66296E-02
 11   15.00   9.00   -.16774E+02   -.23924E+01    -.11529E-01     .51378E-02
 12   15.00   9.90   -.23669E+02   -.14669E+01    -.58498E-02     .26069E-02
 13   15.00  10.80   -.24057E+02   -.79020E+00    -.19032E-02     .84816E-03
 14   15.00  11.70   -.20938E+02   -.34787E+00    -.46774E-04     .20844E-04
 15   15.00  12.60   -.16268E+02   -.10069E+00     .50839E-03    -.22656E-03
 16   15.00  13.50   -.11142E+02   -.78000E-03     .54689E-03    -.24372E-03
 17   15.00  14.40   -.60067E+01   -.28403E-06     .51919E-03    -.23137E-03
```

Fig. 3.37 Bending and twisting moments by PLATO

Table 3.4 Girder moments at mid-span

Details of analyses	Moments in kN·m in girder no.							
	1	2	3	4	5	6	7	8
SECAN 5 harmonics	1508	1005	476	140	−14	−51	−40	−17
PLATO 15 harmonics	1419	971	501	171	8	−40	−37	−20

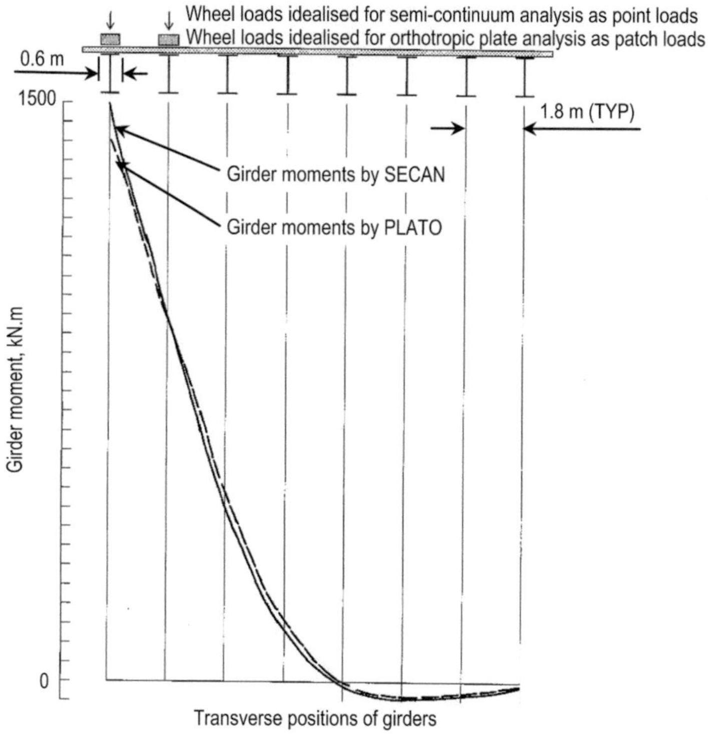

Fig. 3.38 Girder moments at the mid-span calculated by SECAN and PLATO

because the loads used for the orthotropic plate analysis are wide loads, whereas the semi-continuum analysis assumes the load to be point loads. The fact that two methods based on entirely different idealisations and formulisations give very nearly the same results attests to their accuracy.

References

AASHTO (1989) Standard specifications for highway bridges. American Association of State Highway and Transportation Officials, Washington, DC

Bakht B, Jaeger LG (1985) Bridge analysis simplified. McGraw-Hill, New York

Bakht B, Jaeger LG (1986) Analysis of bridges with intermediate supports by the semi-continuum method. Centennial Conference, Canadian Society for Civil Engineering

Bakht B, Mufti AA (1991) Behaviour of steel girder bridge with timber decking, Structural Research Report SRR-90-06. Ministry of Transportation, Ontario

CAN/CSA-S6-00 (2006) Canadian highway bridge design code. Canadian Standards Association, Rexdale

Cusens AR, Pama RP (1975) Bridge deck analysis. Wiley, London

Hambley EC (1994) Bridge deck behaviour. Chapman and Hall, London

Hendry AW, Jaeger LG (1955) General method for the analysis of grid frameworks. Proceedings of Institution of Civil Engineers, London, Part III(4), pp 939–971

Jaeger LG (1957) The analysis of grid frameworks of negligible torsional stiffness by means of basic functions. Proceedings of Institution of Civil Engineers, London, Part III(6), pp 735–757

Jaeger LG, Bakht B (1982) The grillage analogy method in bridge analysis. Can J Civ Eng 9(2):224–235

Jaeger LG, Bakht B (1985) Bridge analysis by the semi-continuum method. Can J Civ Eng 12(3):573–582

Jaeger LG, Bakht B (1989) Bridge analysis by microcomputer. McGraw-Hill, New York

Jaeger LG, Bakht B (1990) Semi-continuum analysis of shear-weak bridges. Can J Civ Eng 17(3):294–301

Jaeger LG, Bakht B (1993) Handling of transverse diaphragms by the semi-continuum method of analysis. Proceedings of the CSCE annual conference, vol II, Fredericton, pp 1–10

Jaeger LG, Bakht B, Mufti AA, Zhu GP (1998) Analysis of girder bridges with diaphragms by the semi-continuum method. Proceedings of the CSCE annual conference, Halifax

Mufti AA, Tadros G, Agarwal AC (1994) On the use of finite element programs in structural evaluation and developments of design charts. Can J Civ Eng 21(5):797–804

Mufti AA, Bakht B, Zhu GP (1998) Calculation of flexural stiffness of diaphragms in girder bridges. Proceedings of the CSCE annual conference, Halifax

Sawko F (1968) Recent developments in the analysis of steel bridges using electronic computers. Conference on steel bridges, British Steelwork Association, London, pp 1–10

Timoshenko S, Woinowsky-Krieger S (1959) Theory of plates and shells. McGraw-Hill, New York

Chapter 4
Arching in Deck Slabs

4.1 Introduction

The term *deck slab* is typically used in North America to describe the concrete slab of a girder bridge which supports the vehicle loads directly before transmitting their effects to the girders. In this chapter and elsewhere in this book, the term is used with this same meaning, rather than to describe a slab bridge as is done in some countries.

Until about four decades ago, deck slabs were designed throughout the world as if they were in pure flexure under the vehicular loads; this assumption necessitated a plate bending type of analysis. Such analyses led to envelopes of design transverse moment intensities which varied from maximum positive midway between the girders to maximum negative directly above the girders. A typical envelope of transverse bending moment intensities due to live loads can be seen in Fig. 4.1. For many years, there had been an almost universal acceptance of this assumption of flexure and the deck slabs so designed had been performing satisfactorily from the point of view of strength. Because of this, there was no reason to suspect that the deck slabs were, in general, grossly over-designed.

An extensive laboratory and field research program undertaken in Ontario, Canada, about four decades ago concluded that most deck slabs, instead of being in pure flexure, develop an internal arching system under live loads, and that because of this arching action they fail under concentrated loads in a punching shear mode, at a much higher load than the failure load corresponding to a purely flexural behaviour. The two modes of failure are shown schematically in Fig. 4.2a, b, respectively. The Ontario research program led to the conclusion that concrete slabs with only nominal steel reinforcement have more than adequate strength to sustain modern commercial heavy vehicles safely. This conclusion is at the basis of the empirical method of design specified in the Ontario Highway Bridge Design Code (1979, 1983, 1992) and in the Canadian Highway Bridge Design Code (2000, 2006) which has to date been used to design hundreds of deck slabs, all of which

B. Bakht, A. Mufti, *Bridges*, DOI 10.1007/978-3-319-17843-1_4

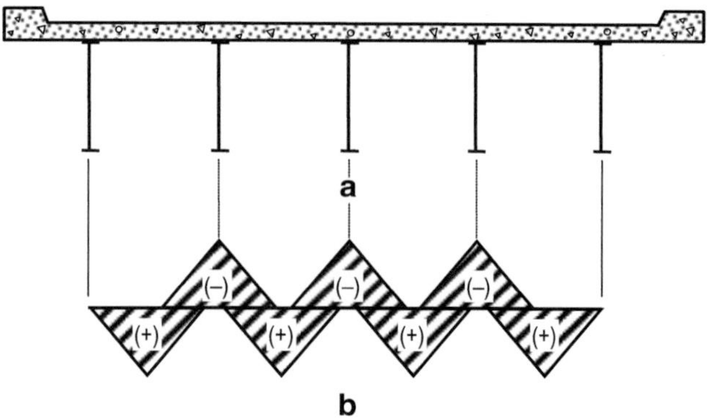

Fig. 4.1 Bending moments in deck slabs, obtained by plate bending analysis: (**a**) cross-section, (**b**) envelope of transverse moment intensities

Fig. 4.2 Failure modes of deck slab under a concentrated load: (**a**) punching shear failure mode; (**b**) flexural failure mode

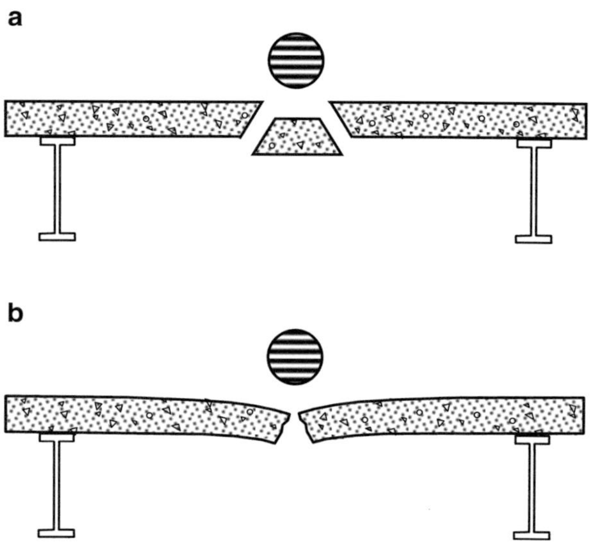

have been performing well. Details of this empirical method are provided in Sects. 4.3.5 and 4.3.7.

The purpose of this chapter is to explain the mechanics of internal arching in deck slabs, to provide background information, and to introduce two sets of design provisions. One of these can be used to design deck slabs with considerably reduced steel reinforcement, whilst the other is for the design of deck slabs which may be entirely free of tensile reinforcement.

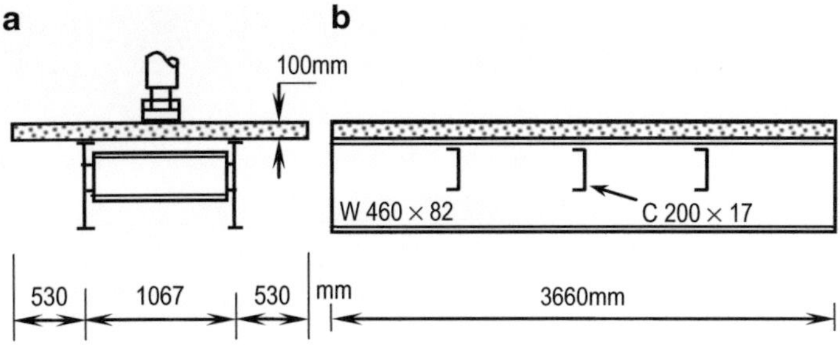

Fig. 4.3 Details of a half-scale model: (**a**) cross-section, (**b**) longitudinal section

4.2 Mechanics of Arching Action

For the authors of this book and probably for others as well, the internal arching system of concrete deck slabs, alluded to in the preceding section, was for some time more of an article of faith than a wholehearted acceptance of a scientific fact. That the arching action does indeed develop in the deck slabs was truly appreciated by the authors with the help of some laboratory tests which are reported by Mufti et al. (1993) and Bakht and Agrawal (1995); these tests are described in the following Subsections.

4.2.1 Model that Failed in Bending

A half-scale model of a two-girder bridge was constructed with a concrete deck slab without tensile reinforcement. For controlling shrinkage cracks, the concrete was mixed with low-modulus polypropylene fibres. Details of this model are given in Fig. 4.3 in which it can be seen that the model had no transverse end diaphragms. The fibres used in the deck slabs have so low a modulus of elasticity that their inclusion does not alter the tensile strength of concrete significantly.

When the deck slab of the model of Fig. 4.3 was tested under a central concentrated load, failure occurred at a load of 173 kN and the failure mode was that of flexure under which the damage covered the entire extent of the slab, as can be seen in Fig. 4.4.

Figure 4.5a shows a latent arch within the cross-section of the deck slab. It is because of such latent arches in both the longitudinal and transverse directions that the deck slab under loads normal to its plane is predominantly in compression and not in pure flexure.

As illustrated in Fig. 4.5b, in deck slabs with conventional steel reinforcement the tie to the transverse arch within the cross-section is provided by transverse reinforcement near the bottom face of the slab (the role of bottom transverse reinforcement on the strength of the slab is discussed later in detail); the tie in the longitudinal direction is clearly provided by the top flanges of the girders. It was realized that the deck slab of the model of Fig. 4.3 failed in bending because its latent arch in the transverse direction, lacking steel reinforcement, did not have

Fig. 4.4 The deck slab after failure in the flexural mode

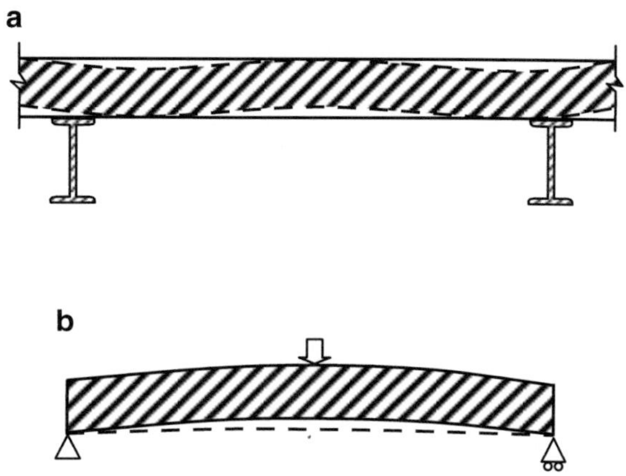

Fig. 4.5 Illustration of the arching action in deck slab, (**a**) latent arch within the cross-section of the deck slab, (**b**) bottom transverse reinforcement serves as tie to the transverse arch

sufficient lateral restraint at its supports, i.e. over the girders. It can be seen in Fig. 4.3 that the three diaphragms within the span were made of steel channels connected through their webs to the girders. This conventional arrangement of diaphragms permitted lateral flexure of the webs above their connections so that the lateral restraint, which was necessary to develop the transverse arch, was absent in the model.

4.2.2 Model that Failed in Punching Shear

It was realized that complete restraint in both the longitudinal and transverse directions is necessary for the development of the internal arching system in the deck slab. With this realization, another half-scale model of a two-girder bridge was built. This model also had a concrete deck slab reinforced only by polypropylene fibres, and was very similar to the previous one, the main difference being that the top flanges of the girders were now interconnected by transverse steel straps lying outside the deck slab. A view of the steel work of this model can be seen in Fig. 4.6. The steel straps were provided so as to serve as transverse ties to the internal arch in the slab.

Fig. 4.6 Top flanges of girders connected by transverse steel straps

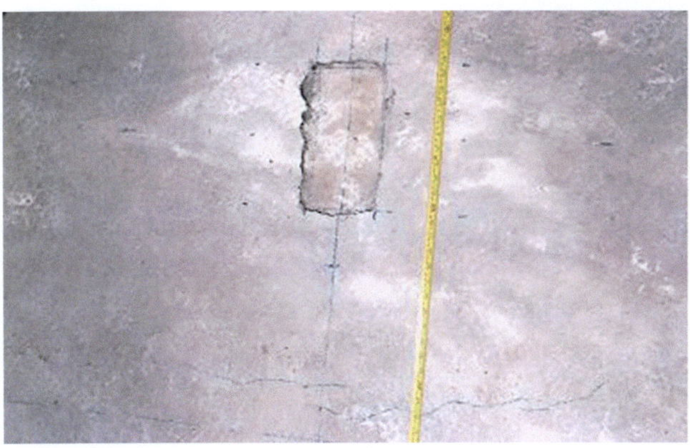

Fig. 4.7 Failure of the deck slab in punching shear

The 100 mm thick concrete slab of the model with transverse straps failed under a central load of 418 kN in a punching-shear failure mode. As can be seen in Fig. 4.7, the damaged area of the slab was highly localized. It can be appreciated that with such a high failure load, the thin deck slab of the half-scale model could have easily withstood the weights of even the heaviest wheel loads of commercial vehicles.

The model tests described above and in Sect. 4.2.1 clearly demonstrate that an internal arching action will indeed develop in a deck slab, but only if it is suitably restrained.

4.2.3 *Edge Stiffening*

A further appreciation of the deck slab arching action is provided by tests on a scale model of a skew slab-on-girder bridge. As will be discussed in Sect. 4.4.2, one transverse free edge of the deck slab of this model was stiffened by a composite steel channel with its web in the vertical plane. The other free edge was stiffened by a steel channel diaphragm with its web horizontal and connected to the deck slab through shear connectors. The deck slab near the former transverse edge failed in a mode that was a hybrid between punching shear and flexure. Tests near the composite diaphragm led to failure at a much higher load in punching shear (Bakht and Agarwal 1993).

The above tests confirmed yet again that the presence of the internal arching action in deck slabs induces high in-plane force effects which in turn demand stiffer restraint in the plane of the deck than in the out-of-plane direction.

4.3 Internally Restrained Deck Slabs

Deck slabs which have embedded reinforcement for strength will now be referred to as *internally restrained deck slabs*. The state-of-art up to 1986 relating to the quantification and utilization of the beneficial internal arching action in deck slabs with steel reinforcement has been provided by Bakht and Markovic (1986). Their conclusions complemented with up-to-date information are presented in this Section in a generally chronological order which, however, cannot be adhered to rigidly because of the simultaneous occurrence of some developments.

4.3.1 *Static Tests on Scale Models*

About four decades ago, the Structures Research Office of the Ministry of Transportation of Ontario (MTO), Canada, sponsored an extensive laboratory-based research program into the load carrying capacity of deck slabs; this research program was carried out at Queen's University, Kingston, Ontario. Most of this research was conducted through static tests on scale models of slab-on-girder bridges. This pioneering work is reported by Hewitt and Batchelor (1975) and later by Batchelor et al. (1985), and is summarized in the following.

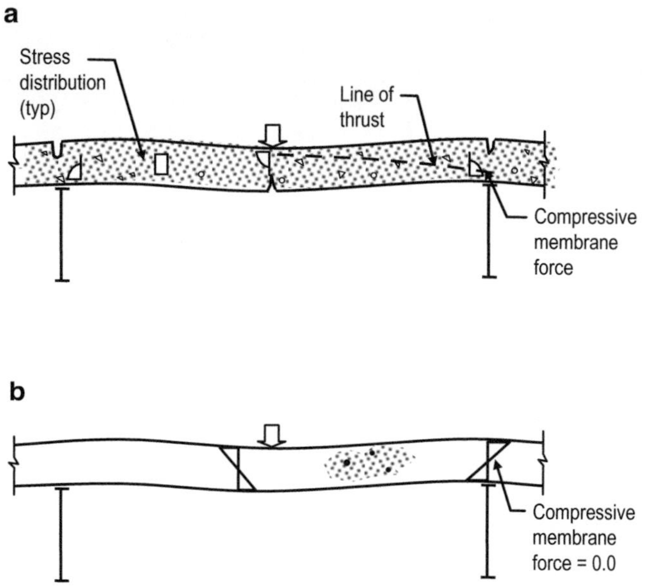

Fig. 4.8 Deck slabs with and without internal arching: (**a**) deck slab which can crack, (**b**) deck slab with the same stress-strain relationships in compression and tension

The inability of the concrete to sustain large tensile strains, which leads to cracking, has been shown to be the main attribute which causes the compressive membrane forces to develop. This phenomenon is illustrated in Fig. 4.8a which shows the part cross-section of a slab-on-girder bridge with a concentrated load on the deck slab. The cracking of the concrete, as shown in the figure, results in a net compressive force near the bottom face of the slab at each of the two girder locations. Midway between the girders, the net compressive force moves towards the top of the slab. It can be readily visualized that the transition of the net compressive force from near the top in the middle region, to near the bottom at the supports corresponds to the familiar arching action. Because of this internal arching action, the failure mode of a deck slab under a concentrated load becomes that of punching shear.

If the material of the deck slab has the same stress-strain characteristics in both tension and compression, the slab will not crack and, as shown in Fig. 4.8b, will not develop the net compressive force and hence the arching action.

In the punching shear type of failure, a frustum of cone separates from the rest of the slab, as shown in schematically in Fig. 4.9. It is noted that in most failure tests, the diameter of the lower end of the frustum extends to near the girders.

From analytical and confirmatory laboratory studies, it was established that the most significant factor influencing the failure load of a concrete deck slab is the confinement of the panel under consideration. It was concluded that this confinement is provided by the expanse of the slab beyond the loaded area; its degree was found difficult to assess analytically. A restraint factor, η, was used as an empirical

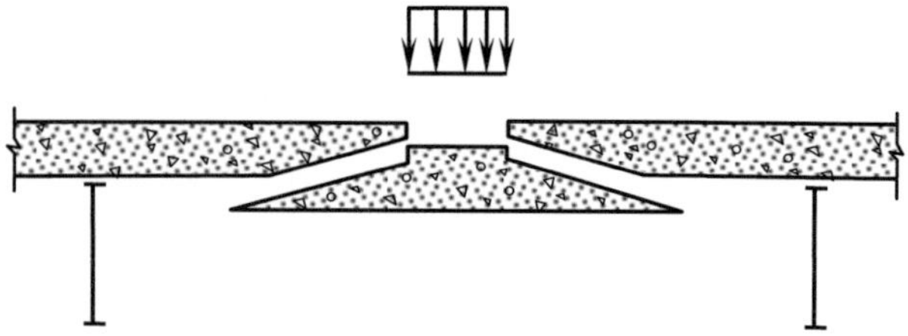

Fig. 4.9 Failure in punching shear mode under a concentrated load

Table 4.1 Effect of various parameters on failure load

Slab thickness, mm	Steel ratio (%)	η	Failure load, kN
200	1.0	0.0	690
		0.5	1600
180	0.2	0.0	110
		0.5	930

measure of the confinement; its value is equal to zero for the case of no confinement and 1.0 for full confinement.

The effect of various parameters on the failure load can be seen in Table 4.1, which lists the theoretical failure loads for various cases. It can be seen that an increase of the restraint factor from 0.0 to 0.5 results in a very large increase in the failure load. The table also emphasizes the fact that the neglect of the restraint factor causes a gross underestimation of the failure load.

It was concluded that design for flexure leads to large amounts of unnecessary steel reinforcement in the deck slabs, and that even the minimum amount of steel required for crack control against volumetric changes in concrete is adequate to sustain modern-day, and even future, highway vehicles of North America.

It was recommended that for new construction, the reinforcement in a deck slab should be in two layers, with each layer consisting of an orthogonal mesh having the same area of reinforcement in each direction. The area of steel reinforcement in each direction of a mesh was suggested to be 0.2 % of the effective area of cross-section of the slab. This empirical method of design was recommended for deck slabs with certain constraints.

4.3.2 Pulsating Load Tests on Scale Models

To study the fatigue strength of deck slabs with reduced reinforcement, five small scale models with different reinforcement ratios in different panels were tested at the Queen's University at Kingston. Details of this study are reported by Batchelor et al. (1978).

Experimental investigation confirmed that for loads normally encountered in North America deck slabs with both conventional and recommended reduced

Table 4.2 Average values of the restraint factor η

Type of slab-on-girder bridge	No. of bridges	No. of tests	Average value of η
Steel girders with non-composite slab	9	15	0.41
Steel girders with composite slab	9	14	0.93
Concrete T-beam	8	17	0.78
Prestressed concrete girders with composite slab	2	2	0.83

reinforcement have large reserve strengths against failure by fatigue. It was confirmed that the reinforcement in the deck slab should be as noted in Sect. 4.3.1. It is recalled that the 0.2 % reinforcement requires that the deck slab must have a minimum restraint factor of 0.5.

The work of Okada et al. (1978) also deals with fatigue tests on full-scale models of deck slabs and segments of severely cracked slab removed from 8 to 10 year old bridges. The application of these test results to deck slabs of actual bridges is open to question because test specimens were removed from the original structures in such a way that they did not retain the confinement necessary for the development of the arching action.

4.3.3 *Field Testing*

Along with the studies described in the preceding Subsection, a program of field testing of the deck slabs of in-service bridges was undertaken by the Structures Research Office of the MTO. The testing consisted of subjecting deck slabs to single concentrated loads, simulating wheel loads, and monitoring the load-deflection characteristics of the slab. The testing and details of the testing equipment are given by Bakht and Csagoly (1979).

Values of the restraint factor, η, were back-calculated from measured deflections. A summary of test results, given in Table 4.2, shows that the average value of η in composite bridges is greater than 0.75, while that for non-composite bridges is 0.42. It was concluded that for new construction, the restraint factor η can be assumed to have a minimum value of 0.5.

Bakht (1981) reports that after the first application of a test load of high magnitude on deck slabs of existing bridges, a small residual deflection was observed in most cases. Subsequent applications of the same load did not result in further residual deflections. It is postulated that the residual deflections are caused by cracking of the concrete which, as discussed earlier, accompanies the development of the internal arching action. The residual deflections after the first cycle of loading suggest that either the slab was never subjected to loads high enough to cause cracking, or the cracks have 'healed' with time.

4.3.4 An Experimental Bridge

The research described above led to the conclusion that deck slabs should be designed by taking account of the internal arching action. There was so much confidence and potential saving inherent in this approach that the MTO decided to build an experimental slab-on-girder bridge which, along with other innovations, incorporated a new design approach for the deck slab. The experimental bridge, called the Connestogo River Bridge, was designed in 1973 and constructed in 1975. The design and testing of the bridge is reported by Dorton et al. (1977). Satisfactory behaviour of the bridge observed by testing further proved the validity of the punching shear design approach for deck slabs.

A second test on the bridge in 1984, so far unreported, confirmed that the performance of the deck slab designed by the new method had not deteriorated with time.

4.3.5 Ontario Code, First Edition

Following the construction and testing of the experimental Connestogo River Bridge, bridge engineers in the MTO were presented with the new method of deck slab design, which simply required that the slab thickness be a minimum of 1/15 of the girder spacing, and that the reinforcement should comprise two orthogonal meshes one near the top of the slab and one near the bottom, with a certain minimum area of reinforcement in each direction of each mesh. There was a general reluctance amongst the designers to adopt the empirical method, mainly because it was not specified or permitted by an acceptable design code.

Until that time, highway bridges in Ontario, like the most of the rest of North America, were designed by the American AASHTO specifications. The process of incorporating any new provisions in that code was so tedious and time-consuming that trying to incorporate the new deck slab design method in the AASHTO specifications appeared to be a futile exercise. The idea of the province of Ontario having its own bridge design code was conceived at that time. The first edition of the Ontario Highway Bridge Design Code (OHBDC) was published in 1979.

In the first edition of the code, an empirical method of deck slab design was permitted, which required two layers of orthogonal isotropic mesh reinforcement with the area of cross-section in each direction in each mesh to be a minimum of 0.3 % of the effective cross-sectional area of concrete. It is noted that the 'effective depth' for obtaining the effective area is measured from the top of the slab to the centroid of bottom reinforcement. The method could be applied only when the following conditions were met:

(a) Deck slab concrete strength is at least 30 MPa;
(b) the deck slab span, i.e. girder spacing, does not exceed 3.7 m;
(c) the deck slab extends at least 1.0 m beyond the exterior girder, or has a curb of equivalent area of cross-section;

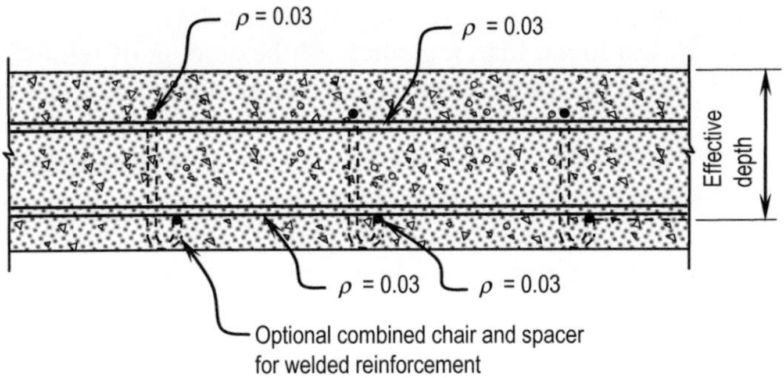

Fig. 4.10 Minimum reinforcement ρ required in deck slab by the Ontario Highway Bridge Design Code (1979)

(d) the spacing of reinforcing bars does not exceed 300 mm;
(e) the bridge has intermediate diaphragms or cross-frames spaced at 8 m;
(f) the bridge has support diaphragms or cross-frames at all supports;
(g) the skew angle of the bridge does not exceed 20°;
(h) the girder spacing to slab thickness ratio is a maximum of 15, with a minimum slab thickness of 190 mm; and
(i) there are at least three girders in the bridge.

Some existing bridges do, indeed, have thinner deck slabs and reinforcements which do not conform to these requirements. For evaluation of deck slabs of existing bridges, the OHBDC provided charts which give failure loads for slabs with various parameters. These charts are derived from the analytical method of Hewitt and Batchelor (1975). The computer program incorporating this method is given by Batchelor et al. (1985), and the charts are also reproduced by Bakht and Jaeger (1985).

Although the research had indicated that 0.2 % steel in each direction of a mesh was sufficient, the writers of the OHBDC (which included some of the researchers cited earlier) decided to be somewhat conservative and specified ρ of 0.3 % as the minimum reinforcement. Minimum reinforcement requirements are shown in Fig. 4.10, which also shows the optional combined chair and spacers, using which the two meshes can be welded to form a prefabricated reinforcement arrangement. It is noted that, with available technology, it is not feasible to protect this prefabricated reinforcement arrangement with the epoxy coating which was required in Ontario to protect the slab from the damaging effects of de-icing salts.

4.3.6 *Research in Other Jurisdictions*

To confirm the conclusions regarding the presence of internal arching in deck slabs, tests on small scale reinforced concrete bridge deck slabs were performed in the State of New York, USA. From the tests, which are reported by Beal (1982), it was again concluded that the large reserve strengths in concrete deck slabs are due to internal arching. It was found that, regardless of the amount of reinforcement, the failure loads of the deck slabs were always more than six times the design wheel load.

A test was also conducted in New York on a deck slab model with only one layer of orthogonal mesh reinforcement placed midway in the slab thickness. Until it developed cracks, this slab behaved similarly to those with two layers of reinforcement. The strains in the reinforcement, however, increased dramatically after the development of cracks, thus making the concept unsuitable for practical applications.

The same type of laboratory testing, which had been carried out in Ontario and New York, was later repeated in the UK. Results and conclusions drawn from this study are reported by Kirkpatrick et al. (1984). This study also led to the conclusion that deck slabs subjected to concentrated loads fail in punching shear, and that the current British code, like several other codes around the world, leads to substantial over-design of the deck slab.

Jackson and Cope (1990) have reported the results of an extensive experimental study conducted in the UK on the behaviour of deck slabs. They have concluded that, contrary to previous belief, the load carrying capacity of the deck slab under a single concentrated load is affected significantly by the presence of other concentrated loads in close proximity. This aspect of deck slab behaviour is of particular concern in the case of the abnormal HB loading which comprises 16 closely-spaced patches of loads, as shown in Fig. 2.17. Jackson and Cope (1990) have concluded that, despite this concern, the Ontario empirical method yields satisfactory designs.

A full-scale model of a 190 mm thick deck slab designed by the Ontario empirical method was tested under cyclic loading at the University of Texas at Austin, USA. The cyclic testing on the deck slab supported on three steel girders was conducted by creating the positive and negative moment conditions. For the positive moment condition, the loading consisted of four concentrated loads simulating two axles, 6.1 m apart, of a commercial heavy vehicle; for the negative moment condition the axle spacing was reduced to 1.2 m. Even after five million cycles of loads varying between 22 kN and 116 kN, applied in each case, the capacity of the deck slab to sustain static concentrated loads was not affected significantly. Thus, it was further confirmed that the deck slab with the reduced amount of reinforcement required by the empirical method is not prone to fatigue damage. Fatigue of deck slabs is discussed more extensively later in this chapter.

4.3.7 Ontario Code, Second and Third Editions

The empirical method of deck slab design was extended in the second and third editions of the OHBDC (1983, 1992) as noted in the following.

The empirical method could now be applied to bridges with concrete girders without intermediate diaphragms. The method was also applicable to bridges with

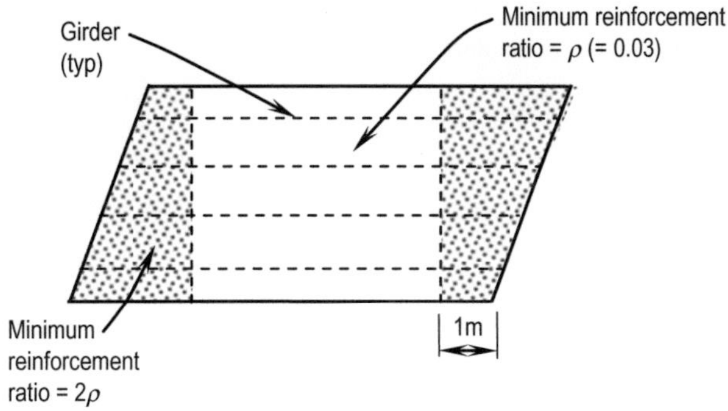

Fig. 4.11 Minimum reinforcement in deck slabs of bridges with large skew angle as required by OHBDC (1992)

skew angles larger than 20°. In these cases, however, the deck slab span was measured along the skew direction, and the minimum reinforcement requirement in the end regions, identified in Fig. 4.11, was increased to 0.6 % isotropic reinforcement in each layer as compared with 0.3 % in middle regions.

The minimum thickness of the deck slab for new construction was raised from 190 mm to 225 mm because of durability considerations which required larger than usual depths of cover over the steel reinforcement for protection against the damaging effects of de-icing salts.

The typical reinforcement for a 190 mm thick deck slab over girders spaced at 2.13 m designed by the conventional design method of AASHTO specifications consists of 16 mm diameter top and bottom bars spaced at 305 mm centres in the transverse direction. The longitudinal reinforcement, which is usually referred to as the distribution steel, is provided mainly near the bottom face; it consists of 16 mm diameter bars at 230 mm centres. This arrangement of steel corresponds to an average volume of steel of about $2.89 \times 10^6\ \mathrm{mm^3/m^2}$ area of the deck slab.

Reinforcement details for the same deck slab designed by the empirical method of the OHBDC (1979) are shown in Fig. 4.12. The reinforcement consists of top and bottom 13 mm diameter straight bars at 220 mm centres in each of the longitudinal and transverse directions. This arrangement of steel corresponds to an average volume of about $2.32 \times 106\ \mathrm{mm^3/mm^2}$ of the deck slab area, and constitutes a saving of 20 % in the steel weight.

It should be noted that for the empirical design, the same slab thickness and reinforcement would suffice for girder spacings of up to 2.83 m, whereas the conventional design method would require larger amounts of steel, and therefore the percentage saving of reinforcement weight affected by the empirical design method would increase with the girder spacing. This can be observed in Fig. 4.13, which plots the total reinforcement ratios for deck slabs of some existing bridges in Ontario designed by the AASHTO method for flexure for HS-20 loading, and by the empirical method of the OHBDC. The figure also shows the mean reinforcement savings achieved by the new method of design plotted against the girder spacings.

It can be seen in Fig. 4.13 that the amount of reinforcement in deck slabs designed by the flexural method is always more than that in slabs designed by the

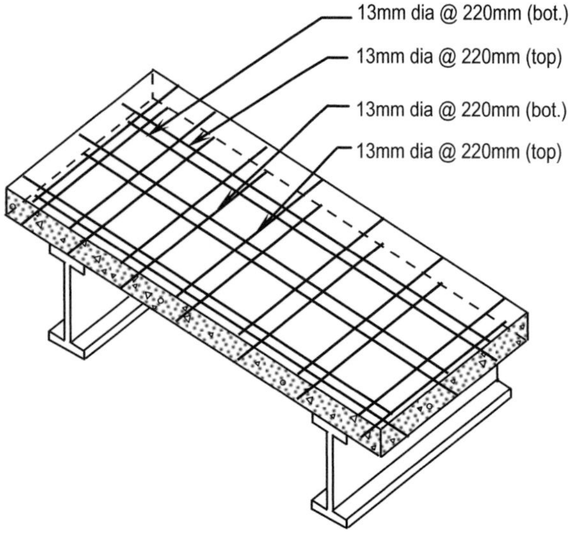

Fig. 4.12 Arrangement of reinforcement in the deck slab designed by the OHBDC (1992) for a girder spacing of 2 m

empirical method. The savings range between 40 % for shorter spans and 35 % for longer spans. The figure also shows the deck slab thicknesses of existing deck slabs plotted against the girder spacings. A clear pattern does not seem to emerge because in Ontario the deck slab thickness is often governed by durability rather than strength criteria.

Between 1980 and 1995, about 1,445,000 m^2 of concrete deck slab were constructed in Ontario, Canada, on 437 Ministry-owned and a much larger number of municipally-owned bridges. Designed by the old method these deck slabs would have required about 13,000 tonnes of extra steel. In 1993 dollars, this savings in steel was estimated to be approximately $14 million, or about $1.0 million per year. A smaller magnitude of savings is achieved in deck slab replacements.

It is worth noting that this saving of about $1 million per year in material cost does not include the very substantial savings resulting from the increased durability of the deck slab, which is the consequence of its reduced steel reinforcement.

4.3.8 Rolling Load Tests on Scale Models

Perdikaris and Beim (1988) have shown that pulsating loads do not cause the same damage in deck slabs as that due to the effect of rolling wheel loads. They have conducted a series of tests on scale models of deck slabs by subjecting them to rolling heavy wheels with pneumatic tires. Notwithstanding the difficulties in extrapolating the results from small scale models to full size structures, Perdikaris and Beim (1988) have concluded that the deck slabs reinforced with smaller amounts of steel have higher fatigue resistance than those of more heavily reinforced slabs.

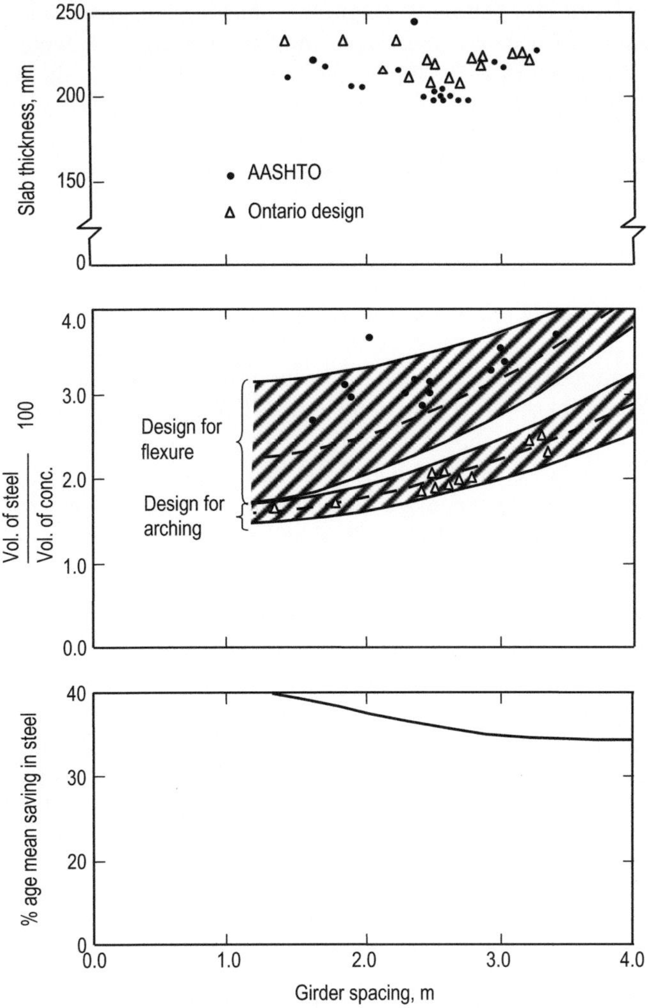

Fig. 4.13 Comparison of reinforcements in deck slabs designed for flexure and by a method which takes account of the internal arching action

4.3.9 Miscellaneous Recent Research

Recent technical literature contains extensive references to both analytical and experimental work on the punching shear resistance of confined concrete slabs. Even though some of the references are not directly related to deck slabs, they are discussed briefly in the following for the sake of completeness. Although the proposed design provisions given in Sect. 4.5 have not benefited directly from the references discussed in this sub-section, it is nevertheless the case that these references implicitly support the concept of internal arching in suitably-confined deck slabs.

Arching in Negative Moment Regions Johnson and Arnaouti (1980) have presented results of tests on three scale models of deck slabs which represent a segment of a girder bridge in the vicinity of the intersection of a girder and a transverse diaphragm. Both of these components were supported at their intersection, thus subjecting the deck slab to biaxial tension. It was found that the presence of this biaxial tension affected significantly neither the mode of failure nor the magnitude of failure load.

Tests on a Full-scale Model Fang et al. (1990) have presented the results of a test on a full-scale model of a slab-on-girder bridge. A part of the deck slab was cast in place and the remainder was constructed by a combination of precast panels and cast-in-place topping. It was found that both segments of the deck slab, which contained reinforcement according to the empirical method of OHBDC, being about 40 % less than that required by AASHTO, performed satisfactorily even under loads that were about three times the AASHTO design loads. Fang et al. (1990) seem to concur with Jackson and Cope (1990) in concluding that, despite the limited extent of damage due to failure under one concentrated load, the load carrying capacity of the deck slab under one concentrated load is affected by the presence of other loads in close proximity.

Instrumented Deck Slabs in New York The state of New York in the USA has several experimental deck slabs which were designed by the empirical method of OHBDC. A number of these deck slabs have instrumented steel bars which are periodically monitored under test loads. Alampalli and Fu (1991) have presented results of tests on three of these decks. The results indicate their continued satisfactory performance. Other tests on deck slabs in New York are reported by Fu et al. (1992).

Tests on Restrained Slab Panels The tests of Kuang and Morley (1992) on 12 square slabs with peripheral beams of different stiffnesses have provided a quantitative assessment of the influence of edge beam stiffness on the failure load of the slab. It is expected that these tests will be found useful in developing a rational criterion for the design of edge stiffening of deck slabs.

Test on a Pier Deck Model Malvar (1992) has presented the results of tests on a reinforced concrete pier deck model which failed in a punching shear mode under a concentrated load. This reference is of particular importance because it also contains the results of an extensive finite element (FE) study which seems to show the promise of solving the difficult problem of deck slab behaviour; further discussion on the FE analysis and its formulation is presented in Sect. 4.8.

4.3.10 Role of Reinforcement on Deck Slab Strength

A series of tests on a full-scale model of a 175-mm thick deck slab containing four different patterns of embedded reinforcement is reported by Khanna et al. (2000).

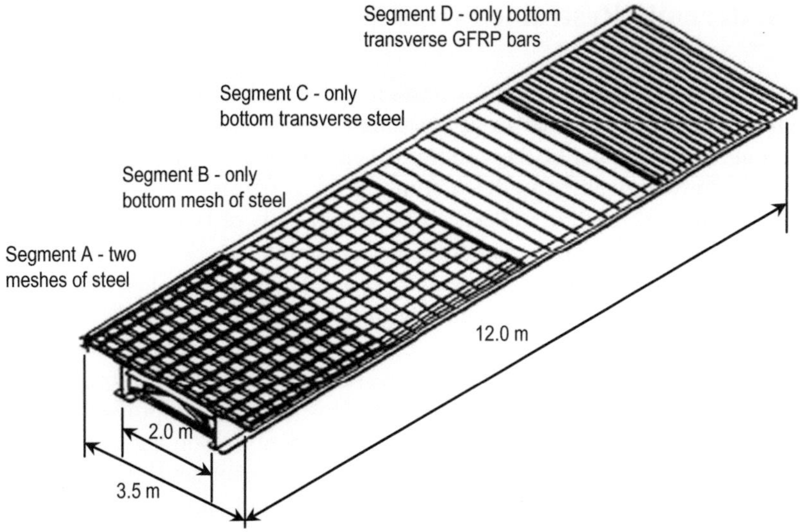

Fig. 4.14 Details of RC deck slab with four segments

The slab was supported by two steel plate girders at a spacing of 2 m. As shown in Fig. 4.14, the slab was conceptually divided into four segments. Segment A of the model was reinforced with top and bottom orthogonal meshes of 15 mm dia. steel bars at a spacing of 300 mm in each direction. Segment B contained only the bottom orthogonal mesh of 15 mm dia. steel bars at a spacing of 300 mm. Segment C was provided with only 15 mm dia. bottom transverse steel bars at a spacing of 300 mm. Segment D contained 25 mm dia. glass fibre reinforced polymer (GFRP) bottom transverse bars at a spacing of 150 mm. The GFRP bars were selected so that their axial stiffness remains the same as that of the steel bars in Segment C. As discussed in Chap. 8, GFRP has a significantly lower modulus of elasticity than that of steel; however, its tensile strength is several times larger than that of steel, with the result that GFRP bars having the same axial stiffness as mild steel bars can have up to 9 times the axial strength of the steel bars.

When tested under a central patch load, Segments A, B, C and D of the 175 mm thick model deck slab failed in punching shear at 808, 792, 882 and 756 kN, respectively. Despite the fact that the axial strength of the GFRP bars in Segment D was about 8.6 times the strength of the bottom transverse steel bars in Segment C, the failure loads of the Segments C and D were similar. This observation confirmed that only the transverse bottom reinforcement of a deck slab governs the load carrying capacity of a deck slab. The tests also confirmed that the axial stiffness of the bottom transverse reinforcement – and not its axial strength – governs the load carrying capacity of the slab. The top mesh of the reinforcement and the bottom longitudinal bars were found to have no influence on the strength of the slab.

4.4 Externally Restrained Deck Slabs

Deck slabs which rely on external restraints, and not embedded reinforcement, for strength will now be referred to as 'externally restrained deck slabs.' It is well-known that de-icing salts or the vicinity of salt-laden sea water cause the steel in concrete to corrode, leading to the spalling of the concrete. The detrimental effect of such spalling, in the bridge deck slabs of hundreds of thousands of bridges in North America and elsewhere, is common knowledge.

The problem of corrosion of steel in concrete deck slab has, to date, been addressed in a number of ways including (a) coating the steel reinforcing bars by epoxy and other protective materials; (b) increasing the depth of concrete cover over the steel bars, thereby increasing the overall thickness of the slab; and (c) using dense concrete mixes. These costly measures have certainly improved the durability of concrete deck slabs; however, the problem of corrosion has not been eliminated completely and; its partial solution leads to a thicker deck slab.

Corrosion in a concrete deck slab can be eliminated entirely by replacing the steel reinforcement by bars or grids of carbon fibre reinforcement polymers (CFRPs), glass fibre reinforced polymers (GFRPs) or other similar new chemically-inert materials, which are discussed in Chap. 8. However, two factors have so far been inhibiting the widespread use of these new materials in bridges: (a) those fibres with a modulus of elasticity close to that of steel are still too expensive for such widespread use; and (b) conversely, those fibres which are relatively inexpensive have so low a modulus of elasticity that they are unsuitable as efficient tensile reinforcement in concrete.

The predominantly compressive forces induced due to the internal arching system in suitably-confined deck slabs subjected to concentrated loads have prompted the authors and their colleagues to develop a concrete deck slab that is entirely devoid of embedded tensile reinforcement. For crack control the concrete is provided with bars of low modulus GFRPs. In some earlier deck slabs, the concrete cracks were controlled by adding low-modulus polypropylene to concrete; however, as discussed later, the addition of fibres alone to concrete had its own problems.

The feasibility of an externally restrained deck slab containing only the inexpensive polypropylene fibres was studied experimentally. The experimental studies, consisting of tests to failure under concentrated loads, on a number of half-scale and full-scale models, are reported by Mufti et al. (1993), Bakht and Agarwal (1993), Selvadurai and Bakht (1995), and Thorburn and Mufti (1995). Some details of these studies have already been presented in Sect. 4.2; others will be presented in this Subsection, which also contains the details of several bridges that incorporate the steel-free deck slab.

4.4.1 First Experimental Study

The experimental program for studying the suitability of ferrous-free concrete deck slabs was somewhat unconventional in that it did not follow the usual practice of planning all the experiments in advance. In this study, each experiment, except the first, was conceived after studying the results of the preceding ones. This procedure led to an acceptable solution after a relatively small number of experimental iterations. Only four models were tested; these are described below, along with the lessons learnt from each (Mufti et al. 1993).

First Model Some details of the first model are presented in Fig. 4.3 and the test results are partly discussed in Sect. 4.2.1 with the help of Fig. 4.4.

The deck slab concrete contained 38 mm long fibrillated polypropylene fibres (FORTA Corporation). These fibres were added to the ready-mixed concrete just prior to placement in the amount of 0.34 % by weight (or 0.88 % by volume). Immediately prior to placement, the necessary degree of workability of concrete to cast the slab was achieved by adding water rather than by the use of the customary super-plasticizer. The concrete did not contain any steel reinforcement.

The deck slab was tested under a central rectangular patch load measuring 257 $\times$127 mm, with the latter dimension being in the longitudinal direction of the bridge. The load was applied through a thick steel plate and a thin neoprene pad to represent the dual tires of a heavy commercial vehicle. The deck slab of the first model failed at a load of 173 kN. Disappointingly, as noted earlier, the mode of failure was not that of punching shear, as is observed in deck slabs with conventional steel reinforcement; it was flexural and similar to that observed by Beal (unpublished report) in an unreinforced deck slab.

Second Model Realizing that the deck slab of the first model lacked lateral restraint at the bridge supports, the collapsed deck slab was carefully removed and end diaphragms added to the steel frame work. With the addition of these end diaphragms, which consisted of two channels, and a new deck slab, the second model resulted. The deck slab of the second model, having the same dimensions as that of the first, was cast in exactly the same way. This deck slab was also tested under a central rectangular patch load. Once again, disappointingly, the deck slab of the second model did not fail in punch shear. At 222 kN, the failure load was somewhat higher, but the mode of failure was practically the same as that of the deck slab of the first model.

Review of the results of the first two tests led to the realization that in conventionally-reinforced deck slabs, the transverse steel reinforcement participates in restraining the lateral movement of the top flanges of the girders. This restraint permits the development of the arching system which is responsible for the enhanced strength of the slab and the punching shear mode of failure. The diaphragms of the first two models, had been lightly welded to the webs of the girders and could not restrain effectively the lateral movement of the girders above their points of connection at the webs. This lateral movement was obviously enough to prevent the arching action from developing in the first two models.

Third Model The transverse steel in a conventional deck slab provides the necessary lateral restraint to the deck slab for it to fail in punching shear under concentrated loads. The same restraint can alternatively be provided by attaching transverse steel straps to the top flanges of the girders. To test the validity of this hypothesis the third model was constructed by using the steel-work of the second model, the only change being the addition of the straps and lower channels at the intermediate diaphragms.

Shown in Fig. 4.6, these additional steel straps comprised bars of 64 × 10 mm cross-section spaced at 457 mm centres, welded to the underside of the upper flanges of the girders. The deck slab of the third model failed under a central load of 418 kN in a punching shear failure mode, thus confirming the hypothesis that the necessary lateral restraint to the deck slab can be provided by the steel straps. It can be seen in Fig. 4.7 that deck slab of the third model, unlike those in the first two models, the failure was highly localized with the rest of the slab remaining virtually undamaged.

Taking advantage of the localized failure under the central load (location 1), the deck slab was tested at two other locations. Locations 2 and 3 were a distance 914 and 457 mm from the closer transverse free edge, respectively, leading to failure loads of 316 and 209 kN; these failure loads are respectively 0.76 and 0.50 times the failure load at the centre. It was obvious that as the load moved towards the unstiffened transverse free edge of the deck slab, the longitudinal restraint declined and the failure mode degenerated towards a flexural one. Contrary to the requirements of the OHBDC (1992), the transverse edges of the deck slab of the third model were not stiffened. As discussed later, the demands for edge-stiffening of a steel-free deck slab are different from those of a slab reinforced with steel bars.

Fourth Model Despite the encouraging results of the tests on the third model, there remained a crucial uncertainty about the ability of the externally restrained deck slab to sustain a pair of concentrated loads which straddle an internal girder and could presumably cause tensile stresses in the concrete above it. A fourth model was constructed to study the behaviour of the slab under pairs of loads, one on each side on an internal girder. As shown in Fig. 4.15, the fourth model was practically the same as the third model except for an additional girder and a larger overall width of the deck slab. The deck slab of the fourth model was cast by using a superplasticizer in the same way as the deck slab of the third model.

The deck slab of the fourth model was first tested under a pair of rectangular patch loads straddling the middle girder at the mid-span of the model. This test location is identified as location 1 in Fig. 4.15. The test at this location resulted in simultaneous punching shear failure under the two loads, with each loading pad carrying a load of 418 kN. Of particular note is the fact that the failure under the two loads occurred simultaneously and in identical patterns, with the punch out area at the top surface being of the same shape and size as the patch loads. It is highly

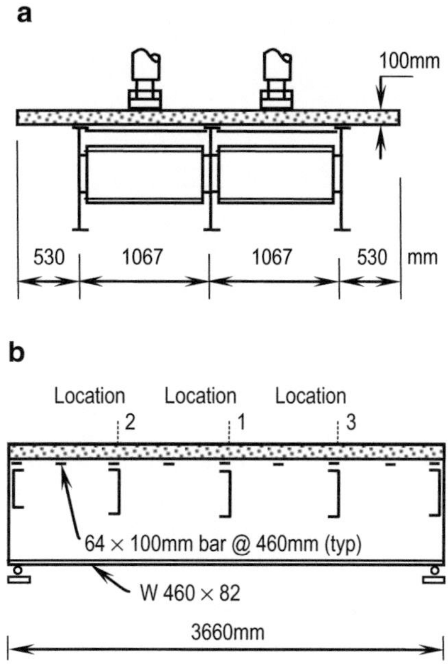

Fig. 4.15 Details of the fourth model: (**a**) cross-section, (**b**) longitudinal section

significant, although somewhat fortuitous, that this failure load per loading pad was exactly the same as the failure load for the deck slab of the third model at location 1. This observation confirmed that the externally restrained deck slab with restrained top flanges of the girders could develop the necessary internal arching system even when subjected to concentrated loads straddling transversely on either side of an internal girder.

The test at location 2 led to simultaneous punching shear failure under the two loads at a load of 373 kN per loading pad; this failure load is about 0.89 times the failure load at location 1. The failure at location 3, which was a mirror image of location 2, occurred under only one loading pad and at 0.84 times the failure load at location 1, i.e. at 352 kN. The mode of failure was again that of punching shear. It is noted that although the mode of failure at locations 2 and 3 was that of punching shear, the punched out area of the deck slab in these cases was slightly larger than at location 1, indicating somewhat reduced in-plane restraint.

Load-Deflection Curves Deflections under the points of load application were measured by means of dial gauges; these deflections included the relatively small deflections of the girders as well. Load-deflection curves constructed from the measured deflections at different load levels are reproduced in Fig. 4.16 corresponding to the tests at three locations on the deck slabs of each of the third and fourth models.

The load-deflection curves corresponding to the three tests on the fourth model have the same pattern but slightly different inclinations; they simply confirm that

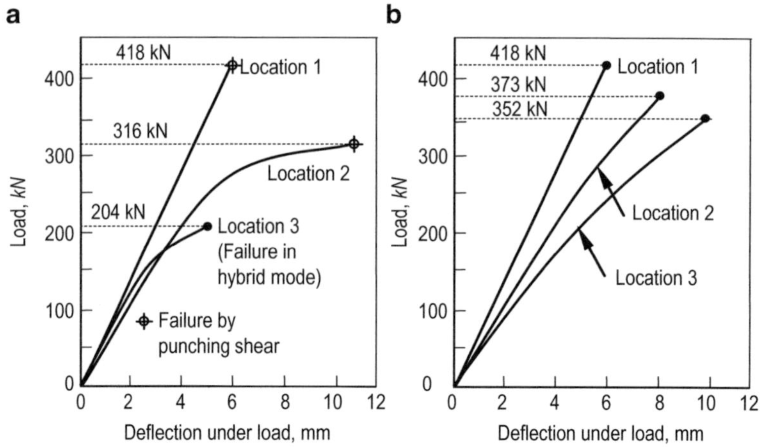

Fig. 4.16 Load-deflection curves for (**a**) deck slab of 3rd model, (**b**) deck slab of 4th model

the degree of internal restraint in the three-girder model diminished slightly, and somewhat irregularly, as the point of load application moved towards a transverse free edge of the deck slab.

The load-deflection curves corresponding to the three tests on the third model tell a somewhat different story. In the initial stages of loading the three curves had similar, nearly-linear, inclinations indicating that the slab had similar stiffness of in-plane restraint at all the three locations. As the load increases, however, the load-deflection curve corresponding to location 2 flattens towards the deflection axis. This softening behaviour, indicating a loss of rigidity, did not however lead to a different failure mode than that observed at locations 2 and 3 of the fourth model.

4.4.2 Second Experimental Study

The OHBDC (1992) implicitly requires that deck slabs at their free edges be stiffened by edge beams that are deeper than the slab. The purpose of such edge beams is obviously to provide a stiff enough component that can sustain the compressive forces developed due to the arching action. In a deck slab with steel reinforcement, part of the in-plane restraint is provided by the steel reinforcement itself, which is effectively tied to the girders. In externally restrained deck slabs, such restraint would have to be provided entirely by the edge beam. It can be appreciated that, for the restraint to be effective, the edge beam should have its major flexural rigidity in the horizontal plane. In addition, there should be some mechanical connection between the deck slab and the edge beam.

In light of the above discussion, it is proposed that, as shown in Fig. 4.17, the edge stiffening for the externally restrained deck slab would be provided by a steel

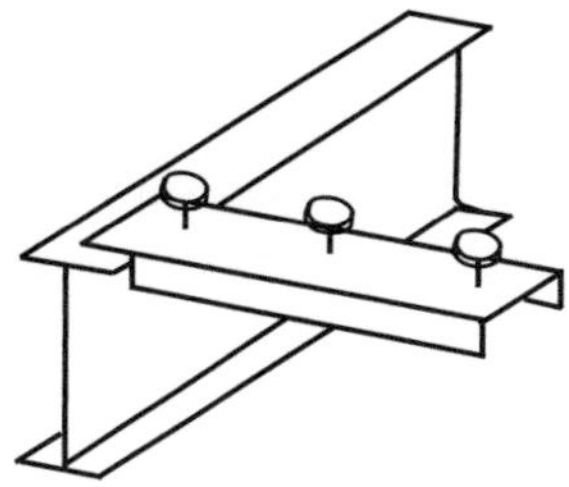

Fig. 4.17 Proposed detail of edge stiffening in deck slabs

channel with its web in the horizontal plane, just above the upper flanges of the girders. The usual shear studs installed on the web, which are also shown in this figure, are to ensure that the horizontal in-plane forces developed in the deck are transferred through the edge beam to the girders. Other permissible edge stiffening details are given in Sect. 4.5.2. The tests on the model are reported by Bakht and Agarwal (1995).

Effect of Angle of Skew The empirical method of the OHBDC is permissible only when the skew angle of the girder bridge is less than 20°. It was postulated that because of the arching action, the load carrying capacity of the deck slab is unlikely to be affected by aberrations in its vertical support system such as those occurring in a skew bridge. It was further postulated that the load carrying capacity of the deck slab near its skew edges can be enhanced significantly by stiffening the edges by the system of Fig. 4.17. To test the latter postulate, a scale-model of a slab-on-girder bridge was constructed. Similarly to the models tested by Mufti et al. (1993), the concrete of the deck slab of this model was mixed with chopped polypropylene fibres.

The externally restrained deck slab was chosen for the study at hand because, as discussed earlier, its mode of failure is highly dependent upon the degree of external in-plane restraint. If the proposed concept could be shown to work for deck slabs without internal reinforcement, then it could be predicted confidently that it would also be effective in conventionally-reinforced deck slabs which have additional in-plane restraint provided by the internal steel reinforcement.

Details of the Model Details of the steel work for the 2/5th scale model of a composite skew bridge are shown in Fig. 4.18a. As can be seen in this figure, the model had a skew angle of 45° and comprised three girders, the top flanges of which were interconnected by straps, and end diaphragms of channel sections. At one end, the channel diaphragm had its web in the vertical plane; it was devoid of shear connectors and had grease on the flange face which was in contact with the slab. The channel diaphragm at the other end had its web in the horizontal plane with frequently-spaced shear connectors on it. An 80 mm thick concrete slab having its concrete mixed with 38 mm long fibrillated polypropylene fibres (FORTA Corporation) was cast on the girders. As shown in Fig. 4.18b, the deck slab had a 400 mm long overhang beyond one outer girder and no overhang beyond the other.

Test Results About 2 months after its casting, the deck slab was tested to failure under a pair of concentrated loads placed successively at each of nine locations.

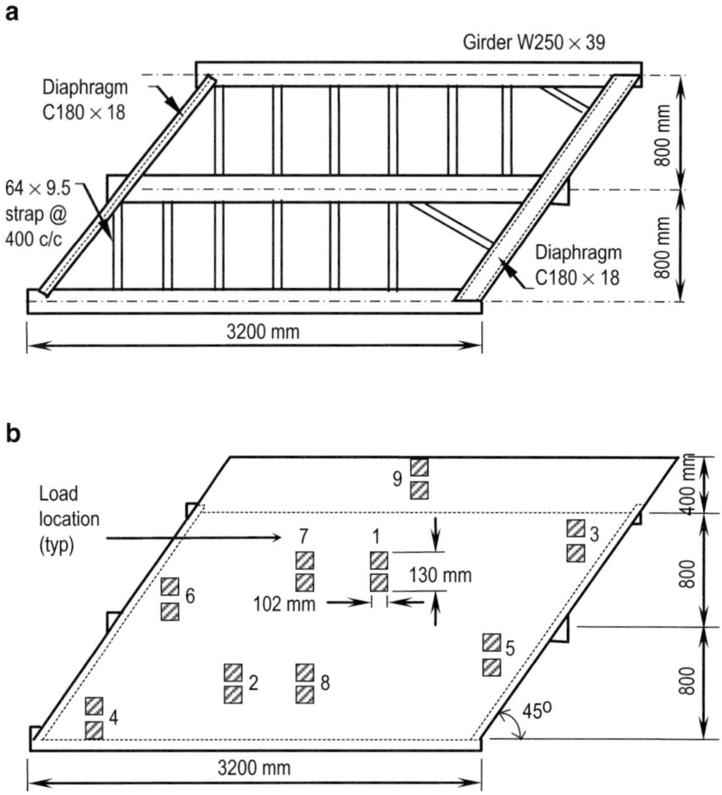

Fig. 4.18 Details of the model of a skew slab-on-girder bridge: (a) plan of steel frame, (b) plan of concrete deck slab showing test locations

These locations are identified in Fig. 4.18b which also shows the contact sizes of the loads which were selected to represent the dual tires of typical heavy commercial vehicles.

As can be seen in Fig. 4.18b load locations 1 and 2 are well away from the skew supports; accordingly, the strength of the deck slab at these locations is treated as the datum strength, which is expected to be free from the effects of skew supports. The slab failed at both these locations in a punching shear mode at loads of 323 and 352 kN, respectively. Locations 3 and 5 are close to the composite end diaphragm; the slab at these locations also failed in a punching shear mode at slightly higher loads, being 363 and 386 kN, respectively. It was found that except for some softening at higher loads, the behaviour of the deck slab in the vicinity of the composite end diaphragm was not much different from that at locations remote from the skew supports.

Locations 4 and 6 near the non-composite end diaphragm are mirror images of locations 3 and 5, respectively. Tests at locations 4 and 6 led to failure in a hybrid

Fig. 4.19 The skew deck slab after failure tests at nine locations

mode at 180 and 232 kN, respectively. It is obvious that the composite diaphragm is far more effective in enhancing the load carrying capacity of the deck slab than the non-composite diaphragm.

The punching shear mode of failure at each of locations 1, 2, 3 and 5 was the familiar one in which the damage, highly localized at the top surface, extends to a much wider area at the bottom surface. The damage at the bottom surface in tests at locations 1 and 2 encroached close to, although not up to, locations 7 and 8, respectively.

Despite such damage, the slab at locations 7 and 8 failed in punching shear mode at somewhat lower load levels, being 252 and 251 kN, respectively. The top surface of the deck slab after tests at all the nine locations can be seen in Fig. 4.19.

Consistent with their intended function as ties to the arch within the slab, the transverse straps in the vicinity of the applied load and beyond were subjected to very high tensile strains as failure load was approached.

Effect of Overhangs The commentary to the first edition of the OHBDC (1979) states that in order to provide lateral restraint to the deck slab, it is necessary to have a minimum of 1.0 m long overhang beyond each outer girder. This contention is negated by the fact that in the model described herein the failure load at location 2 was somewhat higher than that at location 1, it being noted that the former lies in a panel which does not have an overhang. Arguably, one can contend that the overhang is necessary for providing a development length for the bottom reinforcement which develops high tension above the outer girder. If this were the basis for requiring a minimum of 1.0 m long overhang, then the provision which permits an integral curb with an equivalent area of cross-section becomes open to question. It is obvious that at least some of the prerequisites for the applicability of the OHBDC empirical method for deck slab design were formulated from considerations of flexural components; these prerequisites should be reviewed and revised suitably.

Observed Low Strains in Bottom Reinforcement As claimed earlier, the transverse steel bars in the bottom mesh of a conventionally reinforced deck slab serve

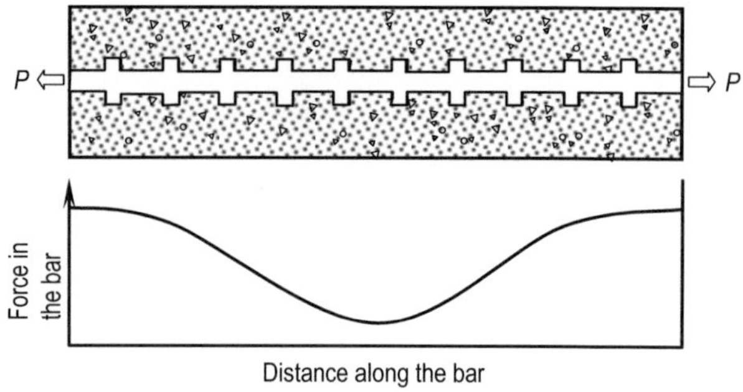

Fig. 4.20 Variation of tensile force in a reinforcement bar embedded in concrete

the same function as the transverse straps of the model. Despite this claim, and contrary to the observation made immediately above, the steel reinforcement has been known to experience very small strains (Dorton et al. 1977; Beal 1982). This apparent discrepancy can be explained with the help of Fig. 4.20, which shows a deformed steel bar embedded in concrete and subjected to a tensile force *P* at its ends. The protrusions of the bar gradually transfer forces to the concrete so that, until the concrete has developed a stable system of cracks, the net tensile force in the bar decreases towards a minimum at the middle. Conditioned by the behaviour of flexural components, in which the flexural strains are highest near the middle, the researchers investigating strains in deck slabs had attempted to measure the strains in the steel reinforcement midway between the girders. Had they measured strains in the bottom reinforcement over the girders, they would probably have recorded high tensile strains. By citing experimental evidence, Bakht (1996) has shown that, during the early life of deck slabs, the tensile strains in the bottom transverse reinforcement of reinforced concrete deck slabs have, indeed, the same pattern as shown in Fig. 4.20.

Conclusions from Second Experimental Study In light of the test results reported in this sub-section, it was concluded that the OHBDC empirical method is also applicable to the deck slabs of skew bridges provided that their end diaphragms are made composite with the deck slab in such a way that they provide a suitably high restraint to in-plane axial forces developed in the slab; this conclusion has already been incorporated in the Canadian Highway Bridge Design Code (CHBDC 2000, 2006).

It is interesting to note that the bridge designers of the MTO, appreciating the need for such restraint, have already designed composite diaphragms; the details of one such diaphragm employed in a right bridge are presented in Fig. 4.21. It is obvious that the composite end diaphragms discussed earlier can, and should, also be used with advantage in deck slabs of right bridges.

The CHBDC (2000) has also removed from the empirical method of deck slab design several other conditions of applicability. For example, for this method to be applicable, it is no longer necessary to have a minimum of three girders.

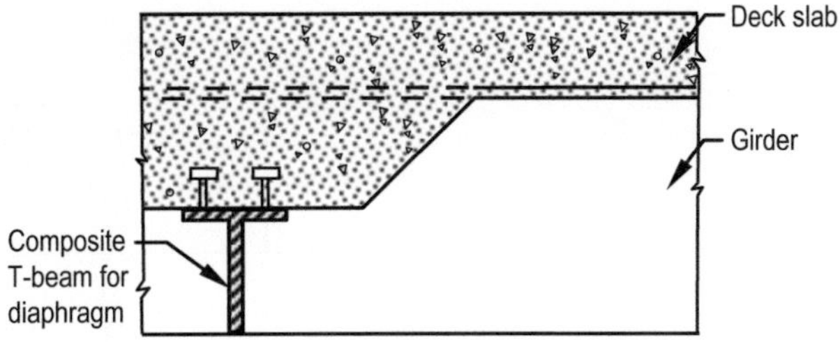

Fig. 4.21 Detail of edge-stiffening employed in bridges designed by the MTO

4.4.3 Reinforcement for Negative Transverse Moments

The concept of the externally restrained deck slab with external steel straps, discussed in Sects. 4.4.1 and 4.4.2, is only applicable to that portion of the deck slab which lies transversely between the two outer girders and which is not subjected to negative, i.e. hogging, transverse moments due to loads on the cantilever overhangs. When a deck slab has either or both cantilever overhangs and barrier wall upstands, certain portions of it are subjected to significant transverse negative moments. To sustain these negative moments, the deck slab has to contain some kind of tensile reinforcement. In addition, tensile reinforcement is also needed to connect the barrier wall with the deck slab.

A new kind of connection between the barrier wall and the deck slab has been developed in Ontario, Canada. Brief details of this development, reported by Maheu and Bakht (1994), are presented in this sub-section which also describes the application of non-ferrous tensile reinforcement for transverse negative moments in both the deck slab and barrier walls.

Barrier Wall Connection The cross-section of a New Jersey type barrier, used commonly in North America, is shown in Fig. 4.22. Cast-in-place New Jersey type barrier walls are generally reinforced with a combination of vertical and horizontal steel bars.

The vertical bars in the barrier wall typically follow the bi-planar profile of the wall face, and a sufficient number extend from the toe of the wall into the deck slab where they are tied to the slab reinforcement, ensuring flexural continuity across the joint. In new construction, these bars are placed in the deck slab before it is cast. In bridge deck rehabilitation, it is often necessary to provide the connection by means of deformed dowel bars secured by epoxy into holes drilled in the deck.

It was postulated that the connection between the barrier wall and deck slab can be provided by double-headed tension bars placed as shown in Fig. 4.22. This connection was foreseen to compel the barrier wall under horizontal loads to behave like a cantilever with its root at the top level of the bars. In this case, the tensile reinforcement for the barrier wall below the root is needed only to provide a sufficient development length.

The validity of the above concept was established through a test on a full-scale model representing a mirror-image arrangement of the deck slab overhang and the barrier wall. The half cross-section of the model is shown in Fig. 4.22. It will be

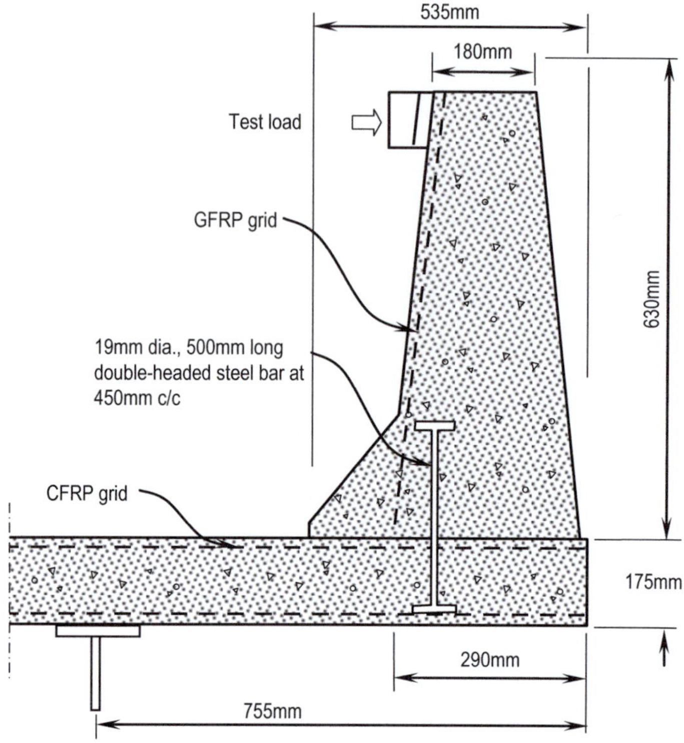

Fig. 4.22 Part cross-section of a full-scale model tested to verify the validity of a proposed connection between the barrier wall and the deck slab

Fig. 4.23 Barrier wall with the design of Fig. 4.22 just after being hit by a heavy pendulum

appreciated that the double-headed tension bar can be made of stainless steel if it is found necessary to have only corrosion free reinforcement in the deck. In later designs, the vertical double-headed tension bars of Fig. 4.22 were inclined, making them parallel to the grid.

Carbon Fibre Reinforced Polymer Reinforcement The test specimen shown in Fig. 4.22 also contained CFRP grids as the tensile reinforcement for negative moments in the deck slab overhangs and GFRP in the barrier walls. Taking advantage of the double-headed tension bar connections, the barrier walls were reinforced with only planar grids, known by the trade name of NEFMAC.

As illustrated in Fig. 4.22, the deck slab overhangs near their top face were also reinforced with a CFRP grid. The test specimen was able to withstand safely the factored design loads thus confirming the validity of both the proposed connection and the use of CFRP grids as tensile reinforcement for negative moments in the deck slab. Pendulum crash tests by the Structures Research Office of the MTO have confirmed that the design of Fig. 4.22 has ample capacity to withstand impact loading from stray commercial vehicles as well (Road Talk 1995). Figure 4.23 shows the new barrier just after it has been hit by the heavy pendulum. The pendulum impacted the barrier through a pair of inflated truck wheels to simulate the truck impact of the barrier wall.

By using the concept presented in this sub-section, in conjunction with the externally restrained deck slab with external transverse steel straps presented in Sects. 4.4.1 and 4.4.2, it is feasible to construct the entire deck slab with corrosion-free reinforcement.

4.4.4 *Static Tests on a Full-Scale Model*

Figure 4.24 shows the cross-section of a full-scale model of an externally restrained deck slab. The steelwork for this model included equally-spaced welded steel straps, the cross-sectional area for which was varied in such a way that the 12 m long deck slab can effectively have four longitudinal segments each with a different transverse confining stiffness. The four confining stiffnesses were 299, 150, 113 and 76 MN/m/m length of the deck slab. Each segment of the deck slab was tested to failure under a concentrated load which was equivalent in contact area to that of a dual tire of a typical commercial vehicle. The failure loads corresponding to the confining stiffnesses noted above were 1127, 923, 911 and 844 kN, respectively. As noted by Thorburn and Mufti (1995), the full-scale testing had provided valuable information, using which the transverse confinement to the deck slab can be optimized.

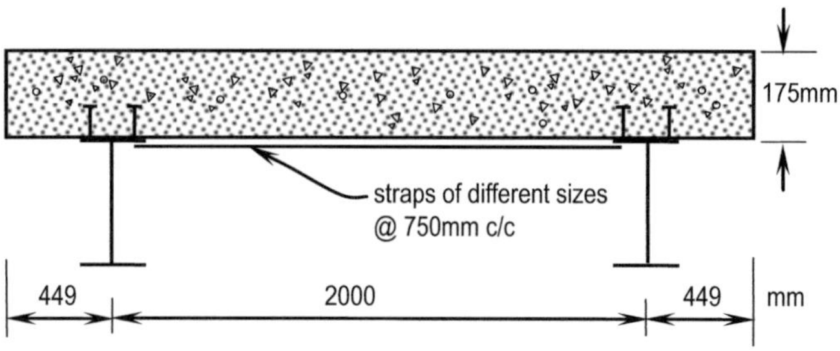

Fig. 4.24 Cross-section of full scale model to investigate the effect of strap stiffness on failure load. Girders are not to scale

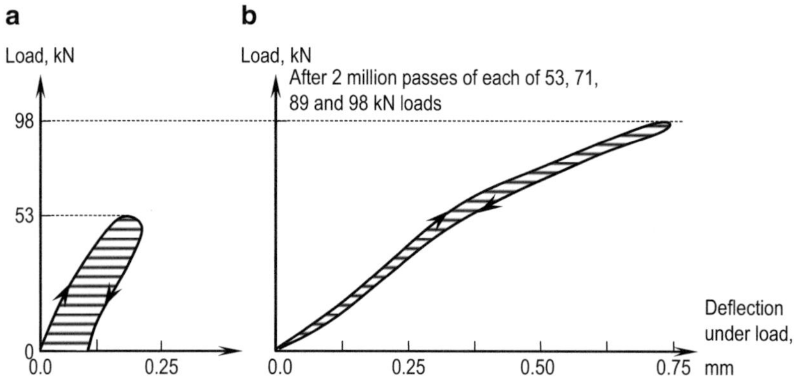

Fig. 4.25 Load-deflection curves (**a**) before dynamic testing, (**b**) after dynamic testing

4.4.5 Rolling Wheel Tests on a Full-Scale Model

A full-scale model with the same cross-section as shown in Fig. 4.24, but with a length of 6.0 m, was constructed to investigate the fatigue resistance of the externally restrained deck slabs under rolling wheels. The transverse straps had an effective axial stiffness of 423 MN/m/m length of the deck slab. As described by Selvadurai and Bakht (1995), the effect of rolling wheels was simulated by means of a number of loading pads at fixed locations. The magnitude of loads on these pads was controlled sequentially in such a way that the load was passed on from one pad to the next according to a predetermined pattern so that during one cycle of loading, the test specimen experienced loads similar to a rolling wheel. The sequential wheel load system employed for testing the full-scale externally restrained deck slab can apply concentrated loads of up to 100 kN at a speed of about 40 km/hr.

The load-deflection curves of the virgin slab are shown in Fig. 4.25a, and with the deflections after the slab had been subjected to two million passes of 53, 71, 89 and 98 kN wheels, respectively are shown in Fig. 4.25b. It can be seen in the latter figure that the externally restrained deck slab, after it had been subjected to a very large number of moving loads, shakes down to an elastic and stable structure.

After the tests noted above, the stiffness of the transverse confinement of the slab was reduced by replacing existing steel straps with those having an axial stiffness of 106 MN/m/m. The slab was then subjected to four million additional passes of a 98 kN wheel. When no deterioration was observed, the deck was saturated under a 25 mm layer of water and subjected to another four million passes of the 98 kN wheel; this was done to explore whether the presence of water has a damaging effect, as reported Matsui (1994), for slab segments which were isolated from the girders. The externally restrained deck slab suffered no noticeable damage even after this test. All the steel straps were then removed so that the transverse confinement was provided by the flexural rigidity of the girder flanges spanning between the end diaphragms and the axial rigidity of the end diaphragms themselves; clearly, the quantification of such confinement is not easy. After the slab without the transverse straps survived additional four million passes of the 98 kN load, it was subjected to a gradually-increasing single concentrated load until it failed in a hybrid mode at about 400 kN.

The maximum wheel load permitted on commercial vehicles in Canada and the rest of the world is less than 60 kN, and the maximum observed wheel load is about 100 kN which is close to the heaviest wheel load of the CL-625 truck of the CHBDC (2006). The externally restrained deck slab was able to sustain several million passes of a 98 kN wheel under both dry and wet conditions and with even minimal transverse confinement. This outcome clearly establishes that the externally restrained deck slab has more than sufficient static strength. The fatigue resistance of deck slabs is further discussed in Sect. 4.5. The deck slab developed a system of cracks during early stages of loading. However, these cracks did not influence the strength and stiffness of the deck slab.

4.5 Fatigue Resistance of Deck Slabs

During its lifetime, a bridge deck slab is subjected to several hundred million truck wheels, the loads on which range from the lightest to the heaviest. Passes of lighter wheels are very large in number, whereas those of the heavier wheels are fewer. By contrast, the laboratory investigation of the fatigue resistance of a deck slab is usually conducted under a test load of constant magnitude. The time available for such investigations is necessarily much smaller than the lifetime of a bridge. Consequently, the test loads are kept large so that the number of passes required to fail the slab in fatigue are manageably small. To the authors' knowledge, no method other than that based on the work of the authors and their colleagues is currently available to correlate the actual wheel loads with the fatigue test loads on deck slabs. The design codes (e.g. AASHTO and CHBDC) are also not explicit with respect to the design fatigue loads on the deck slab. An analytical method, proposed by Mufti et al. (2002), is presented in this section for establishing the equivalence between fatigue test loads and a given population of wheel loads. While the method is general enough to be applicable to all deck slabs of concrete construction, it is developed especially for externally restrained deck slabs, which are relatively new and do not yet have a long track-record of field performance.

4.5.1 Wheel Loads Data

Commentary Clause C3.6.1.4.2 of the AASHTO Specifications (1998) notes that the Average Daily Traffic (ADT) in a lane is physically limited to 20,000 vehicles; and the maximum fraction of trucks in traffic is 0.20. Thus the maximum Average Daily Truck Traffic (ADTT) in a lane is 4,000. When two lanes are available to trucks, the number of trucks per day in a single lane, averaged over the design life, ($ADTT_{SL}$) is found by multiplying ADTT with 0.85, giving $ADTT_{SL} = 3{,}400$. It is assumed that the average number of axles per truck is four (a conservative assumption), and that the life of a bridge is 75 years. The maximum number of axles that a bridge deck would experience in one lane during its lifetime is $3400 \times 4 \times 365 \times 75 = 372$ million. A well-confined deck slab under a wheel load fails in the highly-localized punching shear mode. Accordingly, the consideration of wheel loads in more than one lane is not necessary.

4.5.2 Number of Cycles Versus Failure Load

The Calibration Report in the Commentary to the Canadian Highway Bridge Design Code (CHBDC 2000) is based on vehicle weight surveys in four Canadian provinces; from this report, it can be calculated that the expected annual maximum axle loads in Canada is 314 kN. The expected maximum lifetime axle loads are about 10 % larger than the annual maximum loads (Agarwal 2002, Private communication), thus leading to the maximum lifetime axle load anywhere in Canada being 345 kN. As noted by Matsui et al. (2001), the maximum axle load observed in Japan is 32 t, or 313 kN. The close correspondence between the expected annual maximum axle weight in Canada and the maximum observed axle load in Japan indicates similarity between the axle loads in the two countries. Matsui et al. (2001) have also provided a histogram of axle weights observed on 12 bridges in Japan. In the absence of data on Canadian trucks, this histogram was used to construct the wheel load statistics, which are shown in Table 4.3, it being noted that the wheel load is assumed to be half the axle load. This table also includes the numbers of wheels of various magnitudes, corresponding to a total of 372 million wheels. Any fatigue test load on a bridge deck slab should induce the same damage in the slab as the damage induced by all the wheel loads included in this or a similar table.

A given number of cycles N of a load P can be equated to N_e cycles of an experimental load P_e only on the basis of an established relationship between

Table 4.3 Statistics of wheels loads for a total of 372 million wheels (Adopted from Matsui et al. (2001))

Wheel weight, tonnes	Percentage of total	No. of wheels, in millions
1	21.25	79.05
2	32.06	119.26
3	21.61	80.39
4	12.60	46.87
5	6.48	24.11
6	3.24	12.05
7	1.44	5.37
8	0.54	2.01
9	0.32	1.19
10	0.18	0.67
11	0.11	0.41
12	0.07	0.26
13	0.04	0.15
14	0.02	0.07
15	0.01	0.04
16	0.004	0.01

P and *N*. Matsui and his colleagues in Japan are the only researchers (Matsui et al. 2001) who have provided a *P-N* relationship based on rolling wheel tests on full-scale models of both reinforced concrete and reinforcement-free deck slabs; their conclusions are quantified by the following equation, which is applicable to both reinforced and un-reinforced slabs.

$$\log\,(P/P_s) = -0.07835 \times \log(N) + \log(1.52) \tag{4.1}$$

where P_s is the static failure load. Equation (4.1) gives P/P_s greater than 1.0 for *N* smaller than about 500. Matsui (2001, Private communication) contends that this equation is valid only for *N* greater than 10,000.

North American Researchers (Petrou et al. 1993; Perdikaris and Beim 1988) and Korean Researchers (Youn and Chang 1998) have also presented similar relationships. However, their results are based on tests on models having scales of 1–6.6 and 1–3.3, respectively. The slight differences between the Matsui et al. relationship and those by the above researchers could be attributed to the effect of scaling in the models.

Several standard cylindrical specimens of 35 MPa concrete were tested in the University of Manitoba under compressive fatigue loads (Memon 2005). Notwithstanding the inconclusive nature of results of tests on some specimens, it was observed that the concrete cylinders do fail in fatigue under compressive loads; the fatigue failure loads are smaller than the 'static' failure loads. Cylinders under higher loads fail under smaller number of cycles. Similar observations have also been made by others (Dyduch and Szerszen 1994). An intuitive interpretation of the results of tests by Memon (2005) led to the following variation of the Matsui et al. (2001) equation, i.e. Eq. (4.1).

$$P/P_s = 1.0 - 1n(N)/30 \tag{4.2}$$

For N greater than 10,000, this simple equation gives nearly the same results as Eq. (4.1). It also gives the correct result for $N = 1$. In the absence of a more-reliable relationship, the above equation is used to determine equivalent number of cycles of various loads. The following notation is introduced.

P_1 and P_2 are two different wheel loads; n_1 and n_2 are the corresponding number of passes of P_1 and P_2, respectively, so that the two loads have the same damaging effect; N_1 and N_2 are the limiting number of passes corresponding to P_1 and P_2, respectively; $R_1 = P_1/P_s$; and $R_2 = P_2/P_s$.

It is assumed that for ratio R_i, the cumulative damage to the deck slab is proportional to (n_i / N_i) m, where n_i is the number of passes of the load, N_i is the limiting number of passes, and m is any value larger than or equal to 1. It can be shown that the following relationship holds true for any value of m.

$$N_2/N_1 = n_1/n_2 \tag{4.3}$$

The significance of Eq. (4.3) is that it is equally applicable to linear and non-linear models of damage. Equations (4.2) and (4.3) lead to the following equation for n_2

$$n_2 = n_1 \times e^S \tag{4.4}$$

where

$$S = (R_1 - R_2) \times 30 \tag{4.5}$$

Consider an externally restrained deck slab that has a static failure load (P_s) of 100 t (979 kN). By using Eqs. (4.4) and (4.5), it can be shown that for this deck slab, the wheel loads of Table 4.3 are equivalent to 105 million passes of a 7.5 t (73 kN) wheel, or 49 million passes of a 10 t (98 kN) wheel, or 173,800 cycles of a 25 t (244 kN) wheel load, or 6115 passes of a 40 t (391 kN) wheel, or 510 passes of a 60 t (587 kN) wheel, and so on.

4.5.3 Fatigue Tests on Externally Restrained Deck Slabs

A full-scale externally restrained slab was tested in Dalhousie University under simulated rolling wheel loads. The static failure load (P_s) of a similar slab was found to be 986 kN. This slab has already withstood more than 50,000 passes of a 394 kN load, which is 40 % of P_s. About 6000 passes of this test load are equivalent to all the wheel loads that the deck slab is likely to sustain during its lifetime

(Limaye 2004). It can be concluded that the slab under consideration has considerably more fatigue resistance than required.

Another test, reported by Memon (2005), involved the fatigue testing of three 175 mm thick deck slabs, each on two girders spaced at 2 m. Each slab had different crack control and transverse restraining systems. The first slab contained two orthogonal meshes of 15 mm dia. steel reinforcing bars at a spacing of 300 mm in each direction. The second slab was transversely restrained with external steel straps, and contained one orthogonal crack control mesh of 10 mm dia. CFRP bars with transverse bars at a spacing of 200 mm and longitudinal bars at a spacing of 300 mm; the ratio of the volumes of CFRP bars and concrete was 0.34 %. The third slab was also confined transversely by external steel straps, but contained an orthogonal crack control mesh of 13 mm dia. GFRP transverse and longitudinal bars at spacing of 150 and 250 mm, respectively; the ratio of volumes of GFRP bars and concrete was 0.85 %. Both the crack control meshes were placed near the bottom of the respective slab, each with a clear cover of 40 mm.

Each of the three slabs described above was subjected to successive one million cycles of each of 25 t and 50 t loads. After 200,000 cycles of the 25 t load, the maximum crack widths in deck slabs with steel, CFRP and GFRP bars were nearly 0.35 mm. The study confirmed that the maximum crack widths in all the three tested slabs, after they were subjected to the expected lifetime damage, were well within 0.5 mm, the upper limit proposed to be adopted for externally restrained deck slabs. The crack widths in all the three slabs after one million cycles of the 25 t load increased to nearly 0.4 mm, indicating that the cross-sectional area of crack control meshes provided in the tested externally restrained deck slabs was significantly more than required to keep the crack widths within 0.5 mm.

After they were subjected to one million cycles of the 50 t loads, the maximum crack widths in the slabs with steel, CFRP and GFRP bars grew to about 1.5, 1.2 and 0.6 mm, respectively. By the end of this sequence of loads, each slab is estimated to have been subjected to 40,000 times the fatigue damage that it is likely to receive during its lifetime. Although at the end of this loading sequence, their maximum crack widths were bigger than the proposed limit of 0.5 mm, the slabs showed no sign of impending failure. It is noted that the effect of varying temperatures on the fatigue strength of the deck slabs is yet to be studied.

It was decided to subject each slab to the higher pulsating load of 60 t, and continue testing till failure. The outcome of this last sequence of testing, presented in Fig. 4.26, is instructive in comparing the fatigue resistance of slabs with bars of different materials. The reinforced concrete slab with steel bars failed in punching shear after 23,162 cycles of the 60 t load. The externally restrained deck slab with CFRP bars failed after 198,863 cycles of the same load. The externally restrained deck slab with GFRP bars had the best fatigue resistance, failing at 420,684 cycles.

The observation that slabs with steel bars have the worst fatigue resistance and slabs with GFRP bars the best, might appear surprising. However, after some reflection, it becomes obvious that the fatigue resistance of a concrete slab containing bars with a much higher modulus of elasticity than that of concrete should indeed be inferior to the fatigue resistance of a slab, in which both the concrete and embedded bars have similar moduli of elasticity. The interface

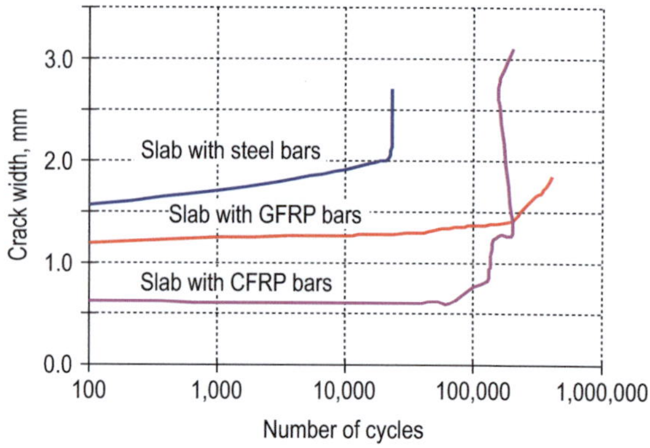

Fig. 4.26 Crack width plotted against number of cycles of a 60 t pulsating load

between a stiff inclusion and a soft surrounding material is clearly subjected to higher fatigue damage than would be the case if the stiffness of both the inclusion and the surrounding material were nearly the same. It is recalled that the modulus elasticity of a steel bar, being about 200 GPa, is more than eight times larger than 24 GPa, the modulus of elasticity of normal weight 35 MPa concrete. On the other hand, the modulus of elasticity of a GFRP bar, lying between 30 and 42 GPa, is much closer to that of concrete.

Perdikaris and Beim (1988) tested several scale models of deck slabs with different amounts of steel reinforcement. The 'orthotropically reinforced specimens' were designed for bending and contained reinforcement that was about 3 5 % by volume of concrete. The 'isotropically reinforced specimens' were designed for arching and contained about 1.2 % by volume of concrete. The following statement by Perdikaris and Beim (1988) confirms the conclusion that reduction of the amount of reinforcement enhances the fatigue resistance of deck slabs: "The fatigue strength of orthotropically reinforced specimens under moving constant wheel loads appears to be lower than that of the isotropically reinforced specimens."

4.6 Bridges with Externally Restrained Deck Slabs

The concept of an externally restrained deck slab was introduced in the technical literature by Mufti et al. 1991. Within 4 years of this introduction, the world's first steel-free deck slab was cast on October 25, 1995, on the Salmon River Bridge on the Trans-Canada Highway in Nova Scotia, Canada. A photograph of the deck slab during casting (Fig. 4.27) shows that the slab was devoid of any embedded reinforcement.

The deck slab of the Salmon River Bridge was transversely restrained by means of steel straps welded to the top flanges of the steel girders (Newhook and Mufti 1996); the contractor for this bridge was to bid separately for reinforced concrete and steel-free (externally restrained) deck slabs. The bid for the latter slab was 6 %

Fig. 4.27 Casting of the steel-free deck slab on the Salmon River Bridge in Nova Scotia, Canada

Table 4.4 Details of Canadian bridges with 1st generation externally restrained deck slabs

Structure	Year of construction	Girder type	Girder spacing, m	Slab thickness, mm
Salmon River Bridge, Kemptown, Nova Scotia	1995	Steel plate	2.7	200
Chatham Bridge, Chatham, Ontario	1996	Steel plate	2.1	175
Crowchild Trail Bridge, Calgary, Alberta	1997	Steel plate	2.0	185
Waterloo Creek Bridge, British Columbia	1998	Precast concrete	2.8	190
Lindquist Bridge, British Columbia	1998	Steel plate	3.5	150

higher than that for the conventional slab. The contractor admitted that the higher bid was a result of perceived difficulties in handling concrete with fibres; his fears were subsequently found to be without foundation.

The externally restrained deck slabs without any crack-control grid are now regarded as the 1st generation steel-free deck slabs. Between 1995 and 1999, five highway bridges in Canada were installed with the 1st generation steel-free deck slabs; some details of these structures are given in Table 4.4. Bakht and Mufti (1998) have provided a summary of the five of these deck slabs. More detailed information can be found in Table 4.5, which also includes the salient features of each of these deck slabs.

Table 4.5 Five salient features of 1st generation externally restrained deck slabs

Structure	Main reference	Salient features
Salmon River Bridge, Kemptown, Nova Scotia	Newhook and Mufti (1996)	1st steel-free deck slab in new construction Welded steel straps
Chatham Bridge, Chatham, Ontario	Aly et al. (1997)	1st steel-free deck slab in rehabilitation Welded steel straps CFRP bars for transverse negative moments Concrete parapet with GFRP bars and double-headed connectors
Crowchild Trail Bridge, Calgary, Alberta	Bakht and Mufti (1998)	1st steel-free deck slab on continuous span bridge Partially studded steel straps GFRP bars for transverse negative moments
Waterloo Creek Bridge, British Columbia		1st steel-free deck slab on concrete girders Partially studded steel straps
Lindquist Bridge, British Columbia	Sargent et al. (1999)	1st application of steel-free deck slab in precast construction Composite action of deck slab with girders through clusters of shear connectors

The design method for 1st generation of externally restrained deck slabs was included in the Canadian Highway Bridge Design Code (CHBDC 2000). Soon after being opened to traffic, all 1st generation externally restrained deck slabs developed about 1 mm wide cracks roughly midway between the girders (Mufti et al. 1999). Through fatigue testing on full-scale models of these slabs, it was confirmed that the presence of even full-depth cracks does not affect the safety of the bridge (Limaye 2004). However, many engineers are not happy with wide cracks in deck slabs without any embedded reinforcement.

Since the presence of cracks in the 1st generation externally restrained deck slabs is considered unacceptable by many engineers, it was decided to include in these slabs nominal crack control grids made preferably of GFRP. Initial estimates of the cross-sectional area and spacing of the GFRP bars in the crack-control grid were established from the results of fatigue testing already conducted (Limaye 2004; Memon 2005). It is expected that the amount of reinforcement in the crack-control grids will be optimized after further tests.

The second edition of the CHBDC, published in 2006, requires that the externally restrained deck slabs be provided with a GFRP crack control grid placed near the bottom of the slab. The externally restrained deck slabs with crack control grids are also referred to as the 2nd generation of externally restrained deck slabs.

The American Concrete Institute (ACI) has prepared a report on bridge decks free of steel reinforcement (ACI 2004); this report, which provides the complete design method with worked examples, also requires that the these deck slabs be provided with crack control grids.

The first externally restrained deck slab of the 2nd generation was cast on North Perimeter Bridge in the Canadian province of Manitoba in 2003. The first externally restrained deck slab in Iowa, USA is also of the 2nd generation; it was cast on a

Fig. 4.28 A 2nd generation externally restrained deck slab in Iowa, USA before casting

bridge in the Tama County, Iowa. As can be seen in the photograph in Fig. 4.28, a crack control grid of GFRP is used in this deck slab.

4.7 Proposed Design Method

Based on the design provisions of AASHTO (1998), CHBDC (2000, 2006), and ACI (2004), a set of design provisions can be formulated which are applicable to the deck slabs of highway bridges anywhere in the world subjected to normal vehicular traffic. These provisions are presented in Sect. 4.7.1 for deck slabs with steel reinforcement, in Sect. 4.7.2 for deck slabs with FRP reinforcement, and in Sect. 4.7.3 for the externally restrained deck slabs.

Throughout most of the world, the maximum permitted load on normal roads for a single axle is not more than 12 tonnes. The design provisions given herein are valid for this maximum permissible axle load, it being noted that allowance has been made for a 100 % exceeding of this limit and also for the dynamic amplification of load effects. A revision to the proposed design provisions may be necessary only in the unusual cases where the deck slab is to be subjected to much heavier axle loads.

It is emphasized that the design provisions given in this section do not cater for the transverse negative moments intensities due to loads on either the deck slab overhangs or on the barrier walls. As will be shown in Chap. 5, these negative moments extend from the overhang into the panels contained between the outermost and immediately adjacent girders.

Unlike the rest of the book the design provisions proposed in Sects. 4.7.1–4.7.3 are given in prescriptive rather than descriptive format. This is so as to permit their ready adoption into design codes. For background information regarding the proposed design provisions, the reader should refer to the material presented in Sects. 4.1–4.6.

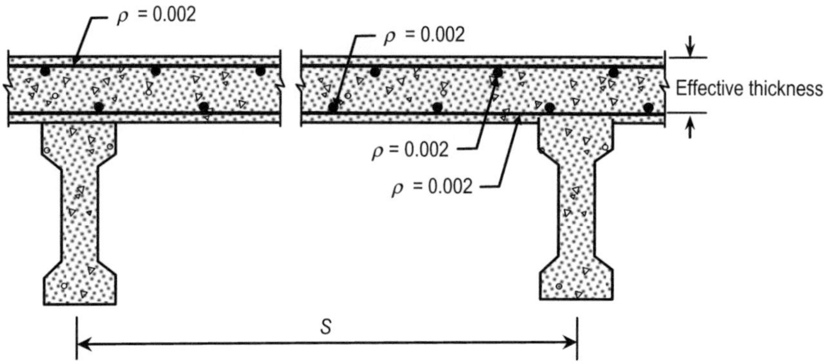

Fig. 4.29 Proposed minimum reinforcement in deck slab

4.7.1 Concrete Deck Slabs with Steel Reinforcement

General The empirical method presented herein is applicable to the design of reinforced concrete deck slabs of composite slab-on-girder bridges in which the centre-to-centre spacing of girders does not exceed 3.7 m. When this method is used, the deck slab need not be analysed, except for the effect of loads on the cantilever overhangs and barrier walls and for the negative longitudinal moment in continuous span bridges, and shall be deemed to have met all the requirements of the relevant design code.

Minimum Deck Slab Thickness Unless a greater thickness is required to provide thicker cover to the reinforcement from considerations of durability, the minimum deck slab thickness shall be the greater of 175 mm and $S/15$, where S is the centre – to-centre spacing of girders. An additional thickness of 10 mm at the top surface of exposed deck slabs shall be provided to allow for wear.

Concrete Strength The concrete used in the deck slab shall have a minimum strength of 30 MPa.

Reinforcement The deck slab shall be reinforced with two meshes of orthogonal steel reinforcement, one near the top surface of the slab and the other near the bottom. The reinforcement ratio ρ, being the ratio of the cross-sectional area of steel and the area of the relevant section of the slab above the centroid of the bottom transverse bars, shall be a minimum of 0.002 for the bars in each direction of each layer. This requirement, which is illustrated in Fig. 4.29, is applicable only when it can be demonstrated that the reinforcement resulting from it is constructible, as for example, through welded wire meshes or cages. If special arrangements cannot be made for the construction and placing of the reinforcement, then the minimum reinforcement ratio in each direction in each mesh shall be taken as 0.003.

Deck slabs of all continuous-span bridges shall have cross-frames or diaphragms extending through the cross-section at all support lines or girders. Steel I-girders

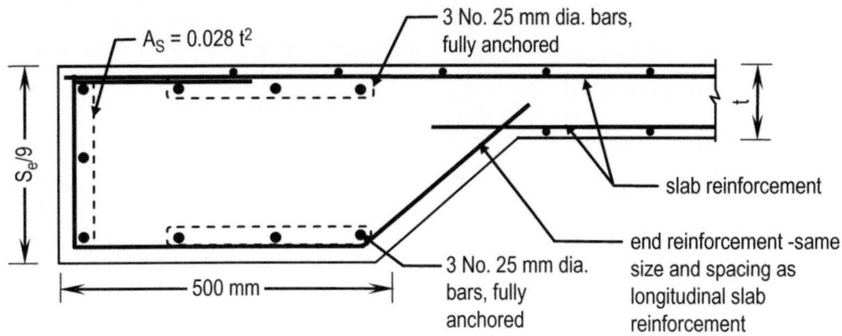

Fig. 4.30 Edge stiffening at transverse free edges

supporting deck slabs designed in accordance with this design method shall have intermediate cross-frames or diaphragms at a spacing of not greater than 8.0 m centre-to-centre.

Except as required in the following, deck slabs on box girders shall have intermediate diaphragms, or cross-frames, at a spacing not exceeding 8.0 m centre-to-centre between the boxes. In lieu of the intermediate cross-frames or diaphragms between the boxes, the deck slab shall contain reinforcement over the internal webs of the boxes, in addition to that required above, to provide for the global transverse bending due to eccentric loads.

Edge Stiffening Free transverse edges of the deck slab at the bridge ends and other discontinuities shall be supported by composite diaphragms either having the details as shown in Fig. 4.30 or with details adopted from CHBDC (2006) Clause 8.18.6. It is noted that t and S_e in this figure are respectively the thickness of the deck slab and the unsupported length of the edge beam.

Overhangs The transverse length of the deck slab overhangs beyond the outermost girders shall be equal to or greater than the development length of the transverse reinforcement in the bottom layer.

4.7.2 Concrete Deck Slabs with FRP Reinforcement

General Fibre reinforced polymers (FRPs), discussed in Chap. 8, are inert to chlorides in concrete. While these building materials are somewhat more expensive than steel reinforcement, they can be cost-effective in corrosive environments. For example, bridge deck slabs exposed to chlorides from either deicing salts or marine environment are prone to rapid corrosion. These slabs can be designed with advantage with FRP bars. The following provisions make use of the arching action in the slab which must be made composite with the supporting beams or girders.

Reinforcement For deck slabs with FRP bars, all design provisions of Sect. 4.7.1 shall apply except that the following conditions shall be satisfied in lieu of those related to steel reinforcement.

The deck slab shall contain two orthogonal assemblies of FRP bars with the clear distance between the top and bottom transverse bars being a minimum of 55 mm.

For the transverse FRP bars in the bottom assembly, the area of cross-section in mm^2/mm shall not be less than $500\, d_s / E_{FRP}$, where d_s is the distance from the top of the deck slab to the centroid of the bottom transverse bars in mm, and E_{FRP} is the modulus of elasticity of the FRP bars in MPa.

Longitudinal bars in the bottom assembly and both the longitudinal and transverse bars in the top assembly shall be of glass fibre reinforced polymer (GFRP) with the minimum reinforcement ratio ρ being 0.0035. As for the steel reinforcement, ρ shall be calculated as the ratio of the area of cross-section of the bars and the area of the relevant section of the slab above the centroid of the bottom transverse bars.

The minimum cover to the FRP bars shall be 35 mm with a construction tolerance of ± 10 mm.

4.7.3 Externally Restrained Deck Slabs

An externally restrained deck slab supported on girders or stringers, being the supporting beams, and satisfying the following conditions need not be analyzed except for negative transverse moments due to loads on the overhangs and barrier walls, and for negative longitudinal moments in continuous span bridges.

Composite Action The deck slab is composite with parallel supporting beams in the positive moment regions of the beams.

Beam Spacing The spacing of the supporting beams, S, does not exceed 3000 mm.

Slab Thickness The total thickness, t, of the deck slab including that of the stay-in-place formwork if present is at least 175 mm and not less than $S/15$.

Diaphragms The supporting beams are connected with transverse diaphragms, or cross-frames, at a spacing of not more than 8000 mm.

Straps The deck slab is confined transversely by means of straps, and the distance between the top of the straps and the bottom of the slab is between 25 and 125 mm.

The spacing of straps, S_l, is not more than 1250 mm, and each strap has a minimum cross-sectional area, A, in mm^2, given by:

$$A = \frac{F_s S^2 S_l}{Et} \tag{4.6}$$

where F_s is 6.0 MPa for outer panels and 5.0 MPa for inner panels, S is the girder spacing in mm, S_l is strap spacing in mm, and E is the modulus of elasticity of the material of the strap in MPa.

The direct or indirect connection of a strap to the supporting beams is designed to have a shear strength in Newtons of at least 200A.

Shear Connectors Either the projection of the shear connectors in the deck slab, t_s, is a minimum of 75 mm, or additional reinforcement with a minimum t_s of 75 mm is provided having at least the same shear capacity as that of the shear connectors.

Cover to Shear Connectors The cover distance between the top of the shear connecting devices and the top surface of the deck slab shall be at least 75 mm when the slab is not exposed to moisture containing chlorides; otherwise, either this cover distance is at least 100 mm, or the shear connecting devices are provided with a coating approved by the authority having jurisdiction on the bridge.

Crack-Control Grid The deck slab is provided with a crack control orthogonal grid of GFRP bars, placed near the bottom of the slab, with the area of cross-section GFRP bars being at least 0.0015 t^2 mm^2/mm. In addition, the spacing of transverse and longitudinal crack control bars is not more than 300 mm.

Fibre Volume Fraction For deck slabs with only one crack control grid, the fibre volume fraction shall be at least 0.002, but shall not exceed 0.005. For deck slabs with two reinforcement grids, no fibre need be added to the concrete.

Edge stiffening The transverse edges of the deck slab are stiffened by one of the composite edge beams shown in Fig. 4.31 or their equivalent in transverse flexural rigidity.

Longitudinal Transverse Negative Moment For continuous span bridges, the deck slab contains longitudinal negative moment reinforcement in at least those segments in which the flexural tensile stresses in concrete due to service loads are larger than 0.6f_{cr}, where f_{cr} is calculated as follows.

$$f_{cr} = 0.4\sqrt{f'_c} \tag{4.7}$$

where both f_{cr} and the specified compressive strength of concrete f'_c are in MPA.

4.8 Analytical Method for Predicting Failure Load

An analytical method has developed by Mufti and Newhook (1998) to predict accurately the failure load of externally restrained deck slabs under concentrated loads. The various assumptions on which the theory of the analytical method is based are as follows.

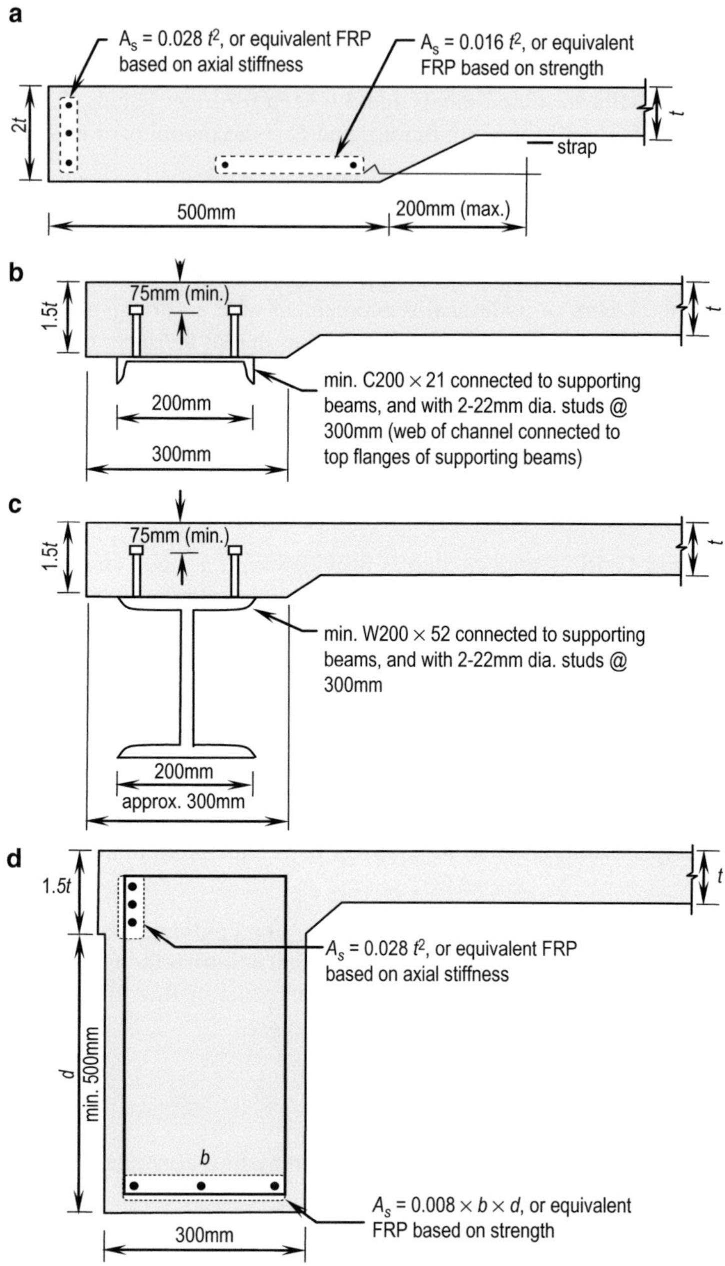

Fig. 4.31 Details of permitted edge stiffening for steel-free deck slabs (**a**) with thickened deck slab, (**b**) with steel channel, (**c**) with steel I-beam, (**d**) with concrete beam

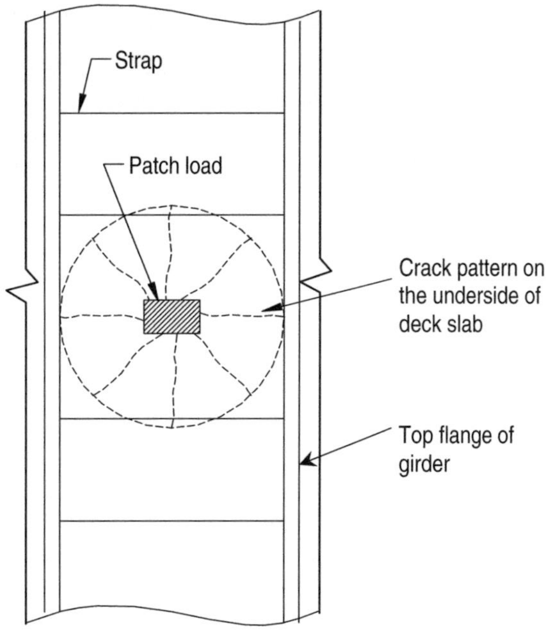

Fig. 4.32 Radial crack pattern

Radial Cracks A concrete deck slab subjected to a central concentrated load, which is well below its failure load, develops a system of radial cracks on its bottom surface. As shown in Fig. 4.32, the extremities of the radial cracks at the bottom surface form a circle, the diameter of which is equivalent to the clear spacing between the top flanges of the girders. As the magnitude of the load level is increased, the radial cracks gradually migrate to the top surface of the slab to eventually become full-depth cracks, but in such a way that the area of the failed surface at the top becomes the same as the area of the patch load. The inclined shear cracks develop at load levels somewhat below the failure load. At punching failure, these inclined cracks form the upper surface of the frustum of a cone, which is punched out, and which will hereafter be referred to as simply the 'cone.' The sections of the cone can be divided into a number of 'wedges' bound by the shear and radial cracks and the outside edge of the slab. At failure, these wedges act as rigid bodies rotating in the radial direction about a centre of rotation.

Compressive Stresses Under Confinement It is assumed that the conical shell region at the intersection of the wedges with the loaded area experiences very high compressive stresses, which the concrete is able to sustain because of its confinement.

Centre of Rotation After the appearance of the shear cracks, the centre of rotation (CR) of the wedges is assumed to be located at the root of the shear crack. As the load increases, the CR moves towards the centre of the patch load. For this model, the CR is assumed to be always at the centre of the load, and located in a plane at a distance *y* from the top surface of the slab, as illustrated in Fig. 4.32. For clarity, the punched cone is not shown in this figure.

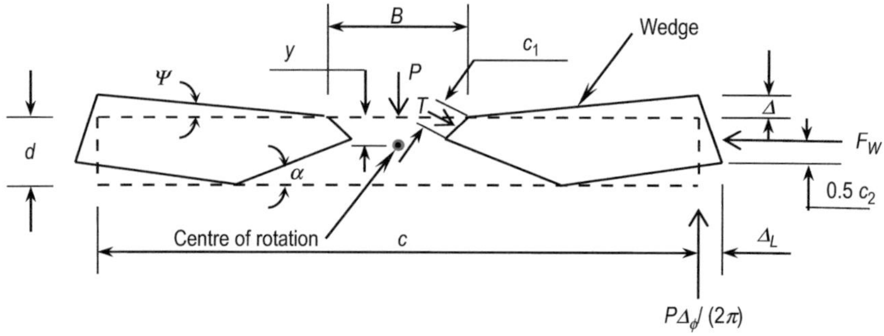

Fig. 4.33 A wedge with an inscribed angle of $\Delta\phi$ under applied and reactive loads

Axi-symmetry The analytical model assumes radial axi-symmetry of both the geometry of the failed area and loading. The loaded area of a deck slab under the dual tire of a vehicle is assumed to be rectangular. For use in the analytical model, the non-circular load contact area is converted, by using the equivalence of the perimeter, into a circular load contact area with diameter B. The equivalent circular slab is defined by the largest circle of diameter C, which can be inscribed between the top flanges of adjacent girders, as illustrated in Fig. 4.31.

4.8.1 Formulation

Following the assumptions discussed above, Eqs. (4.8), (4.9), and (4.10) are developed on the basis of the equilibrium of vertical forces, horizontal forces and moments, respectively.

$$T = \frac{P\Delta\phi}{2\pi\sin(\alpha - \Psi)} \tag{4.8}$$

$$\frac{P}{2\pi}\cot(\alpha - \Psi) + R_r = K - \frac{c}{2}\Psi(d - y) \tag{4.9}$$

$$R_r\left(d - \frac{y}{3} - \frac{c_2}{2}\right) + \frac{P}{2\pi}\cot(\alpha - \Psi)\left[d - \frac{B_1 y}{2} - \frac{c_2}{2} - \Psi\left(\frac{c}{2} - \frac{B}{2}\right)\right] = \frac{P}{2\pi}\left[\frac{c}{2} - \frac{B}{2} + \Psi\left(d - \frac{B_1 y}{2} - \frac{c_2}{2}\right)\right] \tag{4.10}$$

The notation used in the above equations is as shown Fig. 4.33 and as defined in the following.

α = angle of shear crack
Δ_ϕ = angle inscribed by adjacent radial cracks defining a wedge
ψ = angle of rotation of wedge
β_1 = concrete stress block variable
F_w = lateral restraining force for the wedge
T = the confining force

Since these equations are difficult to solve directly, an iterative procedure is used to obtain the three unknowns, being the load P, deflection under the load (Δ), and rotation of the wedge (Ψ). The deflection Δ is increased gradually, and for each assigned deflection, the three equations are solved. The force T gives the confinement stress, from which the confining strain ε is obtained.

Kinnunen and Nylander (1960) established an empirical criterion for the failure of confined concrete under the force T; according to their criterion, the concrete fails when the circumferential strain at the top surface of the deck slab reaches the critical value of 0.0019. Mufti and Newhook (1998) have revised the limiting strain to 0.002, so that when the strain in concrete reaches 0.002, it is assumed that failure has occurred.

Two Closely Spaced Loads The above equations are based on the assumption that the deck slab is subjected to a single patch load. The areas of influence of two closely spaced patch loads overlap, because of which Eqs. (4.8), (4.9), and (4.10) are not strictly applicable; in such cases Mufti and Newhook (1998) propose the following approximate relationship to obtain the failure load $P_u^{'}$ due to loads spaced at a centre-to-centre distance S_u.

$$P_u^{'} = P_u\left(1 + \frac{S_w}{S_g}\right) \tag{4.11}$$

In the above equation, P_u is the failure load due to a single patch load, and S_g is the centre-to-centre spacing of the girders.

4.8.2 *Program PUNCH*

The analytical method discussed above has been incorporated by Newhook and Mufti (2003) into a computer program PUNCH, which can be downloaded from http://extras.springer.com. The followings steps of calculation are used in the program.

(a) Provide input parameters as described later.
(b) Assume an initial value of Δ to be $d/80$.
(c) Calculate ψ wedge rotation from $\Psi = 2\Delta/c$.
(d) Estimate centre of rotation from $y = d/10$.
(e) Calculate α, the shear crack angle.
(f) Calculate the circumferential strain in concrete at the top surface of the deck slab.
(g) Calculate the tensile strain in the steel strap.
(h) When the circumferential strain in concrete reaches 0.002, or the strain in steel strap reaches the specified limiting strain in steel, assume P to be the failure load.

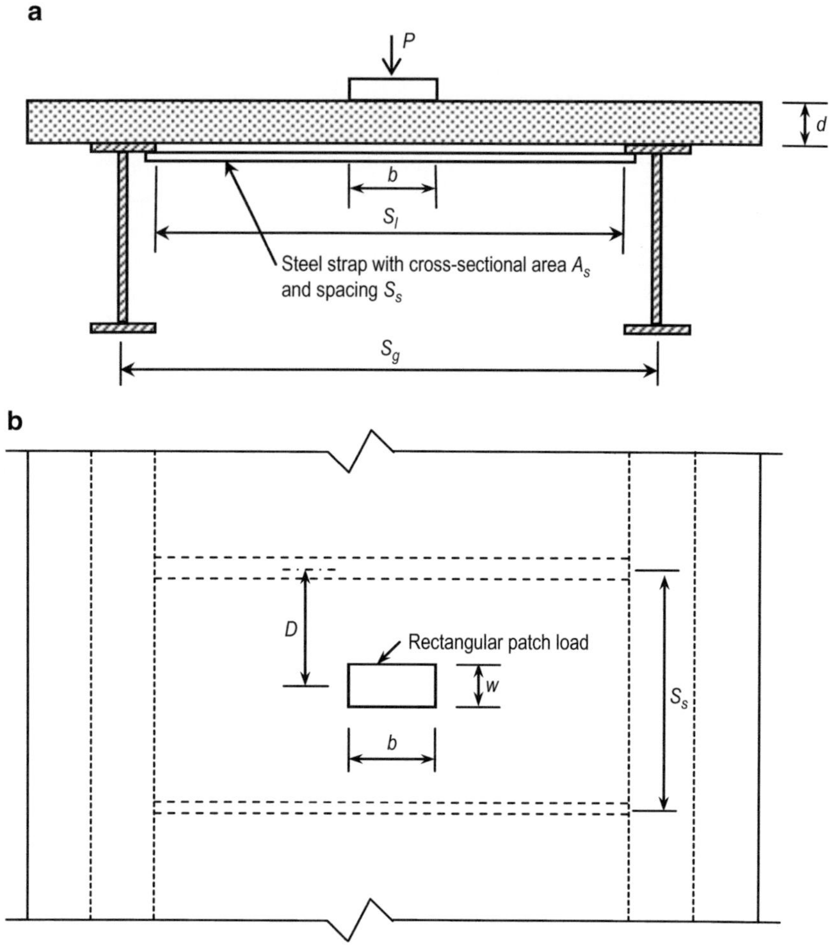

Fig. 4.34 Definition of input parameters used for PUNCH (**a**) cross-section, (**b**) plan

A Triaxial state is assumed for calculating the compressive force at the tip of the wedge using the formula $f'_{cc} = f_c + k\sigma_1$, where σ_1 is the stress under the concentrated load, $= P/A$, in which P is the ultimate load and $A =$ area of tire print; and k is the confinement constant described by Newhook (1997). Complete convergence is assumed if the new value of y is within 0.0001 mm of the previous value. The load corresponding to this step of calculation is taken as the failure load.

Data Input The various input parameters required by program PUNCH are defined in Fig. 4.34 and in the following:

$$B = \frac{2(b + w)}{\pi} \tag{4.12}$$

$$K = \frac{EA_s}{0.5S_l S_s} \tag{4.13}$$

where

K=lateral restraint stiffness per unit length of deck slab
S_ℓ=clear spacing between the top flanges of adjacent girders
S_s = spacing of transverse straps in the longitudinal direction
E=modulus of elasticity of the material of transverse straps
A_s=area of cross-section of each transverse strap
B=diameter of equivalent circular load contact area
b=length of actual load contact area
w=width of actual load contact area

The rectangular stress block parameter β_1 is calculated from the following of the relevant equations which are adopted from CSA A23.3-M84.

$$\beta_1 = 0.85 \quad f_c^{'} \leq 30MPa \tag{4.14}$$

$$\beta_1 = 0.85 - \frac{f_c^{'} - 30}{10} \quad 30MPa < f_c^{'} < 55MPa \tag{4.15}$$

$$\beta_1 = 0.65 \quad f_c^{'} \geq 55MPa \tag{4.16}$$

The order in which the input parameters are required is listed in Table 4.6. A interactive program PUNCHIN can be downloaded from http://extras.springer.com to prepare the input data according to the scheme of Table 4.6. The data is stored in a file named PUNCH.DAT. The program runs with the stored data by clicking on its icon, after which the results are stored in a file named PUNCH.RES.

Example The use of PUNCH is illustrated with the help of the example of an externally restrained deck slab shown in Fig. 4.35. The deck slab has a thickness of 175 mm and the steel straps have a 25 × 10 mm cross-section and 1000 mm centre-to-centre spacing. The modulus of elasticity of steel is 200,000 MPa. The following input data are prepared.

$$S_g = 2000 \text{ mm}$$
$$B = \frac{2(500 + 250)}{\pi} = 477mm$$
$$f_c^{'} = 30MPa$$
$$K = \frac{200,000 \times 2500}{0.5 \times 1670 \times 1000} = 599N/mm$$
$$D = 500 \text{ mm}$$
$$D = 175 \text{ mm}$$
$$\beta_1 = 0.85(\text{from Eq. 4.14})$$
$$k = 10$$
$$A = 500 \times 250 = 125,000 \text{ mm}^2$$
$$\varepsilon_y = 0.002$$
$$\text{Unit indicator} = 1(\text{for metric units})$$

Table 4.6 Details of data input to PUNCH

Input Ref. No.	Input parameter	Notes
1	Girder spacing S_g	It is assumed that all girders in the bridge have the same spacing
2	Diameter of equivalent load circle, B	The actual $b \times w$ rectangular patch area is converted into an equivalent circular area according to Eq. (4.12)
3	Maximum compressive strength of concrete, f_c'	
4	Axial stiffness of strap (or transverse bottom layer of bars), K	The strap stiffness is calculated from Eq. (4.13)
5	Distance of load from the strap measured parallel to the axes of the girder, D	D is calculated from the centre of the strap nearer to the load
6	Thickness of slab, d	For externally restrained deck slabs, d is the actual thickness of the slab; for internally restrained deck slabs, d is the distance between the top of the slab and centroid of bottom transverse reinforcement
7	The rectangular stress block parameter, β_1	β_1 is calculated from the relevant of Eqs. (4.14), (4.15), and (4.16).
8	Confinement factor for triaxial stress condition, k	k is obtained from the relationship provided by Newhook (1997). In the absence of data to justify otherwise, k is assumed to be 10
9	Tire print area, A	$A = b \times w$
10	Yield strain of strap, ε_y	For mild steel, $\varepsilon_y = 0.002$
11	Unit indicator	0 for North American Customary units (in., kips), and 1 for metric units (mm, kN)

The file containing the input data prepared by PUNCHIN for the example deck slab is reproduced in Fig. 4.36.

For the example under consideration, a part of the PUNCH output is reproduced in Fig. 4.37. It can be seen in this figure that the program gives three values for each increment of deflection: α, P and ε (STRN), the last value being the strain in the steel strap. In the case under consideration, the deck slabs fails when the circumferential strain in concrete reaches the limiting strain of 0.002.

Mufti and Newhook (1998) have compared the predictions of PUNCH with the actual failure loads on 15 half- and full-scale models of externally restrained deck slabs; they found that the PUNCH predictions for all deck slabs except 2 were within −7 % to +10 %. For the two exceptional cases, the variation was +20 % and +27 %. Mufti and Newhook (1998) also demonstrated that PUNCH could also give very accurate failure loads for internally restrained, i.e. reinforced concrete, deck slabs, provided that the effective thickness of the deck slab is assumed to be the

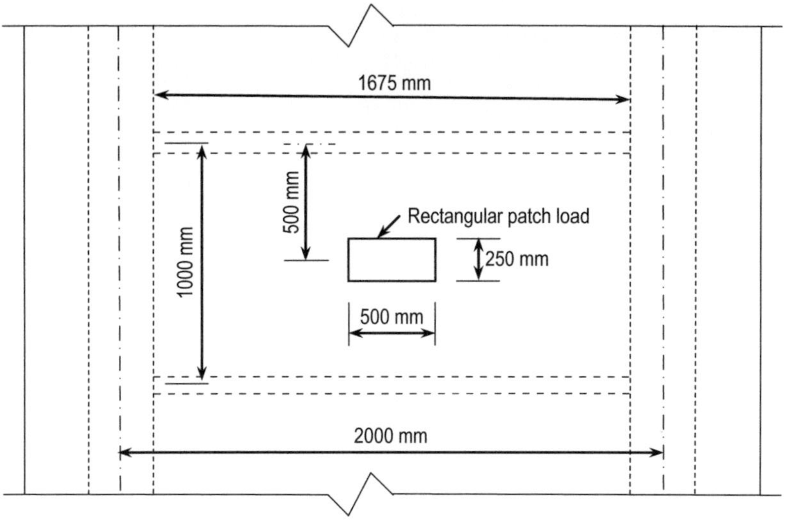

Fig. 4.35 Deck slab example

```
2000.00         477.000         30.0000          599.000
500.00
175.000   0.850000     10.0000        125000.    0.20E-02
1
```

Fig. 4.36 Data input for the example deck slab

distance between the top surface of concrete and the top of the bottom transverse bars as illustrated in Fig. 4.37.

4.9 Other Analytical Method for Predicting Failure Load

Alongside the experimental work reported in Sect. 4.8, analytical studies were also conducted to investigate the behaviour of the steel-free deck slab. The results of preliminary studies using three-dimensional non-linear finite element (FE) analyses are recorded by Wegner and Mufti (1994). While comparisons between the FE and experimental results were not an unqualified success, they were in reasonable correspondence. It was found that the analytical predictions of deck slab behaviour were particularly sensitive to small changes in some of the parameters used in the analysis. The sensitivity of the predicted behaviour to a number of modelling parameters requires the tuning of the analytical modelling scheme. Clearly, such an outcome limits the scope of the FE analysis in the context under consideration.

```
UNITS METRIC  KN, MM, SECOND
Clear Span Between Girders=    2000.00
Diameter of Equivalent Circle for Load=    477.000
Maximum Compressive Stress of Concrete=    30.0000
Elastic Axial Stiffness of Strap=    599.000
Strap to Load Spacing=    500.000
Depth Of Slab=    175.000
Beta to Define Rectangular Stress Block=   0.850000
Concrete Constant used for confinement=    10.0000
Area of Load Patch=    125000.
Yield strain=   0.200000E-02
 DELTA  ALPHA    P        STRN
  0.00  0.00     0.   0.00000
  0.50  19.59    72.   0.00006
  1.00  18.50    138.   0.00012
  1.50  17.67    200.   0.00018
  2.00  17.01    258.   0.00024
  2.50  16.47    314.   0.00029
  3.00  16.01    367.   0.00035
  3.50  15.61    417.   0.00040
  4.00  15.27    466.   0.00045
  4.50  14.95    512.   0.00050
  5.00  14.67    557.   0.00056
  5.50  14.42    601.   0.00061
  6.00  14.18    642.   0.00065
  6.50  13.97    682.   0.00070
  7.00  13.77    721.   0.00075
  7.50  13.58    759.   0.00080
  8.00  13.40    795.   0.00085
  8.50  13.23    829.   0.00089
  9.00  13.08    863.   0.00094
  9.50  12.92    895.   0.00098
 10.00  12.78    926.   0.00103
 10.50  12.64    956.   0.00107
 11.00  12.51    985.   0.00112
 11.50  12.38    1013.   0.00116
 12.00  12.26    1039.   0.00120
 12.50  12.14    1065.   0.00124
 13.00  12.03    1089.   0.00129
 13.06  12.02    1092.   0.00129
*******   Punch Load-Strain in concrete=.002   *******
```

Fig. 4.37 PUNCH output for the example deck slab

References

AASHTO (1998) LRFD bridge design specifications. American Association of State Highway and Transportation Officials, Washington, DC

ACI (2004) Report on bridge decks free of steel reinforcement. ACI-ITG-3-04, Michigan

Alampalli S, Fu G (1991) Influence line tests of isotropically reinforced bridge deck slabs. Client report 54. Engineering Research and Development Bureau, New York State Department of Transportation, New York

Aly A, Bakht B, Schaefer J (1997) Design and construction of a Steel-Free deck slab in Ontario. In: Proceedings of the annual conference of the Canadian Society for Civil Engineering, Sherbrooke

Bakht B (1981) Testing of the Manitou Bridge to determine its safe load carrying capacity. Can J Civ Eng 8(2):218–224

Bakht B (1996) Revisiting arching in deck slabs. Can J Civ Eng 23(4):973–981

Bakht B, Agarwal AC (1993) Deck slabs of skew bridges. In: Proceedings of the annual conference of the Canadian Society for Civil Engineering, vol II

Bakht B, Agarwal AC (1995) Deck slabs of skew bridges. Can J Civ Eng 22(4):514–981

Bakht B, Csagoly PF (1979) Bridge testing, Structural research report SRR-79-10. Ministry of Transportation and Communications, Downsview

Bakht B, Jaeger LG (1985) Bridge analysis simplified. McGraw-Hill, New York

Bakht B, Markovic S (1986) Accounting for internal arching in deck slab design. J Inst Eng (India) 67(CI1):18–25

Bakht B, Mufti AA (1998) Five steel-free bridge deck slabs in Canada. Struct Eng Int (IABSE) 8(3):196–200

Beal BD (1982) Load capacity of concrete bridge decks. ASCE J Struct Div 108(ST4):814–832

CHBDC (2000) Canadian highway bridge design code, CAN/CSA-S6-00. Canadian Standards Association International, Toronto

CHBDC (2006) Canadian highway bridge design code, CAN/CSA-S6-06. Canadian Standards Association International, Toronto

deV Batchelor B, Hewitt BE, Csagoly PF (1978) Investigation of the ultimate strength of deck slabs of composite steel concrete bridges. TRR record no. 664: 162, Transportation Research Board, Washington, DC

deV Batchelor B, Hewitt BE, Csagoly PF, Holowka M (1985) Load carrying capacity of concrete deck slabs. Structural research report SRR-85-03. Ministry of Transportation and Communications, Downsview

Dorton RA, Holowka M, King JPC (1977) The Connestogo River Bridge – design and testing. Can J Civ Eng 4(1):18–39

Dyduch K, Szerszen M (1994) Experimental investigation of the fatigue strength of plain concrete under high compressive loading. Mater Struct 27:505–509

Fang I-K, Worley J, Burns NH, Klinger RE (1990) Behaviour of isotropic R/C bridge decks. ASCE J Struct Eng 116(3)

FORTA Corporation. Fibrous reinforcement Type A-10. 100 Forta Drive, Grove City

Fu G, Alampalli S, Pezze FP III (1992) Long term serviceability of isotropically reinforced bridge deck slabs, Pre-print no. 92-0293. Transportation Research Board, Washington, DC

Hewitt BE, deV Batchelor B (1975) Punching shear strength of restrained slabs. ASCE J Struct Div 101(ST9):1827–1853

Jackson PA, Cope RJ (1990) The behaviour of deck slabs under full global loads. Developments in short and medium span bridge engineering '90. Can Soc Civ Eng 1:253–264

Johnson RP, Arnaouti C (1980) Punching shear strength of concrete slabs subjected to in-plane biaxial tension. Mag Concr Res 32(110)

Khanna OS, Mufti AA, Bakht B (2000) Reinforced concrete bridge deck slabs. Can J Civ Eng 27(3):475–480

Kinnunen S, Nylander H (1960) Punching of concrete slabs without shear reinforcement. Transactions, Royal Institute of Technology, Stockholm, No. 158

Kirkpatrick J, Rankin GIB, Long AE (1984) Strength evaluation of M–beam bridge deck slabs. Struct Eng 62B(3):60–68

Kuang JS, Morley CT (1992) Punching shear behaviour of restrained reinforced concrete slabs. ACI Struct J 89(1):13–19

Limaye VN (2004) Steel-free decks under cyclic loading: a study of crack propagation and strength degradation. Ph.D. thesis, Dalhousie University, Halifax

Maheu J, Bakht B (1994) A new connection between barrier wall and deck slab. In: Proceedings of the annual meeting of the Canadian Society for Civil Engineering, Winnipeg

Malvar LJ (1992) Punching shear failure of a reinforced concrete pier deck model. ACI Struct J 89(5)

Matsui S (1994) New weigh method of axle loads of vehicles and axle weight characteristics of trucks in Japan. In: Proceedings of the 4th international conference on short and medium span bridges, Halifax, pp 533–544

Matsui S, Tokai D, Higashiyama H, Mizukoshi M (2001) Fatigue durability of fiber reinforced concrete decks under running wheel load. In: Proceedings of the third international conference on concrete under severe conditions, Vancouver, vol 1, pp 982–991

Memon AH (2005) Comparative fatigue performance of steel-reinforced and steel-free concrete bridge deck slabs. Ph.D. thesis, University of Manitoba, Winnipeg

Mufti AA, Newhook JP (1998) Punching shear strength of restrained concrete bridge deck slabs. ACI Struct J 95:375–381

Mufti AA, Bakht B, Jaeger LG (1991) FRC deck slabs with diminished steel reinforcement. In: Proceedings of the IABSE symposium, Leningrad, pp 388–389

Mufti AA, Jaeger LG, Bakht B, Wegner LD (1993) Experimental investigation of FRC slabs without internal steel reinforcement. Can J Civ Eng 20(3):398–406

Mufti AA, Newhook JP, Mahoney MA (1999) Salmon river bridge field assessment. In: Proceedings of the 1999 Canadian Society for Civil Engineering annual conference, vol 1, pp 51–61

Mufti AA, Memon AH, Bakht B, Banthia N (2002) Fatigue investigation of steel-free bridge deck slabs. SP-206. In: Concrete: Material Science to Application, a Tribute to Surendra P. Shah. Balaguru P, Naaman A, Weiss W (eds). American Concrete Institute, Farmington Hills, pp 61–70

Newhook JP (1997) The behaviour of steel-free concrete bridge deck slabs under static loading conditions. A thesis submitted to the Faculty of Engineering in partial fulfilment of the requirements for the degree of Doctor of Philosophy, Dalhousie University

Newhook JP, Mufti AA (1996) Steel-free concrete bridge deck -the Salmon River project: experimental verification. In: Proceedings of the annual conference of the Canadian Society for Civil Engineering, Edmonton

Newhook JP, Mufti AA (2003) PUNCH user and theoretical manual. ISIS Canada Research Network, Winnipeg

OHBDC (1979) Ontario highway bridge design code, 1st edn. Ministry of Transportation of Ontario, Downsview

OHBDC (1983) Ontario highway bridge design code, 2nd edn. Ministry of Transportation of Ontario, Downsview

OHBDC (1992) Ontario highway bridge design code, 3rd edn. Ministry of Transportation of Ontario, Downsview

Okada K, Okamura M, Sononoda K (1978) Failure mechanism of reinforced concrete bridge deck slabs. Transportation research record no. 664, Washington, DC

Perdikaris PC, Beim S (1988) RC bridge decks under pulsating and moving load. ASCE J Struct Eng 114(3):591–607

Petrou MF, Perdikaris CP, Wang A (1993) Fatigue behaviour on noncomposite reinforced concrete bridge deck models, Transportation research record 1460. Transportation Research Board, Washington, DC, pp 73–80

Road Talk (1995) Strength without steel. Ont Transp Technol Tran Dig 1(3)

Sargent DD, Mufti AA, Bakht B (1999) Design construction and field testing of steel-free arch panel bridge deck for forestry bridges. In: Proceedings of the 1999 Canadian Society for Civil Engineering annual conference, vol I, pp 95–104

Selvadurai APS, Bakht B (1995) Simulation of rolling wheel loads on an FRC deck slab. In: Proceedings of the 2nd University-Industry workshop on FRC, Toronto, pp 273–287

Thorburn J, Mufti AA (1995) Full-scale testing of externally reinforced FRC bridge decks on steel girders. In: Proceedings of the annual conference of CSCE, Ottawa, vol II, pp 543–552

Wegner LD, Mufti AA (1994) Finite element investigation of fibre-reinforced concrete deck slabs without internal steel reinforcement. Can J Civ Eng 21(2):231–236

Youn S-G, Chang S-P (1998) Behaviour of composite bridge decks subjected to static and fatigue loading. ACI Struct J 95(3):249–258

Chapter 5
Cantilever Slabs

5.1 Introduction

The concrete deck slabs of girder bridges are usually projected transversely beyond the outermost girders. These projections, which are provided for reasons of economy and aesthetics, are referred to in this book as *cantilever slabs* or *cantilever overhangs*.

The internal arching system of the deck slabs, which has been discussed in Chap. 4, is limited to that portion of the deck slab which is contained transversely between the outermost girders and which is subjected to live loads also located within these bounds. Until recently, it was believed that the load effects induced by loads on the cantilever overhangs respond to a purely flexural behaviour; these flexural effects are not limited to only the overhangs, but also extend into the internal panels of the deck slab. While the arching action is also present in certain cantilever slabs, there is no design code in the world which recognises this arching action.

In addition to dealing with the flexural analysis of load effects induced in the deck slab by loads applied to the cantilever overhang, this chapter also discusses arching action in the cantilever slabs in the hope that in future, some design codes will also permit the use of this beneficial arcing action.

5.1.1 Definitions

The terminology and geometry used in this chapter are defined in this section with the help of Fig. 5.1.

Root The support of a cantilever, which provides some measure of restraint against both rotation and vertical deflection, is commonly referred to as the root

B. Bakht, A. Mufti, *Bridges*, DOI 10.1007/978-3-319-17843-1_5

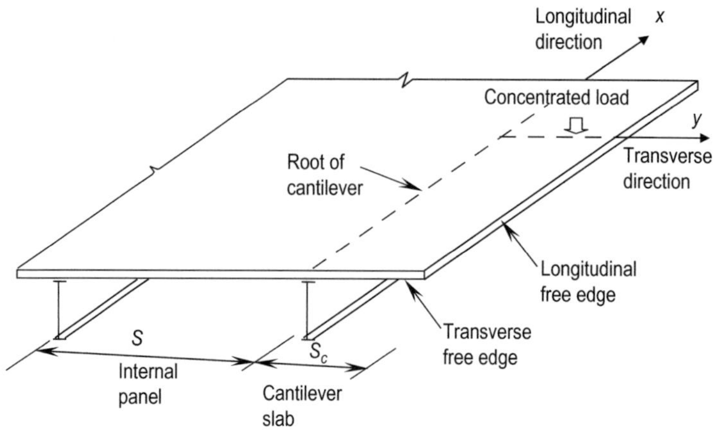

Fig. 5.1 Illustration of notation

of the cantilever; these restraints may individually be infinite or finite, i.e. rigid or semi-rigid.

Directions It is customary in bridge analysis to regard the longitudinal direction as the direction of the flow of traffic on the bridge, which is usually parallel to the axes of the girders, if present. Accordingly, the directions parallel to and perpendicular to the root are taken as longitudinal and transverse, respectively as illustrated in Fig. 5.1. Also, as shown in this figure, the x- and y-axes are parallel to the longitudinal and transverse directions, respectively.

Free Edges The free edges of a cantilever slab parallel to and perpendicular to the root are designated as longitudinal and transverse free edges, respectively.

Cantilever Span The transverse distance between the longitudinal free edge and the root is the span of the cantilever slab. As shown in Fig. 5.1, this span is denoted as S_c. Some design codes require the root to be taken at the outer edge of the girder if it is of concrete. This practice is justifiable only if the cantilever slab is assumed to be fully fixed against rotation at its root.

Moment and Shear Intensities It is customary in slab analysis to refer to moments and shears on a per-unit-length basis; these quantities are called moment intensities and shear intensities and they have the units of force-length/length and force/length, respectively. The moment intensities acting along x- and y- directions are denoted as M_x and M_y, respectively. In cantilever slabs, the latter is usually more significant and is referred to either as transverse moment intensity or cantilever moment intensity. Consistent with the terminology used in textbooks on plate analysis, the moment intensity causing tension in the top fibres of the slab is regarded as negative and, naturally, that causing tension in the bottom fibres as positive.

Cantilever Slab of Infinite Length When the transverse free edges of a cantilever slab are so remote from the applied concentrated load that the resulting load effects are negligible in their vicinity, the slab is regarded for purposes of analysis as being of infinite length. For example, in an unstiffened cantilever slab with its root rigidly restrained, the moment and shear intensities are negligible at a distance $3S_c$ measured longitudinally from the applied concentrated load. Thus, a slab can be regarded as being of infinite length if its transverse free edges are at least a distance $3S_c$ from the nearest load. Edge stiffening and relaxation of restraints at the root cause the longitudinal distributions of load effects to become less peaky. Because of these two factors, the influence of a concentrated load may extend beyond a longitudinal distance $3S_c$ from the load. Clearly, in such cases the transverse free edges must be even farther from the nearest load, if the slab is to be regarded as of infinite length.

Cantilever Slab of Semi-infinite Length For purposes of analysis, a cantilever slab is regarded as being of semi-infinite length when one of its two transverse free edges is close enough to the applied load to have an influence on the distribution of load effects whilst the other is not.

Thickness Ratio The methods of analysis presented in this chapter are applicable to cantilever slabs in which the thickness varies linearly in the transverse direction. This variation in thickness is defined by the thickness ratio t_2/t_1 where t_1 is the thickness at the root and t_2 the thickness at the tip.

Internal Panel As noted earlier, load effects due to applied loads on the cantilever slab are not confined only to the cantilever slab, but are also induced in the internal portions of the deck slab. As will be explained later in the chapter, these load effects in the internal portions of the deck slab are significant only in the slab panel adjacent to the overhang. Such panels are called internal panels. For reference in this chapter, an *internal panel* is defined as that portion of the deck slab, which lies transversely between the outermost girder and the one immediately adjacent to it.

5.1.2 Mechanics of Behaviour

Before conducting the force analysis of a structure, the engineer must have a clear idea of the manner in which the loads are transmitted through the structure. The *feel* for structural behaviour is necessary not only to enable the engineer to seek the relevant information from the analysis, but also to enable him or her to check that the results obtained from the analyses, especially if they are computer-based, are within the expected range.

This subsection describes the pattern of behaviour of the cantilever overhang and the internal panel under loads on the former. The methods of analysis here presented are for the determination of transverse moment intensities in cantilever

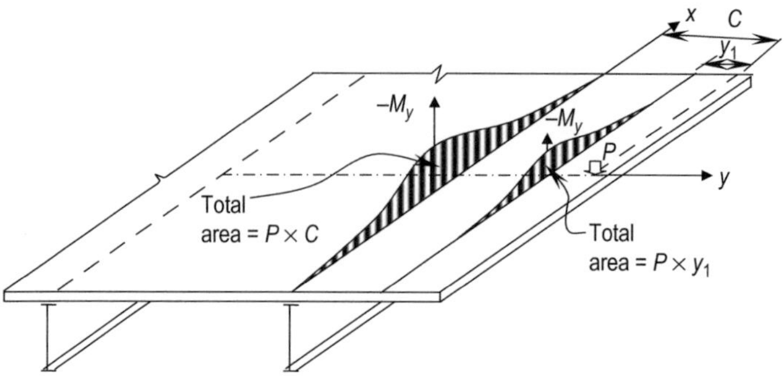

Fig. 5.2 Distribution of transverse moment intensities in the cantilever slab due to a concentrated load

slabs and internal panels of infinite length. Accordingly the discussion presented below is limited in the main to the same components.

Figure 5.2 shows the deck slab overhang of a slab-on-girder bridge subjected to a single concentrated load; this figure also shows the distributions of cantilever moment intensity M_y at two longitudinal sections, one at the root and the other between the load and the root. It can be seen that the patterns of distribution of M_y, which are similar at the two sections, are bell-shaped with well-defined peaks. The intensity of moments drops rapidly from the peak and then gradually reduces to almost zero. It can also be seen in Fig. 5.2 that the peak intensity of M_y at the root is higher than the peak intensity at the other section, and that M_y at the root diminishes to nearly zero at a much larger value of x than is the case at the other section.

Locations where M_y drops to nearly zero define, in a certain sense, the boundary of the zone of influence of the concentrated load. It can be appreciated that this zone of influence spreads out longitudinally as the reference section moves away from the load towards the root.

Of particular note in Fig. 5.2 is the observation that the total areas under the curves for M_y are equal to the total cantilever moments at the respective sections. For example, the total moment at the root is PC where P is the load and C its transverse distance from the root. The total cantilever moment and hence the total area under the M_y curve at a longitudinal section, is determined by overall static equilibrium alone and so is not affected by factors other than the magnitude of the load and its distance from the section under consideration.

It is interesting to note that for corresponding positions of the load and the reference section, the peak intensity of M_y is not affected by the span of the cantilever. This observation may at first glance appear contrary to engineering judgement. However, upon reflection it becomes clear that this phenomenon exists because the cantilever slabs under consideration have infinite lengths, because of which the length of the plate effectively sustaining the load increases with the span length of the cantilever, thereby maintaining the same value of the peak intensity.

There are four factors which affect the pattern of distribution of M_y along a longitudinal section, and consequently the value of the peak cantilever moment intensity; the effects of these factors are discussed briefly in the following.

Edge Stiffening It is intuitively obvious that the stiffening of the longitudinal edge, by spreading the effect of concentrated loads in the longitudinal direction, would help to improve the distribution of M_y, i.e. to reduce its peak intensity. As is required by overall statics, the reduction in the peak value of M_y leads to increase in the values of M_y elsewhere. This observation is significant, since in the case of multiple concentrated loads, edge-stiffening may not reduce the peak value of M_y as much as it does in the case of single concentrated loads.

Thickness Ratio The effect of the thickness ratio on the distribution of M_y can be appreciated by considering the unlikely case in which the slab thickness at the tip is larger than at the root, i.e. when t_2/t_1 is greater than 1.0. It can be visualized readily that the cross-section of such a slab is similar to that of a slab of uniform thickness with edge-stiffening. On the basis of this similarity it can be postulated that an increase in the value of t_2/t_1 leads to a reduction in the peak value of M_y. This postulate is confirmed by analysis. The corollary of this observation is that the distribution of M_y is made 'peakier' by a reduction of t_2/t_1. It is estimated that the maximum value of M_y due to a single concentrated load at the tip in a fully-fixed cantilever slab with $t_2/t_1 = 0.33$, is about 37 % larger than the corresponding value in a slab with $t_2/t_1 = 1.00$.

Restraint Against Deflection An increase in the flexibility of restraint against vertical deflections at the root has the effect of improving the distribution of M_y. It is noted, however, that in most highway bridges the restraint offered by the girders even near the mid-span is high enough for it to be considered rigid without affecting significantly the governing values of M_y.

Restraint Against Rotation The flexibility of the rotational restraint at the root has the same effect on the distribution of M_y as that of the restraint against deflection; its increase also reduces the peak value of M_y. However in slab-on-girder bridges, unlike the restraint against deflections, restraint against rotation has a significant effect on the distribution of M_y and therefore should not be neglected in the analysis. For example, the peak value of M_y due to a single concentrated load on the overhang of a bridge having a girder spacing equal to 2.5 times the cantilever span would be overestimated by up to 40 % if the overhang were analysed by assuming the rotational restraint at the root to be rigid.

5.1.3 Negative Moments in Internal Panel

As discussed earlier, the cantilever moments induced in the overhangs of the deck slab are carried over into the internal panel. To study the distribution of these negative moments in the internal panel it is instructive to recall first the familiar case of a beam having a simple support at one end and a cantilever overhang beyond the other, and with a concentrated load on the overhang. The statically determinate negative moment in this beam varies linearly from zero at the simple support to a maximum moment at the root of the cantilever.

By using this beam analogy, and assuming that negative moment falls to zero at the girder next to the root, it is obvious that the total moment along a longitudinal section of an internal panel of a deck slab, is statically determinate. However, similarities between the beam and the actual deck slab are not valid any further. Contrary to the usual misconception, the negative moment intensity M_y at a transverse section in the internal panel does not vary linearly in the transverse direction, even though the total negative moment does.

Two factors influence the transverse variation of the peak intensity of M_y in the internal panel, these being: (a) the ratio of the spans of the cantilever and the internal panel; and (b) the thickness ratio. The former has more pronounced effect, which is discussed in the following.

Three deck slabs are considered each having internal panels with a span of 10 units, and having cantilever overhangs with spans of 5, 10 and 20 units respectively. As shown in Fig. 5.3 these slabs are respectively subjected to 6.04, 2.21 and 1.00 units of a single concentrated load at the tips of the cantilever. The loads have been chosen so that they lead to the same peak intensity of M_y at the root of the cantilever of the three deck slabs. Since the intensities of M_y at the two supports are the same for the three cases, a direct comparison of the patterns of their transverse variations can be made. It can be seen in Fig. 5.3 that the variation of peak M_y tends to become linear as the length of the cantilever becomes very large with respect to the span of the internal panel, i.e. when the rotational restraint at the cantilever root tends to become infinity.

Conversely, the distribution of peak M_y becomes distinctly non-linear as the cantilever span becomes shorter with respect to the internal span. It is thus concluded that the increase of the span of the internal panel with respect to the cantilever span has two beneficial effects. Firstly, it reduces the degree of rotational restraint at the root, thereby reducing the peak intensity of M_y at the root; and secondly, it causes a rapid diminution of the peak intensity in the internal panel.

5.1.4 Cantilever Slab of Semi-infinite Length

The mechanics of distribution of transverse moment intensity, M_y due to a concentrated load in the vicinity of a transverse free edge is discussed below with the help of Figs. 5.4a, b and c.

Figure 5.4a shows M_y at the root of a slab of infinite length due to a concentrated load P at a distance C from the root. As discussed earlier in conjunction with Fig. 5.1, the total area under the M_y curve, i.e. the total moment, is equal to PC. A transverse section to the left of the concentrated load is now considered which

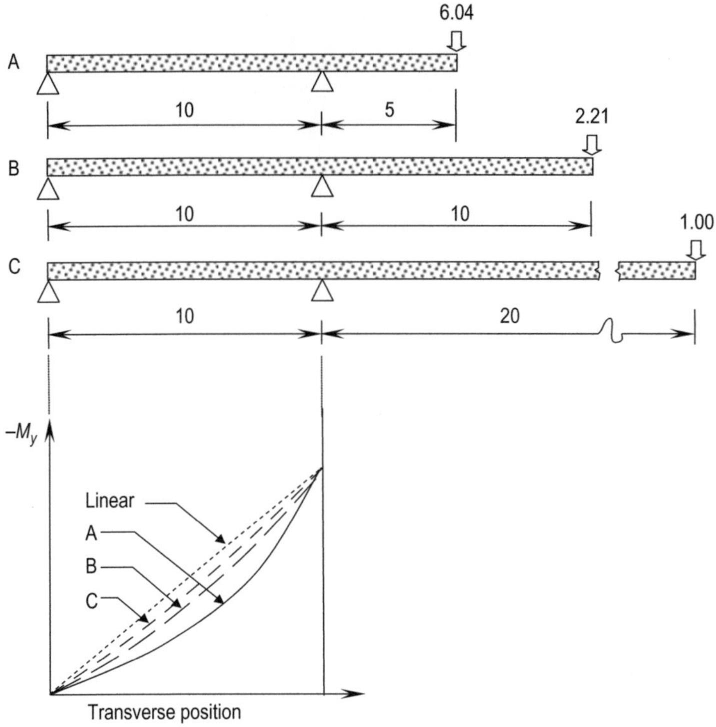

Fig. 5.3 Distribution of peak negative moment intensities in the internal panel

divides the total area under the M_y curve into two areas A_L and A_R with the former being on the left hand side of the section and the latter on the right hand side.

The slab of infinite length is now considered to be cut at the transverse section discussed above thus making it one of semi-infinite length. To satisfy statics, the moments represented by area A_L must be redistributed within the curtailed slab. It is intuitive to postulate that the moments A_L are redistributed into the curtailed slab according to the pattern shown in Fig. 5.4b i.e. that they 'reflect back'; this pattern has, indeed, been confirmed, sufficiently closely for design purposes, by rigorous analysis.

As shown in Fig. 5.4c, the net moments in the slab of semi-infinite length can then be obtained by superimposing the redistributed moments A_L over the moments A_R corresponding to the slab of infinite length. From this discussion, it will be appreciated that the peak values of M_y in a cantilever slab of semi-infinite extent are larger than the peak values of M_y in a corresponding cantilever slab of infinite extent. A method is provided in Sect. 5.2.3 for the analysis of semi-infinite slabs; however, it is noted that the Canadian Highway Bridge Design Code (2006) has simplified the task of designers by stipulating that the cantilever slab within a distance S_c from the transverse free edges should be designed for twice the live load moments for which the rest of the cantilever slab is designed.

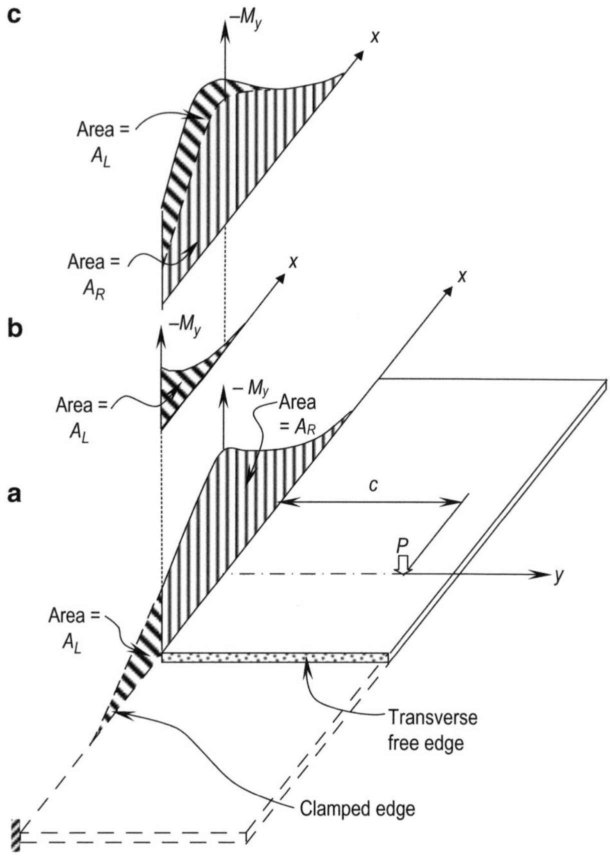

Fig. 5.4 Distribution of cantilever moment intensities in the vicinity of a transverse free edge

5.2 Methods of Analysis

All methods of analysis presented in this section are based on the assumption that both the cantilever slab and the internal panel behave in a linear elastic fashion at all load levels.

5.2.1 Unstiffened Cantilever Slab of Infinite Length

A simplified method was proposed by Bakht and Holland (1976) for determining the cantilever moment intensity M_y in a cantilever slab of infinite length subjected to a concentrated load P. According to this method:

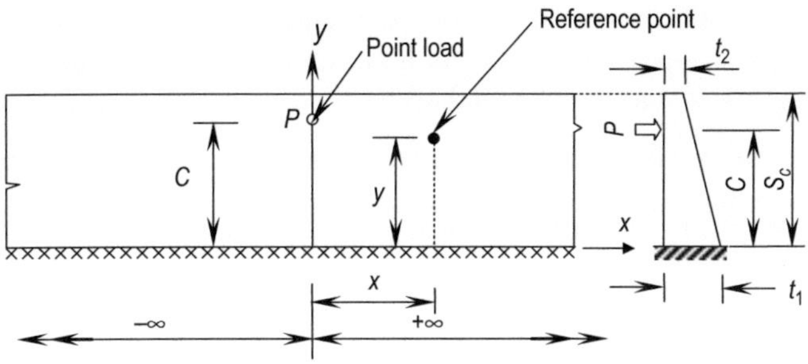

Fig. 5.5 Illustration of notation used for proposed simplified methods of analysis

$$M_y = -\frac{PA'}{\pi}\frac{1}{\cosh\left(\frac{A'x}{C-y}\right)} \tag{5.1}$$

where A' is a coefficient whose values depend upon the positions of the load and the reference point with respect to the root of the cantilever; other notation is as shown in Fig. 5.5. Graphical charts were provided by Bakht and Holland (1976) for the values of A' for different load and reference point locations in cantilever slabs of linearly varying thickness having thickness ratio t_2/t_1 of 1.0, 0.5 and 0.33, where t_1 and t_2 are the thicknesses of the slab at the root and tip, respectively, as is also shown in Fig. 5.5.

The method noted above is for the analysis of cantilever slabs without edge stiffening. Bakht (1981) has shown that Eq. (5.1) is also applicable to edge-stiffened cantilever slabs for which a different set of values of coefficient A' are required depending upon the ratio of the flexural rigidity of the cantilever slab and the edge beam.

Algebraic Equation Jaeger and Bakht (1990) have shown that the cosine hyperbolic (*cosh*) function of Eq. (5.1) be replaced by an algebraic function, so that M_y takes the following form.

$$M_y = -\frac{2PB}{\pi}\frac{(C-y)^4}{\left[(C-y)^2 + (Bx)^2\right]^2} \tag{5.2}$$

where

$$B = \frac{A'}{2} \tag{5.3}$$

Equations (5.1) and (5.2) yield practically the same results. In terms of accuracy, neither equation can be preferred over the other. It is recognized, however, that some engineers, might prefer Eq. (5.2) because of being able to relate more readily to its algebraic function. Both Eqs. (5.1 and 5.2) satisfy the three important conditions discussed earlier: (a) at any longitudinal section, the integration of M_y from $x=\infty$ to $x=-\infty$ is equal to the total applied negative moment, (b) M_y is maximum at $x=0.0$, and (c) $M_y=0.0$ at $x=\infty$.

Cantilever Slabs with Finite Rotational Restraints The solution proposed by Bakht and Holland (1976) is applicable to cantilever slabs with their roots restrained completely against both deflection and rotation. Dilger et al. (1990) have shown that Eq. (5.1), and by inference Eq. (5.2), are also applicable to cantilever slabs in which the rotational restraint at the root is finite; such slabs are encountered in the deck slab overhangs of girder bridges. Graphical charts of the coefficient A' for cantilever slabs having finite rotational restraint at the root are provided by Dilger et al. (1990); they have shown that for the analysis of deck slab overhangs of girder bridges subjected to single concentrated loads, the assumption of full rotational restraint at the root necessarily leads to a significant overestimation of the peak intensity of cantilever moments.

Moment Intensities in Internal Panels The methods proposed by Bakht and Holland (1976) and Dilger et al. (1990) provide the intensities of cantilever, or negative, moments in only the deck slab overhang and do not provide any information regarding the distribution of these negative moments in the internal panels. In the absence of this information, designers usually formulate their own empirical rules for curtailing the negative moment reinforcement in the internal panel. These empirical rules are usually based on the assumption that the peak intensity of the negative moment varies linearly across the internal panel.

Mufti et al. (1993) have shown that the moment intensity M_y in the internal panel can be obtained by any one of Eq. (5.4) and (5.5), which are the counterparts of the *cosh* function of Eq. (5.1) and the algebraic function of Eq. (5.2) respectively.

$$M_y = -\frac{2PB}{\pi}\frac{1}{\cosh\left[\frac{2BSx}{C(S-y)}\right]} \tag{5.4}$$

$$M_y = -\frac{2PB}{\pi}\frac{C^4(S-y)^4}{\left[C^2(S-y)^2 + S^2(Bx)^2\right]^2} \tag{5.5}$$

The notation used for Eqs. (5.4) and (5.5) is illustrated in Fig. 5.6, which shows that the direction of y for the cantilever slab is as defined in Fig. 5.5; however, the direction of y for the internal panel is reversed. In both cases, the origin of y is at the root of the cantilever. The values of the coefficient B used in Eq. (5.4) or (5.5) are different from those which are used in Eq. (5.1) or (5.2).

Similar to Eqs. (5.1) and (5.2), Eqs. (5.4) and (5.5) satisfy the three conditions discussed earlier.

Boundary Conditions It can be shown that the distribution of moment intensities in a cantilever slab subjected to concentrated loads is affected by the degrees of restraint against both rotations and out-of-plane deflections at the root. Several

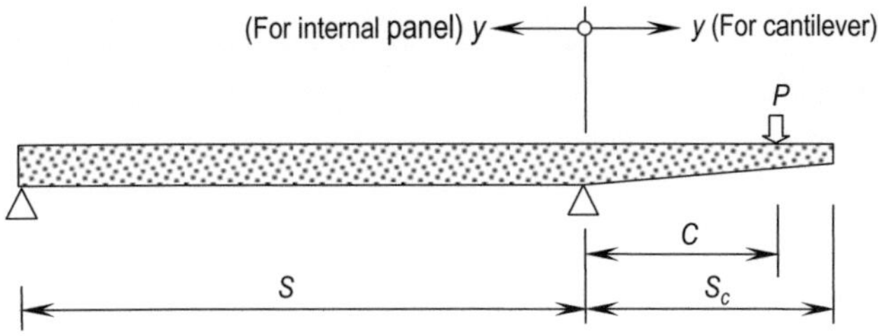

Fig. 5.6 Definition of transverse coordinate y

analyses by the finite-element program SPAST (Ghali and Tadros 1980) have confirmed that the magnitude of restraint against out-of-plane deflections offered by typical girders used in bridges is relatively very large. For purposes of the cantilever slab analysis, the girders can be assumed to be non-deflecting without significant loss of accuracy.

A further simplification can be made by assuming that the deck slab is simply supported over the girder adjacent to the external girder, as shown in Fig. 5.7b. After confirming by rigorous analyses that the neglect of the deck slab beyond the second girder also does not lead to significant errors in the estimation of negative moment intensities beyond the range of design accuracy, the boundary conditions shown in Fig. 5.7b were adopted for developmental analyses. It is noted that Dilger et al. (1990) have also used the same boundary conditions.

Cantilever Slabs with Stiffened Longitudinal Edges Bakht (1981), and later Tadros et al. (1994), have confirmed that Eqs. (5.1), (5.2), (5.4) and (5.5) are also applicable to cantilever slabs with stiffened longitudinal edges provided that the loads applied to the cantilever are at least $3 \times S_c$ away from any transverse free edge; however, the values of coefficient B depend upon the ratio of the flexural rigidity of the edge stiffening and that of the cantilever slab. Tadros et al. (1994) have made analysis of cantilever slabs with stiffened longitudinal edge and the internal panel considerably simpler by proving that the increase of the flexural rigidity of the edge stiffening beyond a certain limit has little effect on the distribution of transverse negative moments in both the cantilever slab and the internal panel. Fortunately, the limiting flexural rigidity corresponds to concrete barrier walls commonly employed in bridge decks. Such barrier walls, commonly referred to as the New Jersey barrier walls, are about 1.4 m tall; at the base, they are nearly 430 mm wide, and at the top their width is nearly 180 mm. The approximate cross-sectional dimensions of the barrier wall considered for analysis of edge-stiffened cantilever slabs are shown in Fig. 5.8.

Cantilever Slabs of Semi-infinite Length Bakht et al. (1979) have shown that M_y at the root of a fully clamped cantilever slab due a concentrated load near a transverse free edge is given by the following equation.

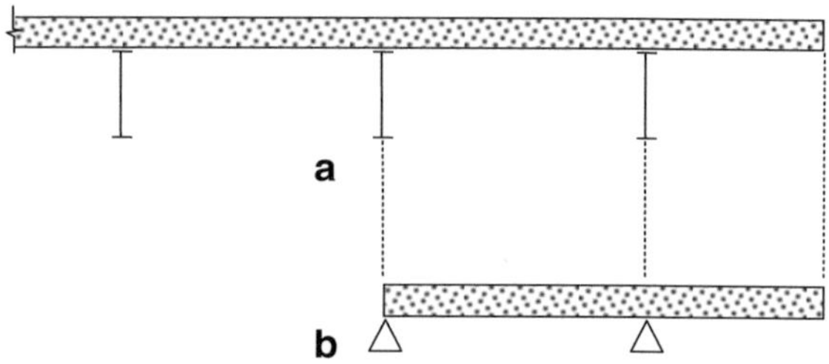

Fig. 5.7 Illustration of boundary conditions: (**a**) actual cross-section, (**b**) assumed cross-section

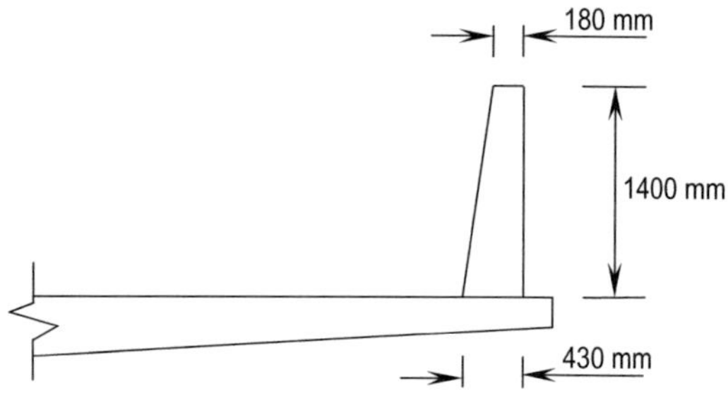

Fig. 5.8 Approximate cross-sectional dimensions of the barrier wall considered in the analysis of edge-stiffened cantilever slabs

$$M_y = -\frac{2PB}{\pi}\left[\frac{1}{\cosh\left(\frac{2PB}{C(S-y)}\right)} + De^{(-KX/S_c)}\right] \tag{5.6}$$

where K is given by Eq. (5.7).

$$K = \frac{BDS_c}{C}\left[\frac{1}{\tan^{-1}\left[\exp\left(-2Bx_{edge}/C\right)\right]}\right] \tag{5.7}$$

In Eqs. (5.6) and (5.7), the following additional notation has been used.

D = a coefficient, which is similar to B and which is obtained from rigorous analysis.

x_{edge} = the distance of the point load from the nearer transverse free edge, as shown in Fig. 5.9.

X = the distance of the reference point at the root from the nearer transverse free edge = $x + x_{edge}$ as shown in Fig. 5.9.

5.2.2 *Proposed Method of Analysis for Slabs of Infinite Length*

It is recommended that Eq. (5.1) or (5.2) be adopted for the analysis of M_y in the cantilever slab, and Eq. (5.4) or (5.5) for the internal panel. The values of B to be

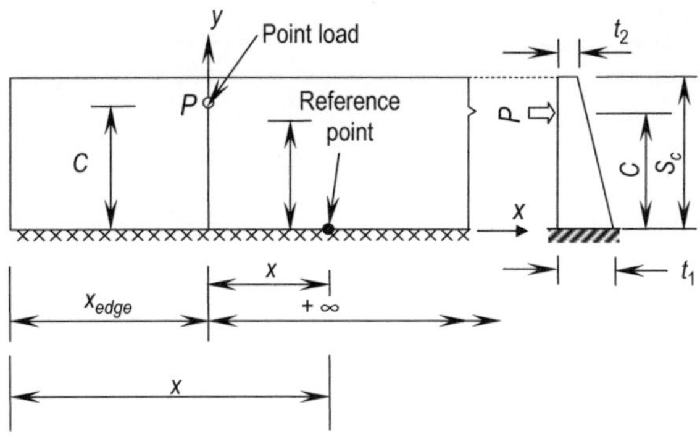

Fig. 5.9 Illustration of notation used for proposed simplified methods of analysis for cantilever slabs of semi-infinite length

used in conjunction with these equations are recommended to be those provided by Mufti et al. (1993) in the form of tables; these tables cover a wide range of cases. As shown in Fig. 5.6, the deck slab in the internal panel has a constant thickness; the thickness ratio for the cantilever can vary between 0.0 and 1.0. The tables of values of B are reproduced as follows: (a) Tables 5.1, 5.2, 5.3, and 5.4 for unstiffened cantilever slabs, (b) Tables 5.5, 5.6, 5.7, and 5.8 for the internal panels corresponding to unstiffened cantilever slabs, (c) Tables 5.9 and 5.10 for cantilever slabs with stiffened longitudinal edges, and (d) Tables 5.11 and 5.12 for internal panels corresponding to edge-stiffened cantilever slabs. It can be seen that these values are given for discrete values of four dimensionless parameters being (a) the thickness ratio, (b) the ratio of the spans of the cantilever and the internal panel, (c) the ratio of the distance of the concentrated load from the root to the cantilever span, and (d) the ratio of the distance of the reference section from the root to the cantilever span.

Tables 5.1, 5.2, 5.3, 5.4, 5.5, 5.6, 5.7, and 5.8 contain values of B pertaining to unstiffened cantilever slabs having $t_2/t_1 = 1.00$, 0.50, 0.33 and 0.00. The values of B pertaining to edge-stiffened cantilever slabs, given in Tables 5.9, 5.10, 5.11, and 5.12, are for $t_2/t_1 = 1.00$ and 0.50. The reason why values of B for edge stiffened cantilevers are not given for $t_2/t_1 < 0.50$ is because in edge-stiffened cantilever slabs it is rare to have $t_2/t_1 < 0.50$.

Validity of Proposed Method Mufti et al. (1993) have confirmed that several comparisons were made of plots of M_y against x obtained by the proposed method and by the finite element analysis. It was found that without exception, the two sets of results were very close to each other, thus confirming the validity and accuracy of the proposed method.

5.2.3 Method of Analysis for Slabs of Semi-infinite Length

It is recommended that the cantilever moments at the root of a cantilever slab of semi-infinite length be calculated by using Eq. (5.6) or (5.7). The proposed method

Table 5.1 Values of B for unstiffened cantilever slabs with $t_2/t_1 = 1.0$

		B for $C/S_c=$				
S_c/S	y/S_c	1.0	0.8	0.6	0.4	0.2
0.4	1.00	0.00				
	0.80	0.21				
	0.60	0.23	0.06			
	0.40	0.31	0.18	0.04		
	0.20	0.40	0.30	0.20	0.08	
	0.00	0.53	0.46	0.39	0.33	0.27
0.5	1.00	0.00				
	0.80	0.21				
	0.60	0.23	0.06			
	0.40	0.31	0.18	0.04		
	0.20	0.41	0.31	0.20	0.09	
	0.00	0.54	0.47	0.40	0.34	0.28
0.67	1.00	0.00				
	0.80	0.21				
	0.60	0.24	0.06			
	0.40	0.32	0.19	0.05		
	0.20	0.42	0.32	0.21	0.09	
	0.00	0.56	0.48	0.41	0.35	0.30
1.0	1.00	0.00				
	0.80	0.22				
	0.60	0.24	0.07			
	0.40	0.33	0.20	0.06		
	0.20	0.43	0.33	0.22	0.10	
	0.00	0.58	0.50	0.43	0.36	0.32
2.0	1.00	0.00				
	0.80	0.22				
	0.60	0.26	0.08			
	0.40	0.35	0.21	0.07		
	0.20	0.47	0.36	0.25	0.12	
	0.00	0.64	0.55	0.47	0.40	0.35
*	1.00	0.00				
	0.80	0.23				
	0.60	0.28	0.10			
	0.40	0.39	0.25	0.10		
	0.20	0.54	0.43	0.31	0.17	
	0.00	0.76	0.67	0.59	0.52	0.52

*Fully clamped cantilever

is applicable only to fully clamped cantilevers. Accordingly, the values of coefficient B should be obtained for fully clamped cantilevers from the relevant Tables 5.1, 5.2, 5.3, and 5.4 for unstiffened plates, and from the relevant Tables 5.9 and 5.10 for edge-stiffened plates. Table 5.13 provides calues of coefficient B for cantilever slabs of semi-infinite length.

Table 5.2 Values of B for unstiffened cantilever slabs with $t_2/t_1 = 0.5$

		B for C/S_c=				
S_c/S	y/S_c	1.0	0.8	0.6	0.4	0.2
0.4	1.00	0.00				
	0.80	0.35				
	0.60	0.42	0.21			
	0.40	0.51	0.34	0.18		
	0.20	0.60	0.47	0.34	0.19	
	0.00	0.73	0.62	0.52	0.43	0.33
0.5	1.00	0.00				
	0.80	0.35				
	0.60	0.42	0.21			
	0.40	0.51	0.34	0.18		
	0.20	0.60	0.47	0.34	0.19	
	0.00	0.73	0.63	0.53	0.44	0.34
0.67	1.00	0.00				
	0.80	0.35				
	0.60	0.42	0.21			
	0.40	0.51	0.35	0.18		
	0.20	0.61	0.48	0.34	0.19	
	0.00	0.75	0.64	0.54	0.45	0.36
1.0	1.00	0.00				
	0.80	0.35				
	0.60	0.42	0.22			
	0.40	0.52	0.35	0.18		
	0.20	0.62	0.49	0.35	0.20	
	0.00	0.77	0.66	0.56	0.46	0.38
2.0	1.00	0.00				
	0.80	0.36				
	0.60	0.43	0.22			
	0.40	0.53	0.36	0.19		
	0.20	0.65	0.51	0.37	0.21	
	0.00	0.82	0.71	0.60	0.49	0.42
*	1.00	0.00				
	0.80	0.36				
	0.60	0.44	0.23			
	0.40	0.56	0.39	0.22		
	0.20	0.71	0.57	0.42	0.26	
	0.00	0.95	0.83	0.72	0.61	0.56

*Fully clamped cantilever

5.2.4 *Program ANDECAS*

Although the proposed method of analysis can be applied manually, its application becomes time-consuming when the number of applied loads is more than one, and

Table 5.3 Values of B for unstiffened cantilever slabs with $t_2/t_1 = 0.33$

		B for C/S_c=				
S_c/S	y/S_c	1.0	0.8	0.6	0.4	0.2
0.4	1.00	0.00				
	0.80	0.43				
	0.60	0.51	0.29			
	0.40	0.60	0.42	0.24		
	0.20	0.69	0.55	0.40	0.23	
	0.00	0.81	0.69	0.58	0.47	0.35
0.5	1.00	0.00				
	0.80	0.43				
	0.60	0.51	0.29			
	0.40	0.61	0.42	0.24		
	0.20	0.70	0.55	0.40	0.23	
	0.00	0.82	0.70	0.59	0.48	0.37
0.67	1.00	0.00				
	0.80	0.43				
	0.60	0.51	0.29			
	0.40	0.61	0.43	0.24		
	0.20	0.70	0.55	0.40	0.24	
	0.00	0.84	0.72	0.60	0.49	0.39
1.0	1.00	0.00				
	0.80	0.43				
	0.60	0.51	0.29			
	0.40	0.61	0.43	0.24		
	0.20	0.71	0.56	0.41	0.24	
	0.00	0.86	0.74	0.62	0.50	0.41
2.0	1.00	0.00				
	0.80	0.43				
	0.60	0.52	0.30			
	0.40	0.62	0.44	0.25		
	0.20	0.74	0.58	0.42	0.25	
	0.00	0.91	0.78	0.66	0.53	0.44
*	1.00	0.00				
	0.80	0.43				
	0.60	0.53	0.30			
	0.40	0.65	0.46	0.27		
	0.20	0.79	0.63	0.47	0.29	
	0.00	1.04	0.90	0.78	0.64	0.58

*Fully clamped cantilever

the various ratios are other than those for which the coefficient B is provided in tabular form. Parr (1993) has incorporated the proposed method into a program which interpolates the value of B for the given set of ratios, and searches for the maximum moment intensity; the tables of coefficients B have been incorporated in this program through a number of data files.

Table 5.4 Values of B for unstiffened cantilever slabs with $t_2/t_1 = 0.0$

		B for C/S_c=				
S_c/S	y/S_c	1.0	0.8	0.6	0.4	0.2
0.4	1.00	0.00				
	0.80	0.66				
	0.60	0.78	0.52			
	0.40	0.87	0.63	0.40		
	0.20	0.94	0.74	0.54	0.33	
	0.00	1.03	0.87	0.71	0.55	0.40
0.5	1.00	0.00				
	0.80	0.66				
	0.60	0.78	0.52			
	0.40	0.87	0.63	0.40		
	0.20	0.94	0.74	0.54	0.33	
	0.00	1.04	0.88	0.72	0.56	0.41
0.67	1.00	0.00				
	0.80	0.66				
	0.60	0.78	0.52			
	0.40	0.87	0.63	0.40		
	0.20	0.94	0.74	0.54	0.33	
	0.00	1.06	0.89	0.73	0.57	0.43
1.0	1.00	0.00				
	0.80	0.66				
	0.60	0.78	0.52			
	0.40	0.87	0.63	0.40		
	0.20	0.95	0.75	0.54	0.34	
	0.00	1.08	0.91	0.75	0.59	0.46
2.0	1.00	0.00				
	0.80	0.66				
	0.60	0.78	0.52			
	0.40	0.88	0.64	0.40		
	0.20	0.97	0.76	0.56	0.35	
	0.00	1.14	0.96	0.78	0.62	0.49
*	1.00	0.00				
	0.80	0.66				
	0.60	0.79	0.52			
	0.40	0.89	0.65	0.41		
	0.20	1.02	0.81	0.59	0.37	
	0.00	1.28	1.09	0.90	0.72	0.61

*Fully clamped cantilever

The program, ANDECAS, derives its acronym from ANalysis and DEsign of CAntilever Slab overhangs of slab on girder bridges. In its current version, the program analyses and designs slabs subjected to Canadian design loadings and user-defined design trucks. The program ANDECAS can be downloaded from http://extras.springer.com.

Table 5.5 Values of B for internal panels corresponding to unstiffened cantilever slabs with $t_2/t_1 = 1.0$

		B for C/S_c=				
S_c/S	Y/S_c	1.0	0.8	0.6	0.4	0.2
0.4	0.00	0.53	0.46	0.39	0.33	0.27
	1.00	0.00	0.00	0.00	0.00	0.00
0.5	0.00	0.54	0.47	0.40	0.34	0.28
	0.20	0.33	0.27	0.22	0.16	0.09
	0.40	0.21	0.17	0.13	0.09	0.05
	0.60	0.03	0.10	0.08	0.05	0.03
	0.80	0.06	0.05	0.04	0.03	0.01
	1.00	0.00	0.00	0.00	0.00	0.00
0.67	0.00	0.56	0.48	0.41	0.35	0.30
	0.10	0.45	0.38	0.31	0.24	0.15
	0.20	0.37	0.31	0.25	0.18	0.10
	0.30	0.30	0.25	0.20	0.14	0.08
	0.40	0.24	0.20	0.16	0.11	0.06
	0.50	0.19	0.16	0.12	0.09	0.05
	0.60	0.15	0.12	0.10	0.07	0.03
	0.70	0.11	0.09	0.07	0.05	0.02
	0.80	0.07	0.06	0.05	0.03	0.02
	0.90	0.14	0.03	0.02	0.02	0.01
	1.00	0.00	0.00	0.00	0.00	0.00
1.0	0.00	0.58	0.50	0.43	0.36	0.32
	0.10	0.49	0.42	0.35	0.28	0.19
	0.20	0.42	0.35	0.29	0.22	0.13
	0.30	0.35	0.29	0.24	0.18	0.10
	0.40	0.29	0.24	0.19	0.14	0.08
	0.50	0.24	0.20	0.15	0.11	0.06
	0.60	0.18	0.15	0.12	0.09	0.05
	0.70	0.14	0.11	0.09	0.06	0.03
	0.80	0.09	0.07	0.06	0.04	0.02
	0.90	0.04	0.04	0.03	0.02	0.01
	1.00	0.00	0.00	0.00	0.00	0.00
2.0	0.00	0.64	0.55	0.47	0.40	0.35
	0.10	0.56	0.48	0.41	0.33	0.25
	0.20	0.49	0.42	0.35	0.28	0.19
	0.30	0.42	0.36	0.30	0.24	0.16
	0.40	0.36	0.30	0.25	0.20	0.12
	0.50	0.29	0.25	0.21	0.16	0.10
	0.60	0.23	0.20	0.16	0.12	0.08
	0.70	0.17	0.15	0.12	0.09	0.06
	0.80	0.12	0.10	0.08	0.06	0.04
	0.90	0.16	0.05	0.04	0.03	0.02
	1.00	0.00	0.00	0.00	0.00	0.00

Table 5.6 Values of B for internal panels corresponding to unstiffened cantilever slabs with $t_2/t_1 = 0.5$

		B for C/S_c=				
S_c/S	Y/S_c	1.0	0.8	0.6	0.4	0.2
0.4	0.00	0.73	0.62	0.52	0.43	0.33
	1.00	0.00	0.00	0.00	0.00	0.00
0.5	0.00	0.73	0.63	0.53	0.44	0.34
	0.20	0.41	0.34	0.26	0.19	0.10
	0.40	0.25	0.20	0.15	0.10	0.05
	0.60	0.14	0.12	0.09	0.06	0.03
	0.80	0.07	0.05	0.04	0.03	0.01
	1.00	0.00	0.00	0.00	0.00	0.00
0.67	0.00	0.75	0.64	0.54	0.45	0.36
	0.10	0.58	0.49	0.40	0.30	0.18
	0.20	0.46	0.38	0.31	0.22	0.12
	0.30	0.37	0.30	0.24	0.17	0.09
	0.40	0.29	0.24	0.19	0.13	0.07
	0.50	0.23	0.19	0.14	0.10	0.05
	0.60	0.18	0.14	0.11	0.08	0.04
	0.70	0.13	0.10	0.08	0.05	0.03
	0.80	0.08	0.07	0.05	0.03	0.02
	0.90	0.04	0.03	0.03	0.12	0.01
	1.00	0.00	0.00	0.00	0.00	0.00
1.0	0.00	0.77	0.66	0.56	0.46	0.38
	0.10	0.64	0.54	0.45	0.35	0.22
	0.20	0.53	0.45	0.36	0.27	0.16
	0.30	0.44	0.37	0.29	0.21	0.12
	0.40	0.36	0.30	0.23	0.17	0.09
	0.50	0.29	0.24	0.19	0.13	0.07
	0.60	0.22	0.18	0.14	0.10	0.05
	0.70	0.16	0.13	0.10	0.07	0.04
	0.80	0.11	0.09	0.07	0.05	0.02
	0.90	0.05	0.04	0.03	0.02	0.01
	1.00	0.00	0.00	0.00	0.00	0.00
2.0	0.00	0.82	0.71	0.60	0.49	0.42
	0.10	0.72	0.62	0.52	0.41	0.30
	0.20	0.63	0.53	0.44	0.35	0.23
	0.30	0.54	0.45	0.37	0.29	0.18
	0.40	0.45	0.38	0.31	0.24	0.14
	0.50	0.37	0.31	0.25	0.19	0.11
	0.60	0.29	0.25	0.20	0.15	0.09
	0.70	0.22	0.18	0.15	0.11	0.06
	0.80	0.14	0.12	0.10	0.07	0.04
	0.90	0.07	0.06	0.05	0.04	0.02
	1.00	0.00	0.00	0.00	0.00	0.00

Table 5.7 Values of B for internal panels corresponding to unstiffened cantilever slabs with $t_2/t_1 = 0.33$

		B for C/S_c=				
S_c/S	y/S_c	1.0	0.8	0.6	0.4	0.2
0.4	0.00	0.81	0.69	0.58	0.47	0.35
	1.00	0.00	0.00	0.00	0.00	0.00
0.5	0.00	0.82	0.70	0.59	0.48	0.37
	0.20	0.44	0.36	0.28	0.20	0.11
	0.40	0.26	0.21	0.16	0.11	0.06
	0.60	0.15	0.12	0.09	0.06	0.03
	0.80	0.07	0.06	0.04	0.03	0.01
	1.00	0.00	0.00	0.00	0.00	0.00
0.67	0.00	0.84	0.72	0.60	0.49	0.39
	0.10	0.64	0.54	0.43	0.33	0.19
	0.20	0.50	0.41	0.33	0.23	0.13
	0.30	0.40	0.32	0.25	0.18	0.09
	0.40	0.31	0.25	0.20	0.14	0.07
	0.50	0.24	0.20	0.15	0.10	0.05
	0.60	0.19	0.15	0.11	0.08	0.04
	0.70	0.03	0.11	0.08	0.06	0.03
	0.80	0.09	0.07	0.05	0.04	0.02
	0.90	0.04	0.03	0.03	0.02	0.01
	1.00	0.00	0.00	0.00	0.00	0.00
1.0	0.00	0.86	0.74	0.62	0.50	0.41
	0.10	0.71	0.60	0.49	0.37	0.24
	0.20	0.58	0.49	0.39	0.29	0.17
	0.30	0.48	0.39	0.31	0.23	0.13
	0.40	0.39	0.32	0.25	0.18	0.10
	0.50	0.31	0.25	0.20	0.14	0.07
	0.60	0.24	0.20	0.15	0.11	0.06
	0.70	0.17	0.14	0.11	0.08	0.04
	0.80	0.11	0.09	0.07	0.05	0.03
	0.90	0.06	0.05	0.04	0.02	0.01
	1.00	0.00	0.00	0.00	0.00	0.00
2.0	0.00	0.91	0.78	0.66	0.53	0.44
	0.10	0.80	0.68	0.56	0.45	0.31
	0.20	0.69	0.58	0.48	0.37	0.24
	0.30	0.59	0.50	0.41	0.31	0.16
	0.40	0.49	0.41	0.34	0.25	0.15
	0.50	0.40	0.34	0.27	0.20	0.12
	0.60	0.32	0.27	0.21	0.16	0.09
	0.70	0.24	0.20	0.16	0.12	0.07
	0.80	0.16	0.13	0.10	0.08	0.04
	0.90	0.08	0.06	0.05	0.04	0.02
	1.00	0.00	0.00	0.00	0.00	0.00

Table 5.8 Values of B for internal panels corresponding to unstiffened cantilever slabs with $t_2/t_1 = 0.0$

		B for C/S_c=				
S_c/S	y/S_c	1.0	0.8	0.6	0.4	0.2
0.4	0.00	1.03	0.87	0.71	0.55	0.40
	1.00	0.00	0.00	0.00	0.00	0.00
0.5	0.00	1.04	0.88	0.72	0.56	0.41
	0.20	0.50	0.41	0.31	0.22	0.11
	0.40	0.28	0.23	0.17	0.12	0.06
	0.60	0.16	0.13	0.10	0.06	0.03
	0.80	0.07	0.06	0.04	0.03	0.01
	1.00	0.00	0.00	0.00	0.00	0.00
0.67	0.00	1.06	0.89	0.73	0.57	0.43
	0.10	0.77	0.64	0.51	0.37	0.21
	0.20	0.58	0.48	0.37	0.26	0.14
	0.30	0.45	0.36	0.28	0.19	0.10
	0.40	0.35	0.28	0.21	0.15	0.07
	0.50	0.27	0.21	0.16	0.11	0.06
	0.60	0.20	0.16	0.12	0.08	0.04
	0.70	0.14	0.12	0.09	0.06	0.03
	0.80	0.09	0.07	0.06	0.04	0.02
	0.90	0.05	0.04	0.03	0.02	0.01
	1.00	0.00	0.00	0.00	0.00	0.00
1.0	0.00	1.08	0.91	0.75	0.59	0.46
	0.10	0.86	0.72	0.57	0.43	0.26
	0.20	0.69	0.57	0.45	0.32	0.18
	0.30	0.56	0.45	0.35	0.25	0.14
	0.40	0.44	0.36	0.28	0.19	0.10
	0.50	0.35	0.28	0.22	0.15	0.08
	0.60	0.27	0.22	0.17	0.11	0.06
	0.70	0.19	0.16	0.12	0.08	0.04
	0.80	0.13	0.10	0.08	0.05	0.03
	0.90	0.06	0.05	0.04	0.03	0.01
	1.00	0.00	0.00	0.00	0.00	0.00
2.0	0.00	1.14	0.96	0.78	0.62	0.49
	0.10	0.98	0.82	0.67	0.51	0.34
	0.20	0.84	0.70	0.56	0.42	0.26
	0.30	0.71	0.59	0.47	0.35	0.21
	0.40	0.59	0.49	0.39	0.28	0.16
	0.50	0.48	0.40	0.31	0.23	0.13
	0.60	0.38	0.31	0.25	0.18	0.10
	0.70	0.28	0.23	0.18	0.13	0.07
	0.80	0.19	0.15	0.12	0.08	0.05
	0.90	0.09	0.08	0.06	0.04	0.02
	1.00	0.00	0.00	0.00	0.00	0.00

Table 5.9 Values of B for edge-stiffened cantilever slabs with $t_2/t_1 = 1.0$

		B for $C/S_c=$				
S_c/S	y/S_c	1.0	0.8	0.6	0.4	0.2
0.4	1.00	0.00				
	0.80	−0.05				
	0.60	0.02	−0.07			
	0.40	0.08	0.03	−0.07		
	0.20	0.14	0.11	0.07	−0.01	
	0.00	0.21	0.20	0.21	0.23	0.24
0.5	1.00	0.00				
	0.80	−0.05				
	0.60	0.02	−0.07			
	0.40	0.08	0.03	−0.07		
	0.20	0.15	0.12	0.08	−0.01	
	0.00	0.21	0.21	0.22	0.23	0.25
0.67	1.00	0.00				
	0.80	−0.05				
	0.60	0.02	−0.07			
	0.40	0.09	0.034	−0.06		
	0.20	0.15	0.12	0.08	−0.00	
	0.00	0.22	0.21	0.22	0.24	0.25
1.0	1.00	0.00				
	0.80	−0.04				
	0.60	0.03	−0.06			
	0.40	0.10	0.05	−0.06		
	0.20	0.16	0.13	0.09	0.01	
	0.00	0.23	0.23	0.24	0.25	0.26
2.0	1.00	0.00				
	0.80	−0.04				
	0.60	0.04	−0.05			
	0.40	0.11	0.06	−0.04		
	0.20	0.18	0.15	0.11	0.03	
	0.00	0.25	0.25	0.26	0.28	0.28
*	1.00	0.00				
	0.80	−0.03				
	0.60	0.05	−0.04			
	0.40	0.13	0.08	−0.01		
	0.20	0.21	0.19	0.16	0.09	
	0.00	0.29	0.31	0.36	0.43	0.48

*Fully clamped cantilever

The computer program, ANDECAS uses the methods proposed in Sect. 5.2.2 including Tables 5.1, 5.2, 5.3, 5.4, 5.5, 5.6, 5.7, 5.8, 5.9, 5.10, 5.11, and 5.12 to calculate design hogging moments due to Canadian, or user-defined, design loads at 21 reference sections, as shown in Fig. 5.10 (Parr 1993; Parr and Bakht 1993; Bakht et al. 2006).

Table 5.10 Values of B for edge-stiffened cantilever slabs with $t_2/t_1 = 0.5$

		B for C/S_c=				
S_c/S	y/S_c	1.0	0.8	0.6	0.4	0.2
0.4	1.00	0.00				
	0.80	−0.01				
	0.60	0.05	−0.02			
	0.40	0.11	0.08	−0.01		
	0.20	0.17	0.16	0.14	0.06	
	0.00	0.22	0.25	0.29	0.31	0.30
0.5	1.00	0.00				
	0.80	−0.01				
	0.60	0.05	−0.02			
	0.40	0.11	0.08	−0.01		
	0.20	0.17	0.16	0.15	0.06	
	0.00	0.23	0.25	0.29	0.32	0.31
0.67	1.00	0.00				
	0.80	−0.01				
	0.60	0.06	−0.02			
	0.40	0.12	0.09	−0.00		
	0.20	0.17	0.17	0.15	0.07	
	0.00	0.23	0.26	0.30	0.32	0.31
1.0	1.00	0.00				
	0.80	−0.00				
	0.60	0.06	−0.01			
	0.40	0.12	0.09	0.00		
	0.20	0.18	0.18	0.16	0.08	
	0.00	0.24	0.27	0.32	0.34	0.32
2.0	1.00	0.00				
	0.80	0.00				
	0.60	0.07	−0.01			
	0.40	0.13	0.10	0.02		
	0.20	0.19	0.19	0.18	0.10	
	0.00	0.26	0.29	0.35	0.37	0.35
*	1.00	0.00				
	0.80	0.00				
	0.60	0.07	0.01			
	0.40	0.14	0.12	0.04		
	0.20	0.21	0.22	0.23	0.16	
	0.00	0.28	0.35	0.46	0.53	0.55

*Fully clamped cantilever

In addition to analysing the cantilever slab and the internal panel, ANDECAS can also design the reinforcement for the calculated negative moment intensities according to the limit states design method prescribed by the CHBDC (2006). The CHBDC (2006) design inequality is as follows.

Table 5.11 Values of B for internal panels corresponding to edge-stiffened cantilever slabs with $t_2/t_1 = 1.0$

		B for $C/S_c=$				
S_c/S	y/S_c	1.0	0.8	0.6	0.4	0.2
0.4	0.00	0.21	0.20	0.21	0.23	0.24
	1.00	0.00	0.00	0.00	0.00	0.00
0.5	0.00	0.21	0.21	0.22	0.23	0.25
	0.20	0.16	0.14	0.12	0.10	0.06
	0.40	0.12	0.10	0.08	0.06	0.03
	0.60	0.08	0.06	0.05	0.04	0.02
	0.80	0.04	0.03	0.03	0.02	0.01
	1.00	0.00	0.00	0.00	0.00	0.00
0.67	0.00	0.22	0.21	0.22	0.24	0.25
	0.10	0.20	0.18	0.17	0.16	0.12
	0.20	0.17	0.15	0.14	0.12	0.08
	0.30	0.15	0.13	0.11	0.09	0.06
	0.40	0.12	0.11	0.09	0.07	0.04
	0.50	0.10	0.09	0.07	0.06	0.03
	0.60	0.08	0.07	0.06	0.04	0.02
	0.70	0.06	0.05	0.04	0.03	0.09
	0.80	0.04	0.03	0.03	0.02	0.01
	0.90	0.02	0.02	0.01	0.01	0.01
	1.00	0.00	0.04	0.00	0.00	0.00
1.0	0.00	0.23	0.23	0.24	0.25	0.26
	0.10	0.21	0.20	0.19	0.18	0.15
	0.20	0.18	0.17	0.16	0.14	0.10
	0.30	0.16	0.14	0.13	0.11	0.08
	0.40	0.14	0.12	0.11	0.09	0.06
	0.50	0.11	0.10	0.09	0.07	0.05
	0.60	0.09	0.08	0.07	0.06	0.03
	0.70	0.07	0.06	0.05	0.04	0.02
	0.80	0.04	0.04	0.03	0.03	0.02
	0.90	0.02	0.02	0.02	0.01	0.01
	1.00	0.00	0.00	0.00	0.00	0.00
2.0	0.00	0.25	0.25	0.26	0.28	0.28
	0.10	0.23	0.22	0.23	0.23	0.21
	0.20	0.20	0.19	0.20	0.19	0.16
	0.30	0.18	0.17	0.17	0.16	0.12
	0.40	0.15	0.14	0.14	0.13	0.10
	0.50	0.13	0.12	0.12	0.11	0.08
	0.60	0.10	0.09	0.09	0.08	0.06
	0.70	0.08	0.07	0.07	0.06	0.04
	0.80	0.05	0.05	0.05	0.04	0.03
	0.90	0.03	0.02	0.02	0.02	0.01
	1.00	0.00	0.00	0.00	0.00	0.00

Table 5.12 Values of B for internal panels corresponding to edge-stiffened cantilever slabs with $t_2/t_1 = 0.5$

		B for $C/S_c=$				
S_c/S	y/S_c	1.0	0.8	0.6	0.4	0.2
0.4	0.00	0.22	0.25	0.29	0.31	0.30
	0.00	0.00	0.00	0.00	0.00	0.00
0.5	0.00	0.23	0.25	0.29	0.32	0.31
	0.20	0.17	0.16	0.15	0.12	0.07
	0.40	0.12	0.11	0.09	0.07	0.04
	0.60	0.08	0.07	0.06	0.04	0.02
	0.80	0.04	0.03	0.03	0.02	0.01
	1.00	0.00	0.00	0.00	0.00	0.00
0.67	0.00	0.23	0.26	0.30	0.32	0.31
	0.10	0.21	0.21	0.22	0.20	0.14
	0.20	0.18	0.17	0.17	0.14	0.09
	0.30	0.16	0.15	0.13	0.11	0.07
	0.40	0.13	0.12	0.11	0.09	0.05
	0.50	0.11	0.10	0.09	0.07	0.04
	0.60	0.09	0.08	0.07	0.05	0.03
	0.70	0.06	0.06	0.05	0.04	0.02
	0.80	0.04	0.04	0.03	0.02	0.01
	0.90	0.02	0.02	0.02	0.01	0.01
	1.00	0.00	0.00	0.00	0.00	0.00
1.0	0.00	0.24	0.27	0.32	0.34	0.32
	0.10	0.22	0.23	0.25	0.24	0.18
	0.20	0.19	0.19	0.20	0.18	0.12
	0.30	0.17	0.16	0.16	0.14	0.09
	0.40	0.14	0.14	0.13	0.11	0.07
	0.50	0.12	0.11	0.11	0.09	0.05
	0.60	0.09	0.09	0.08	0.07	0.04
	0.70	0.07	0.07	0.06	0.05	0.03
	0.80	0.05	0.04	0.04	0.03	0.02
	0.90	0.02	0.02	0.02	0.02	0.01
	1.00	0.00	0.00	0.00	0.00	0.00
2.0	0.00	0.26	0.29	0.35	0.37	0.35
	0.10	0.23	0.26	0.30	0.30	0.25
	0.20	0.20	0.22	0.25	0.25	0.19
	0.30	0.18	0.19	0.21	0.20	0.15
	0.40	0.15	0.16	0.18	0.17	0.12
	0.50	0.13	0.13	0.14	0.13	0.09
	0.60	0.10	0.11	0.11	0.10	0.07
	0.70	0.08	0.08	0.08	0.08	0.05
	0.80	0.05	0.05	0.06	0.05	0.03
	0.90	0.03	0.03	0.03	0.03	0.02
	1.00	0.00	0.00	0.00	0.00	0.00

Table 5.13 Values of D for stiffened and unstiffened cantilever slabs of semi-infinite length

		D for C/S_c=				
t_1/t_2	X/S_c	1.0	0.8	0.6	0.4	0.2
1.0	0	2.24	2.23	2.25	2.25	1.94
	2	1.55	1.49	1.35	1.18	0.46
	4	1.01	0.91	0.68	0.37	0.08
	6	0.64	0.49	0.30	0.12	0.04
	8	0.37	0.25	0.14	0.04	0.02
	12	0.14	0.10	0.04	0.03	0.01
0.5	0	2.33	2.33	2.40	2.36	2.04
	2	1.55	1.47	1.32	1.00	0.35
	4	0.96	0.79	0.59	0.27	0.01
	6	0.49	0.34	0.15	0.07	0.00
	8	0.32	0.15	0.10	0.00	0.00
0.33	0	2.28	2.33	2.36	2.36	1.98
	2	1.49	1.40	1.27	0.96	0.37
	4	0.84	0.74	0.51	0.22	0.00
	6	0.41	0.29	0.14	0.00	0.00
	8	0.15	0.10	0.00	0.00	0.00

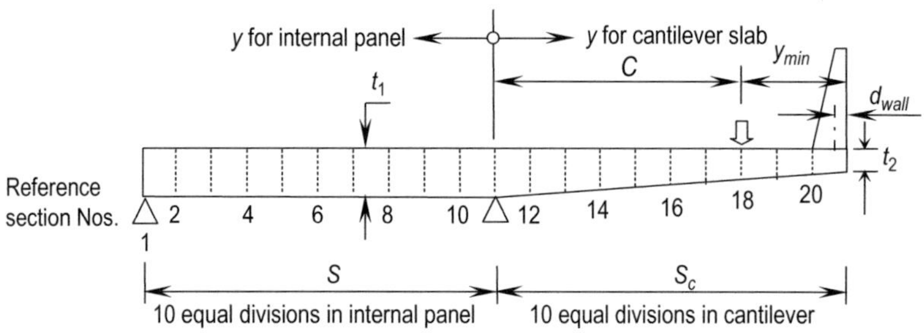

Fig. 5.10 Notation used in computer program ANDECAS6

$$\phi R_n \geq \sum_{i=l}^{N} \alpha_{Di} D_{ni} + \alpha_L (l + DLA) L_n \tag{5.8}$$

where

ϕ = resistance factor, which is $\leq$1.0.
R_n = the nominal failure strength of the component.
N = the number of dead loads, e.g. cast-in-place concrete components and wearing course.
α_{Di} = the dead load factor for dead load number i, which is $\geq$1.0 for loads adding to critical load effects and is $\leq$1.0 for loads reducing the critical load effects.
D_{ni} = the load effect due to nominal dead load number i.
α_L = the load factor for live loads, which is $\geq$1.0.
DLA = the dynamic load allowance, or the impact factor.
L_n = the load effect due to the nominal design live loading.

Table 5.14 Details of data input to ANDECAS as prepared by INPUT

Input Ref. No	Input parameter	Notes
0.	File name	The file name should be one word without spaces; this word is used for the data and results files
1.	Number of cases to be analysed, N_{case}	Data for each case is provided in input ref. Nos. 27–31
2.	Live load factor, α_L	Any number can be given if the run is only for analysis; $\alpha_L = 1.70$ for CHBDC (2006)
3.	Load factor for dead load No. 1 (weight of concrete of cantilever slab and barrier wall), α_{D1}	Any number can be given if the run is only for analysis; $\alpha_{D1} = 1.20$ for CHBDC (2006)
4.	Load factor for dead load No. 2 (weight of wearing course), α_{D2}	Any number can be given if the run is only for analysis; $\alpha_{D2} = 1.50$ for CHBDC (2006)
5.	Load factor for dead load No. 3 (weight of concrete in internal panel), α_{D3}	Any number can be given if the run is only for analysis; $\alpha_{D3} = 0.90$ for CHBDC (2006)
6.	Dynamic load allowance (DLA)	Any number can be given if the run is only for analysis; DLA = 0.4 for single axles, 0.3 for two axles, and 0.2 for three or more axles for CHBDC (2006)
7.	Resistance factor for steel, ϕ_s	Any number can be given if the run is only for analysis; $\phi_s = 0.9$ for CHBDC (2006)
8	Resistance factor for concrete, ϕ_c	Any number can be given if the run is only for analysis; $\phi_c = 0.75$ for CHBDC (2006)
9.	Minimum ratio of steel to concrete, ρ_{min}	Any number can be given if the run is only for analysis; $\rho_{min} = 0.003$ for CHBDC (2006)
10.	Yield stress of steel, f_y, MPa	Any number can be given if the run is only for analysis; typically $f_y = 250$ MPa
11.	Specified strength of concrete f'_c, MPa	Any number can be given if the run is only for analysis; typically $f'_c = 30$ MPa
12.	y_{min}, m	Distance of centreline of wheel from the nearest free edge (see Fig. 5.10)
13.	d_{wall} m	Distance of the c.g. of barrier wall from the nearest free edge (see Fig. 5.10); enter zero if there is no barrier wall
14.	Weight of barrier wall per metre length, kN/m	Enter zero if there is no barrier wall
15.	Moment per metre length due to impact loading on barrier wall, kN.m/m	Enter zero if there is no barrier wall
16.	Thickness of slab of internal panel, m	Any number can be given if the run is only for analysis
17.	Thickness of wearing course, m	Any number can be given if the run is only for analysis

(continued)

Table 5.14 (continued)

Input Ref. No	Input parameter	Notes
18.	Cover to the top reinforcement, m	Any number can be given if the run is only for analysis
19.	Control No., N_{load}, for design truck. Enter 1 for OHBDC design truck, 2 for CSA design truck, 3 for CHBDC (2006) design truck, and 4 for the user-defined truck, which is defined in Fig. 5.11	The OHBDC and CSA design trucks are obsolete. The 5-axle CHBDC truck is defined as CL-W truck, in which the load W is variable and is assumed by the program to be 625 kN. The user-defined truck, shown in Fig. 5.11, can have any number of lines of wheels (N_{line}) and any number of axles (N_{axle})
20.	Control number, N_x, for options for searching for maximum moment intensity	Enter 1 if the program is to search for maximum moment intensity, and 2 if the moment intensity is to be calculated at a specified point
21.	Control number, N_{des}, for analysis and design options	Enter 1 if design is required, and 2 if only analysis is required
22.	Control number, N_{edge}, for edge stiffening	Enter 1 if the longitudinal edge of the cantilever slab is stiffened by a barrier wall shown in Fig. 5.8; enter 2 if the cantilever slab does not have edge stiffening
23.	No. of lines of lines, N_{line}, of wheels and number of axles, N_{axle}, in user-defined truck	This data is not required for OHBDC, CSA and CHBDC trucks
24.	(N_{line}-1) values of successive distances of lines of wheels, starting from the 1st line, m	This data is not required for OHBDC, CSA and CHBDC trucks
25.	(N_{axle}-1) values of successive distances of axles, starting from the 1st axle, m	This data is not required for OHBDC, CSA and CHBDC trucks
26.	N_{axle} values of magnitudes of loads of wheel loads, kN	This data is not required for OHBDC, CSA and CHBDC trucks. It is assumed that all wheels in an axle have the same loads
27.	Title of the case	Up to 60 characters
28.	S_c, S and t_2/t_1	
29.	Weight of CSA design truck, kN	Skip this line if the design truck is not the CSA truck
30.	x location of predefined reference point, m	Skip this line if the control number in input ref. No. 20 is 1
31.	Control number for the kind of analysis	Enter N if the maximum moment intensity is to be searched numerically; enter M if the maximum moment intensity is to be searched mathematically. Both methods give nearly the same results

Note: Input ref. Nos. 27–31 are to be repeated N_{case} number of times

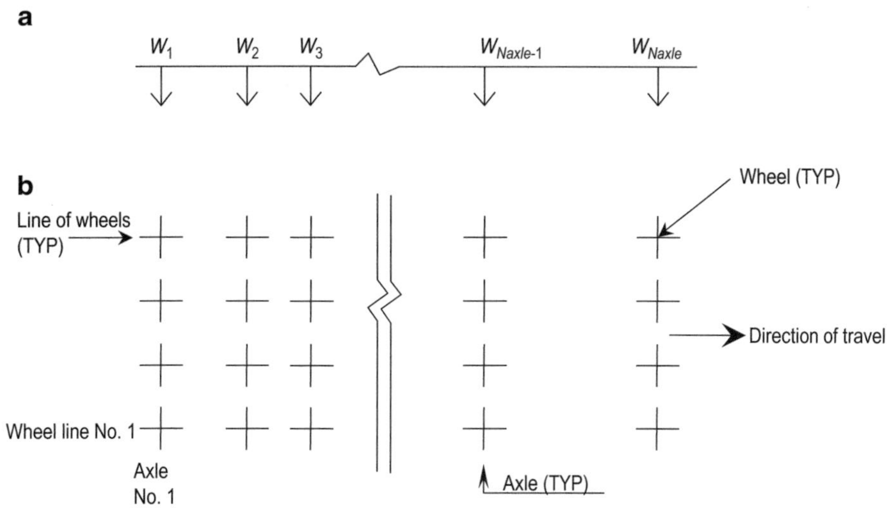

Fig. 5.11 User-defined truck shown with four lines of wheels: (**a**) elevation, (**b**) plan

The program calculates R_n, i.e. the ultimate moment of resistance of the reinforced concrete section, by assuming that the section is balanced and contains only the tensile reinforcement.

Data Input An interactive program INPUT creates the input data for ANDECAS according to the scheme outlined in Table 5.14.

The use of ANDECAS for only analysis is illustrated in the following for the cantilever and internal panel of Fig. 5.12, which shows that the deck slab is subjected to a user-defined truck with two lines of wheels and four axles. The wheel loads in a line are 50, 70, 70 and 40 kN, respectively. The following design input parameters are assumed, it being noted that the values of these parameters are irrelevant since the program is being run only for analysis. $\alpha_L = 1.0$; $\alpha_{D1} = 1.0$; $\alpha_{D2} = 1.0$; $\alpha_{D3} = 1.0$; DLA = 0.0; $\phi_s = 1.0$; $\phi_c = 1.0$; $\rho_{min} = 0.1$; $f_y = 300$ MPa; $f'_c = 30$ MPa.

In addition to the above, it is assumed that (a) the cantilever slab has no edge stiffening, (b) the distance of the lines of wheels closest to the free edge of the cantilever slab is 0.5 m, (c) the distance of the c.g. of the barrier wall is zero, (d) the weight of the barrier wall is zero, (e) moment induced by impact on the barrier wall is zero, (f) the thickness of the internal panel is 0.3 m, (g) thickness of the wearing course is zero, (h) the cover to the top reinforcement is zero. Other input parameters are shown in Fig. 5.12.

The input data created by the program INPUT is reproduced in Fig. 5.13.

The program ANDECAS is run by clicking on the ANDECAS icon, and the results are stored in file having the same name as the title with the extension of '.RES'.

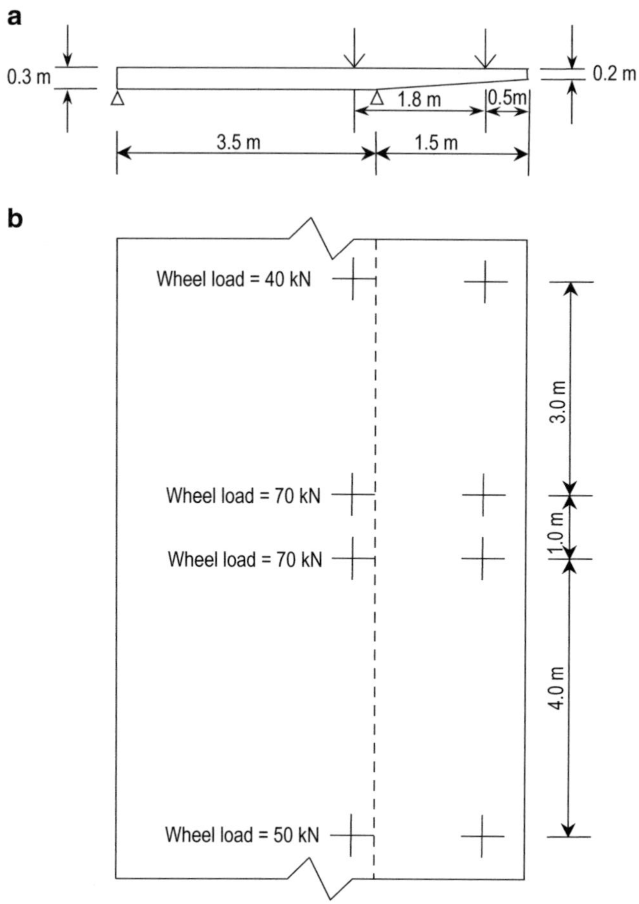

Fig. 5.12 Example cantilever slab and internal panel, (**a**) cross-section, (**b**) plan

```
              1
1.000   1.000   1.000   1.000   0.000   1.000   1.000   0.100
300.000    30.000
0.300   0.000   0.000   0.000   0.300   0.000   0.020   4   1   2
2   4   2
0.000000 4.00000 1.00000  3.00000  0.000000  1.80000
50.0000         70.0000         70.0000         40.0000
4 axle truck
1.500   3.500   0.670 m
```

Fig. 5.13 Input data for the example deck slab created by INPUT

```
ALPHA LL  =   1.000               PHI STELL  =   1.000
ALPHA DL1 =   1.000               PHI CONC.  =   1.000
ALPHA DL2 =   1.000               RHO min.   =   0.100
ALPHA DL3 =   1.000               fy  steel  =   300.000MPa
DLA       =   0.000               f'c conc. =  30.000MPa
    CASE NAME :    4 ax
    Cantilever span, Sc                     =   1.500m
    Girder spacing,   S                     =   3.500m
    Thickness ratio, t2/t1                  =   0.670
    Min. load distance from free edge       =   0.300m
    Concrete thickness                      =   0.300m
    Wearing surface thickness               =   0.000m
    Concrete cover                          =   0.020m
    Wall distance from free edge            =   0.000m
    Weight of the wall                      =   0.000KN/m
    Bending moment due to railing load =   0.000KNm
           4            2
0.000000 4.00000  1.00000   3.00000   0.000000
1.80000  50.0000  70.0000   70.0000   40.0000
Analysis required to determine maximum intensities.
Design NOT required. The slab has no edge stiffening
    ANDECAS4 IS RUNNING MATHMATICAL FORM SOLUTION
```

Fig. 5.14 Echo-printing of input data in the results file for the example deck slab of Fig. 5.12

The echo printing of the input data created by ANDECAS for the example under consideration is reproduced in Fig. 5.14.

ANDECAS gives maximum transverse moment intensities M_y due to the design truck at 11 equally spaced reference points in each of the cantilever and internal panel. The program also gives the longitudinal position of the location of maximum M_y with respect to the first axle of the design truck; this distance is denoted as X in the output file. For the example under consideration, the maximum value of M_y at all the reference points given by ANDECAS are reproduced in Fig. 5.15 along with their longitudinal positions. It is noted that the line wheels on the internal panel is ignored in the analysis, as the wheel loads on this line reduce the negative moments in the internal panel; this procedure is consistent with the CHBDC (2006) requirements.

The envelope of maximum transverse moments due to the user-defined truck of Fig. 5.12 is plotted in Fig. 5.16, in which it can be seen that the variation of maximum negative moments in the internal panel is far from being linear.

5.3 Arching in Cantilever Slabs

Until recently, the arching action of the deck slab discussed in Chap. 4 is believed to take place only in the portion of the slab lying between the outermost girders when the applied concentrated wheel loads are also contained within these two girders. It was believed that the strength of the cantilever deck slab is governed only by the transverse reinforcement near the top of the slab. Some laboratory tests, however,

Fig. 5.15 ANDECAS results for transverse moments for the example deck slab of Fig. 5.12 (Note: negative moments are shown without the – sign)

Cantilever:

Yc (m)	Moment intensity(kNm/m)	X (m)
1.500	0.000	0.000
1.350	0.055	0.000
1.200	0.224	0.000
1.050	6.395	0.795
0.900	12.689	0.795
0.750	18.479	0.826
0.600	24.399	0.847
0.450	30.482	0.861
0.300	36.710	0.872
0.150	43.644	0.858
0.000	50.734	0.849

Internal panel:

Yi (m)	Moment intensity (kNm/m)	X (m)
0.000	50.734	0.849
0.350	37.194	1.137
0.700	27.211	1.594
1.050	19.121	1.859
1.400	12.703	1.938
1.750	7.632	1.969
2.100	4.083	1.969
2.450	1.215	2.000
2.800	0.019	2.000
3.150	-0.572	2.000
3.500	0.000	0.000

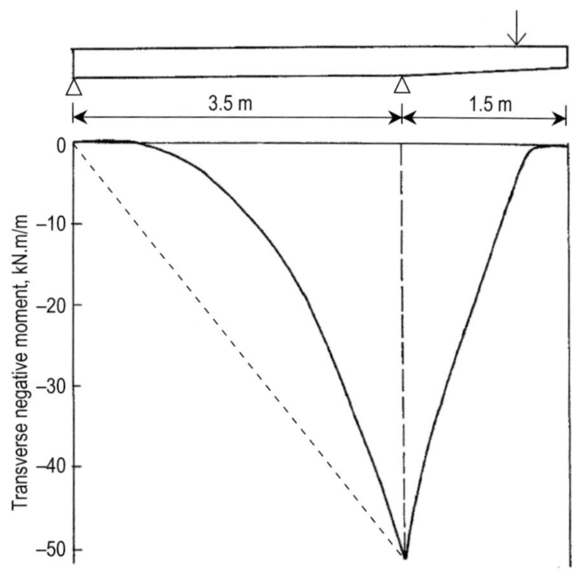

Fig. 5.16 Maximum transverse moment intensities due to the user-defined truck of Fig. 5.12

showed that the failure load of cantilever deck slabs is also influenced by the transverse reinforcement near its bottom surface. For example, Kent and Newhook (1999) found that two segments of the cantilever slab of a full-scale model containing additional bottom transverse reinforcement failed at nearly twice the

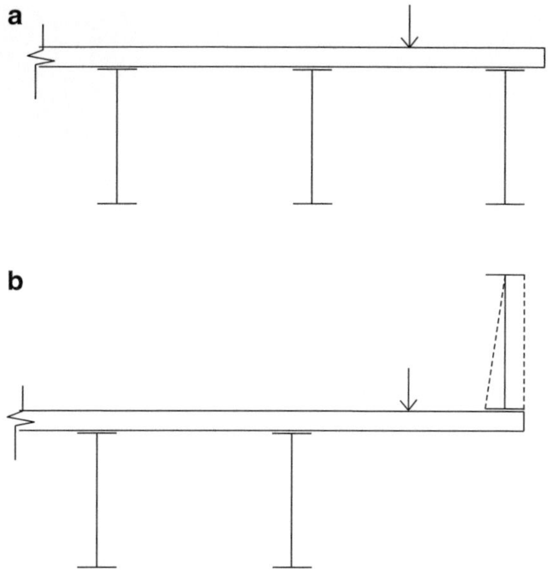

Fig. 5.17 Mechanics of behaviour of an edge-stiffened cantilever slab: (**a**) an external girder under the deck slab, (**b**) external girder moved above the deck slab

load than the failure loads of the segments without the bottom transverse reinforcement. Several tests on full-scale models of concrete deck slabs without edge stiffening conducted by Klowak (2007) also confirmed the presence of arching action in cantilever deck slabs under concentrated wheel loads.

The arching action in unstiffened cantilever slabs is yet to be explored further before it can be made use of in the design. However, there is convincing experimental evidence about the presence of arching action in cantilever slabs with the kind of barrier walls shown in Fig. 5.8. The mechanics of behaviour of an edge-stiffened cantilever slab can be explained with the help of Figs. 5.17a and b. The former figure shows a segment of the deck slab over three girders, one of which is the outermost girder. The deck slab is subjected to concentrated load between the outermost and adjacent girders. In light of the evidence presented in Chap. 4, there is little doubt that the panel of the deck slab below the load will experience arching action.

With the load on the deck slab staying at the same location, the outer girder is now moved above the deck slab, as shown in Fig. 5.17b. It can be appreciated that the behaviour of the deck slab will not be affected by whether the outermost girder is above or below. As shown in Fig. 5.17b, the ‘girder’ above the deck slab could be the barrier wall. This reasoning led to static and fatigue tests on a full-scale model of a deck slab on two steel girders with New Jersey type of barrier walls at its two longitudinal free edges. A partial view of the model with a concentrated load on the cantilever slab can be seen in Fig. 5.18. The 200 mm thick deck slab was supported on two steel girders spaced at 2.5 m, and the cantilever overhang on each side was 2.0 m long. The model was 9.15 m long.

Fig. 5.18 A full-scale model of a cantilever slab stiffened with New Jersey type of barrier walls

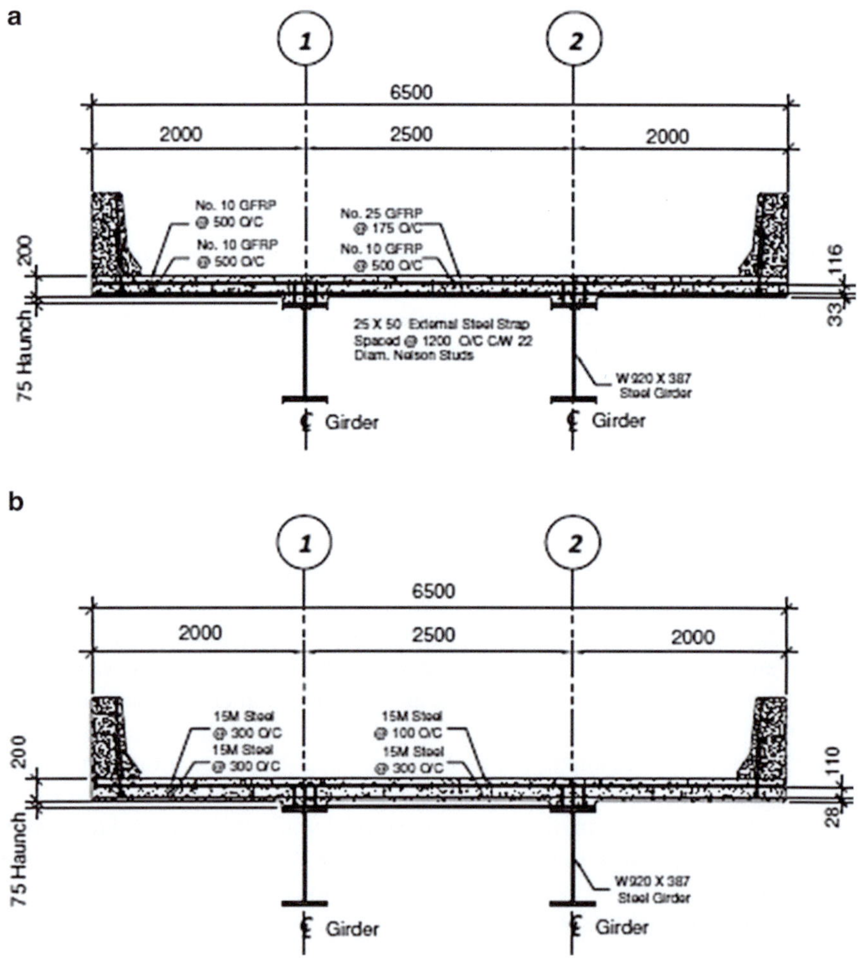

Fig. 5.19 Reinforcement details in deck slab: (**a**), GFRP reinforcement (**b**) steel reinforcement

A 4.575 m length of the deck slab was reinforced with steel bars, and the other 4.575 m length with bars of glass fibre reinforced polymer (GFRP). Details of the reinforcement are shown in Figs. 5.19a and b. The concentrated load, applied through a steel plate and neoprene pad was 610 × 305 mm in plan dimensions.

The model of Fig. 5.18 and results of static and fatigue tests are reported by Klowak et al. (2013). The static failure loads on the cantilever slab with steel and GFRP reinforcement were 772 and 486 kN, respectively. It is recalled that, as noted in Chap. 4, a 175 mm thick deck slab on two girders at a spacing of 2.0 m failed at loads between 750 and 882 kN. The failure loads of the cantilever slab confirm not only the presence of arching action in edge-stiffened cantilever slabs, but also the significant effect of the axial stiffness of the bottom transverse reinforcement on the strength of these slabs.

References

Bakht B (1981) Simplified analysis of edge-stiffened cantilever slabs. ASCE J Struct Eng 103(3):535–550

Bakht B, Holland DA (1976) A manual method for the elastic analysis of wide cantilever slabs of linearly varying thickness. Can J Civ Eng 3(4):523–530

Bakht B, Aziz TS, Bantusevicius KF (1979) Manual analysis of cantilever slabs of semi-infinite width. Can J Civ Eng 6(2):227–231

Bakht B, Mufti AA, Nath Y (2006) ANDECAS6 user manual – analysis and design of cantilever slabs. ISIS Canada Research Network, Winnipeg

CAN/CSA S6-06 (2006) Canadian highway bridge design code. CSA International, Mississauga

Dilger WH, Tadros GS, Chebib J (1990) Bending moments in cantilever slabs. Developments in short and medium span bridge engineering =90. Can Soc Civ Eng 1:256–276, Montreal

Ghali A, Tadros GS (1980) User manual for computer program SPAST. Research report CE80-10. Department of Civil Engineering, University of Calgary. Calgary

Jaeger LG, Bakht B (1990) Rationalization of simplified methods of analyzing cantilever slabs. Can J Civ Eng 17(5):856–867

Kent G, Newhook J (1999) Wheel load tests on cantilever and internal panels on concrete deck slab on steel girders with and without bottom transverse GFRP reinforcement. Internal testing report. Dalhousie University, Halifax

Klowak C (2007) Static & fatigue behaviour of bridge deck cantilever overhangs subjected to a concentrated load. MSc thesis submitted to University of Manitoba, Winnipeg

Klowak C, Mufti A, Bakht B (2013) ASCE J Compos Constr 9

Mufti AA, Bakht B, Jaeger LG (1993) Moments in deck slabs due to cantilever loads. ASCE J Struct Div 119(6):1761–1777

OHBDC (1992) Ontario highway bridge design code. Ministry of Transportation of Ontario, Downsview

Parr S (1993) Analysis and design of deck slab overhangs of girder bridges. Thesis submitted in partial fulfilment of the requirements for the degree of MASc at the University of Toronto, Toronto

Parr S, Bakht B (1993) Design moments due to loads on deck slab overhangs of girder bridges. In: Proceedings of the annual conference of Canadian Society for Civil Engineering, Fredericton

Tadros G, Bakht B, Mufti AA (1994) On the analysis of edge-stiffened cantilever slabs. In: Proceedings of the fifth colloquium on concrete in developing countries, Cairo

Chapter 6
Wood Bridges

6.1 Introduction

Madsen, a leading expert in his times in the structural use of wood made a clear distinction between the terms *wood* and *timber*. He used the former term for defect-free samples, which are employed for determining the fundamental properties of this building material, and reserved *timber* for that useful construction material, which is produced from logs of trees (Madsen 1992). Notwithstanding these definitions, both the terms are used interchangeably in this book as is done commonly in technical literature dealing with the structural applications of timber.

Because in its untreated state it is susceptible to biodegradation, timber is sometimes considered to be unsuitable for permanent and outdoor structures. It has been established, however, that well-treated wood can prove to be a durable material even in warm and humid climates.

Durability There are many examples of well-treated wood having lasted a long time in the outdoors. An example of the durability of wood is the Sioux Narrows Bridge, a photograph of which is presented in Fig. 6.1. This bridge, which had a record span of 64 m, was built in 1936. After about 70 years of uninterrupted service, this bridge was taken down to make room for a wider bridge, and not because of its lack of durability.

The trusses of the Sioux Narrows Bridge, which was located in the northern part of the Province of Ontario, Canada, were made from Douglas fir, pressure-treated with oil-borne creosote. Douglas fir is a soft wood that receives the preservative treatment easily; it has several varieties, of which the Coastal Douglas Fir growing near the west coast of Canada and USA "forms some of the tallest and most magnificent woods in the world" (Mitchell 2014). The timbers of Sioux Narrows Bridge were so durable that the stringers of this discarded bridge were used recently to conduct lab research on the shear strength of stringers as it is affected by the distance of their supports to their ends (Das 2010).

B. Bakht, A. Mufti, *Bridges*, DOI 10.1007/978-3-319-17843-1_6

Fig. 6.1 The Sioux Narrows bridge in Ontario, Canada which was believed to be the longest-span timber bridge in the world

Hardwoods, which are in any case not recommended for permanent outdoor structural applications, are difficult to treat. The secret of having durable timber bridges is to make them out of species, which are easily treatable.

Recently, water-borne preservatives have come to the fore in the treatment of woods against biodegradation. However, their effectiveness is still being debated. It is argued by some experts that unlike timber treated with oil-borne preservatives, water-borne treatment does not seal the wood against the migration of moisture, thereby leading to rapid drying and thence to the development of longitudinal cracks, or checks, in the timber. Another disadvantage of water-borne preservatives is that when the timber is allowed to be saturated with water, its compressive strength in the direction perpendicular to the grain is reduced considerably. This is a serious shortcoming when timber with water-borne treatment is used in the new forms of bridges presented in this chapter.

New Developments In recent years, significant developments have taken place, mainly in Canada and the USA, in the structural forms of wood bridges. The new structural forms that have emerged as a result of these developments permit the use of wood more effectively and efficiently than had been possible in the past. The main purpose of this chapter is to present in summary several new forms of timber bridges.

6.2 Stress-Laminated Wood Decks

The term *nail-laminated deck* is used for a bridge made of sawn timber planks which are about 30–60 mm thick and 150–250 mm wide in cross-section, and which, are usually referred to as laminates. The laminates, with the longer sides of their cross-section vertical, are successively nailed together to form the solid deck, which constitutes the superstructure of the bridge. The length of the laminates, running along the deck in the longitudinal direction, is much shorter than the span

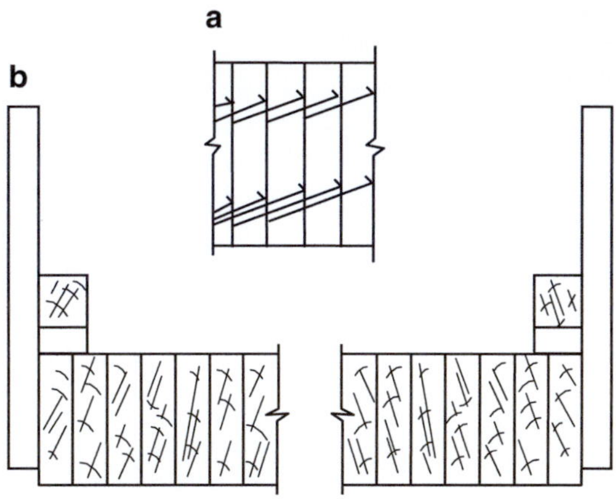

Fig. 6.2 A nail-laminated wood deck: (**a**) nailing detail, (**b**) cross-section

Fig. 6.3 A nail-laminated wood deck bridge in Ontario, Canada

of the deck. Continuity along the span is provided by staggering the butt joints judiciously. The cross-section of a typical nail-laminated deck is shown in Fig. 6.2, and a typical continuous-span nailed-laminated deck bridge in Fig. 6.3.

The transverse distribution of loads in a nail-laminated deck takes place entirely through the nails, which are quite effective in this function during the early life of the bridge. However, under repeated loading, the holes containing the nails tend to

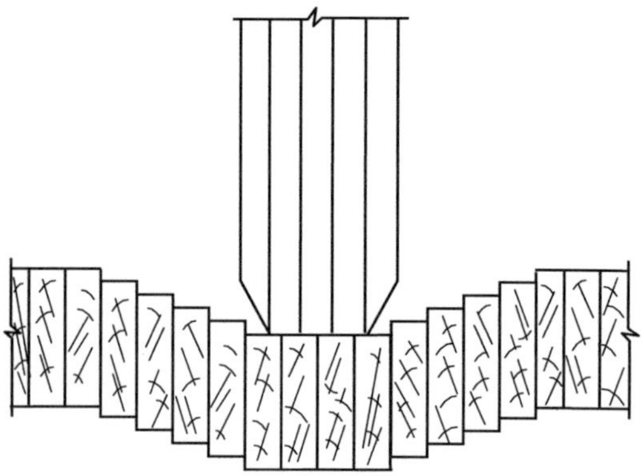

Fig. 6.4 Cross-section of a nail-laminated wood deck with deteriorated nail connections under a wheel load

enlarge, thereby causing the directly loaded laminates to deflect freely a certain distance before engaging the adjacent laminates, as illustrated in Fig. 6.4.

The load distribution characteristics of nail-laminated decks deteriorate with time. Another consequence of the loosening of the nail connections is that longitudinal rotation of the deck at sections containing several butt joints tends to become large, thereby causing the deck to sag excessively. The net result of the loosening of the nail connections under repeated loads is that even under moderate traffic the nail-laminated deck becomes unserviceable after 10–15 years of service.

In the mid 1970s, a technique was developed in Ontario, Canada, for the rehabilitation of deteriorated nail-laminated wood deck bridges; in this technique the laminates are squeezed together by means of substantial lateral pressures applied through high strength steel bars (Csagoly and Taylor 1980). The laminated wood deck stressed laterally in this manner has come to be known as a stress-laminated wood deck (SWD). Although developed primarily for the rehabilitation of deteriorated decks, the concept of SWD has also been applied to new construction in an efficient and imaginative manner (e.g., Taylor and Walsh 1983; Bakht and Tharmabala 1987; Oliva and Dimokis 1988; Gangarao and Latheef 1990; Sarisley and Accorsi 1990). Jaeger and Bakht (1990) have demonstrated analytically that the total effective flexural rigidity of a deck with butt-jointed laminates is enhanced by increasing the lateral pressure on the laminates.

The lateral pressures to the laminated deck can be applied through pairs of bars, one placed at the top of deck and one below it; such a system is called the external post-tensioning system. Alternatively, the lateral pressure can be applied through an internal post-tensioning system in which the post-tensioning bars pass through oversize holes in the deck. The cross-section of a SWD incorporating an external post-tensioning system is shown in Fig. 6.5a and that incorporating an internal post-tensioning system in Fig. 6.5b.

The technique of laterally post-tensioning laminated decks was developed mainly to rehabilitate existing bridges in which the nail connections had deteriorated. The first bridge rehabilitated by this technique was the Herbert Greek Bridge. This bridge had deteriorated so much that practically only laminates directly under

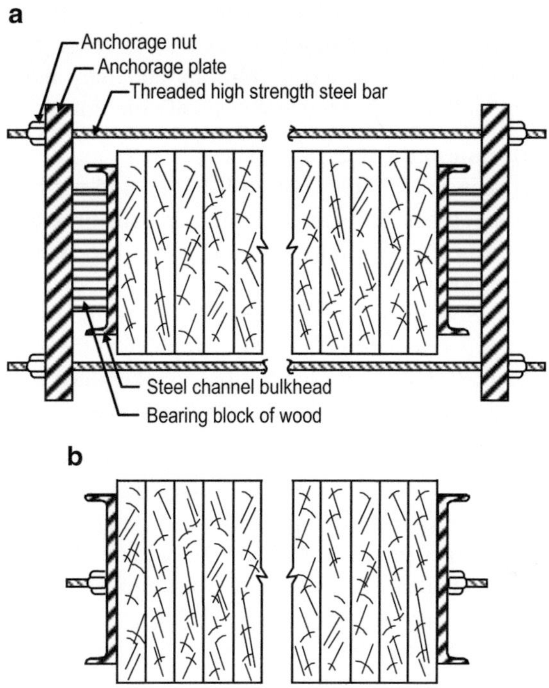

Fig. 6.5 Cross-sections of stress-laminated wood decks: (**a**) bridge with external post tensioning system; (**b**) bridge with internal post-tensioning system

the wheels sustained the applied wheel loads. The post-tensioning operation, which is described by Csagoly and Taylor (1980), permits the whole deck to deform together to such an extent that the maximum deflections of the post-tensioned deck under similar loads were reduced to about 50 % of the corresponding deflections of the bridge before rehabilitation. An example of this behavior is discussed in Sect. 6.3.1.

It is interesting to note that by squeezing the laminates together, the exposed area of wood employed in the deck is reduced by more than 80 %. It is not often appreciated that this reduction of the exposed area of wood not only enhances significantly the durability of the deck but also reduces the amount of preservative needed to treat the deck.

6.2.1 Design Specifications

Until the year 2000, the Ontario Highway Bridge Design Code (OHBDC) was the only available design code dealing with SWDs; the last edition of OHBDC was published in 1992a, b. The AASHTO Specifications in its 13th edition (1992) have also covered SWDs. The OHBDC was superseded by the Canadian Highway Bridge Design Code (CHBDC), the first edition of which was published in 2000, and the second in 2006. Some CHBDC (2006a) provisions for SWD are given in this Subsection. It should be noted, however, that these provisions are relevant to Canadian species and conditions and may have to be revised when applied in conjunction with other species and conditions.

Interlaminate Pressure The CHBDC requires that the average lateral pressure between the laminates of an SWD should not be less than 0.35 MPa, or more than $0.25f_{ql}$, where f_{ql} is the specified limiting pressure perpendicular to the grain for the species of wood under consideration. The smallest value of f_{ql} is 3.5 MPa which corresponds to lodgepole pine and white pine. Through Ontario Provincial Standards Specification (OPSS) 907 (1992), the OHBDC (1992a, b) specified that the full post-tensioning force be reapplied to the deck twice after the initial stressing. The first re-stressing is required to take place about 1 week after the first stressing and the second re-stressing 4–8 weeks later.

The CHBDC (2006a) further specifies that the final lateral pressure between the laminates, after the prestress losses have taken place, shall be assumed to be 0.4 times the corresponding pressure at jacking. While this provision might be construed as a prediction of the magnitude of prestress losses, it should in fact be regarded only as a design aid. The commentary to the code (2006b) notes explicitly that the current state of knowledge is such that the magnitude of prestress loss cannot be predicted with confidence and that provisions should be made for checking the levels of prestress periodically and for re-stressing the deck if the inter-laminate pressure falls below the prescribed minimum level.

Bulkheads Both the external and internal post-tensioning systems require continuous longitudinal bulkheads, which may be formed out of steel channels with their flat sides touching the laminated deck as shown in Fig. 6.5a, b. The spacing of the prestressing bars and the flexural stiffness of the bulkheads should be such that the prestressing forces are distributed at the interface of the bulkheads and the deck.

Stiffness of the Stressing System The CHBDC requires that the cross-sectional area of the stressing steel bars should not be more than 0.0016 times the corresponding area of cross-section of the wood deck. This requirement ensures that the stiffness of the stressing system is not excessively large. As discussed in Sect. 6.3.3, the prestress losses in SWDs are affected significantly by the ratio of the stiffness of the stressing system and that of the wood deck.

Flexural Resistance The design of an SWD is generally governed by its flexural resistance M_r which is determined by the following equation:

$$M_r = \phi k_m k_{sb} f_{bu} S \tag{6.1}$$

where

ϕ = resistance factor which has been discussed in Chap. 1; for the case under consideration, its value is 0.9.

Table 6.1 Load sharing factor, k_m for all species and grades

Number of load sharing components, n	2	3	4	5	10	15	20
k_m k_m	1.10	1.20	1.25	1.25	1.35	1.40	1.40

Table 6.2 Size effect factor for flexure, k_{sb} for all species and grades having the smaller cross-sectional dimension between 64 and 114 mm

Larger dimension of laminate, mm	89	140	184	235	286	337	≥389
k_{sb}	1.7	1.5	1.3	1.2	1.1	1.0	0.9

Table 6.3 Specified flexural strength and mean modulus of elasticity for 38–89 mm wide laminates of various species of wood

Species	Douglas Fir		Hem-Fir		Jack Pine	
Grade	SS	1 & 2	SS	1 & 2	SS	1 & 2
Flexural strength, MPa	11.8	7.1	11.4	7.9	7.6	5.4
Mean modulus of elasticity, MPa	11,200	9,800	10,700	9,800	6,700	6,300

k_m = the load sharing factor whose values are obtained from Table 6.1 in which n is the number of laminates included in a width of 1.75 m.

k_{sb} = the size effect factor whose values depend upon the thickness of the laminates and can be obtained from Table 6.2.

f_{bu} = the 5th percentile strength of the species of wood under consideration the value of which has been discounted suitably for the effect of moisture and live load duration.

S = the elastic section modulus.

The load sharing factors given in Table 6.1 are based on the work of Bakht and Jaeger (1991).

To account for the reduction of the flexural strength due to wet conditions, the actual 5th percentile strengths of the various species of wood are reduced by about 20 % in order to arrive at the specified values of f_{bu}. A selection of these values specified in the CHBDC (2006a) is presented in Table 6.3 for laminates having thickness between 38 and 89 mm. This table also contains the specified mean values of the modulus of elasticity for the various species.

In Table 6.3, *grade* refers to the grading of timber according to the rules of the National Lumber Grading Authority of Canada (NLGA 2014). Select structure (SS) grade timber has minimum defects, and Grade 2 timber has the maximum defects, which could be permitted in lumber that is suitable for structural use.

It is emphasized that the moment of resistance M_r obtained by Eq. (6.1) is applicable to the ultimate limit state, which is discussed in Chap. 1.

Frequency of Butt Joints The total length of a SWD is usually longer than the largest commonly-available lengths of the laminates, requiring the laminates to be

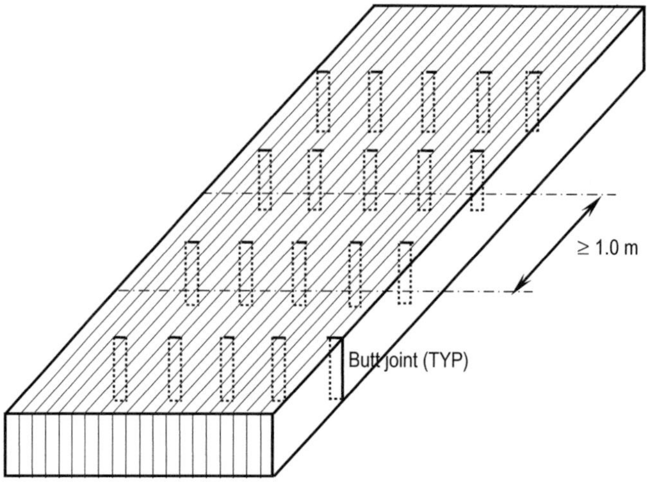

Fig. 6.6 Sketch of a stress laminated wood deck showing butt joints in the laminates

butt-jointed at frequent intervals. It is good practice to stagger the butt joints so that within any band having a width 1.0 m measured along the laminates, a butt joint does not occur in more than one laminate out of any four adjacent laminates. The practice is illustrated in Fig. 6.6.

The CHBDC (2006a) also requires that the stiffness of a SWD be adjusted by a modification factor k_I to account for the effect of butt joints:

$$k_1 = \frac{(N-1)}{N} \tag{6.2}$$

where N is the frequency of butt joints discussed immediately above. Equation (6.2) is based on the recommendation of Bakht and Jaeger (1991). For the example shown in Fig. 6.6, within a width of 1.0 m one of the four laminates has a butt joint, so that, $N = 4$.

Deflection Control Contrary to the usual practice, deflection control has been effectively eliminated in the requirements of both OHBDC (1992a, b) and CHBDC (2006a), except for wood bridges. For these structures, the codes require that the deflection under unfactored dead loads and under live load with a load factor of 1.00 should not exceed 1/400 of the span. It is recalled that the CHBDC design live loads represent directly the maximum permissible truck loads in the jurisdiction where the code is used. It may be noted that the CHBDC (2006a) specifies a measure to control the vibrations of the bridge superstructure with respect to the comfort of the

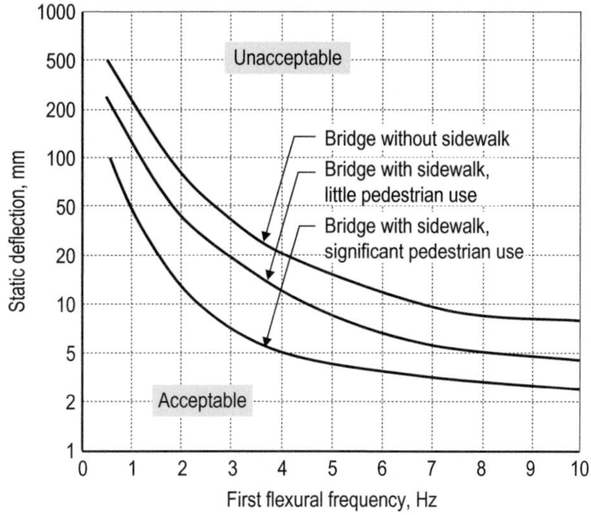

Fig. 6.7 Deflection control criteria of the Canadian Highway Bridge Design Code (2006a) with regard to human discomfort

users of the bridge; this measure, which does require the control of deflections, is noted in the following.

The CHBDC (2006a) requires bridge superstructures to be proportioned so that the maximum deflection due to design live loading with a load factor of 0.80 and without dynamic load allowance does not exceed the limits specified in a chart corresponding to the anticipated degree of pedestrian use; this chart is reproduced in Fig. 6.7. The deflection limit applies to the centre of the sidewalk or, if there is no sidewalk, to the inside face of the barrier wall or railing. It can be seen in Fig. 6.7 that the deflection limit is related to the first flexural frequency, f, of the component. This criterion is based on recognition that human discomfort due to the vibration of a bridge is related more to accelerations than to deflections.

6.3 Examples of SWDs

Since the technique of post-tensioning laminated wood decks was introduced in the mid 1970s, a large number of SWDs have been constructed, mainly in North America; these structures are performing satisfactorily. Some details of the earliest SWDs are provided in this section. All these decks are in Ontario, Canada. Structures with external post-tensioning systems presented in the following sub-section are those bridges that started life as nail-laminated decks and were later turned into SWDs after rehabilitation. Internal post-tensioning, applied exclusively to new structures, is discussed in Sect. 6.3.2.

Fig. 6.8 The Hebert Creek Bridge, which was the first one to be installed with a transverse post-tensioning system

6.3.1 *Decks with External Post-Tensioning*

Hebert Creek Bridge As mentioned in Sect. 6.2, the first nail-laminated deck rehabilitated by post-tensioning was the Hebert Creek Bridge. It is a two-lane, three-span bridge with a total length of about 16.8 m. The spans, which are continuous, have lengths of 5.34, 6.10 and 5.34 m, respectively. The 203 mm deep laminated wood deck of this bridge was rehabilitated in July 1976 by an external post-tensioning system, which incorporates 16 post-tensioning stations along the length of the bridge at a spacing of 910 mm; each station comprised two high strength steel bars, known by the trade name Dywidag bars, each having a diameter of 16 mm.

At the first stressing, applied in July 1976 through a bulkhead of wood, the average inter-laminate pressure was about 0.75 MPa. The deck was re-stressed in October 1976, to an average pressure of about 0.50 MPa. In June 1982, the wood bulkheads were replaced by steel channels whilst ensuring that the pressures in the deck after the replacement were the same as before. The deck has not been re-stressed since. A photograph of the Hebert Creek Bridge is presented in Fig. 6.8, showing the earlier bulkhead of wood.

Kabaigon and Pickerel River Bridges The Kabaigon River Bridge and the Pickerel River Bridge are two-lane structures each with multiple continuous spans; the overall length of the former is about 22.9 m and that of the latter about 42.1 m. Both bridges have 252 mm deep nail-laminated wood decks, which were rehabilitated by external post-tensioning systems incorporating 16 mm diameter Dywidag steel bars.

The Kabaigon River Bridge has five continuous spans at lengths of 4.11, 4.88, 4.88, 4.11 and 4.92 m, respectively; its deck was stressed and re-stressed in September 1980 to an average inter-laminate pressure of about 1.15 MPa; it was re-stressed in September 1982 to the same pressure.

The Pickerel River Bridge is a nine-span continuous bridge with seven inner spans of 4.88 m each and two outer spaces of 3.96 m each; its deck was initially stressed and re-stressed in June 1981 to an average inter-laminate pressure of

Fig. 6.9 The Fox Lake Bridge, which is the first new bridge designed as a stress-laminated wood deck

Fig. 6.10 The Fox Lake Bridge under construction

1.09 MPa; it was re-stressed again in October 1982 to a pressure of about 0.96 MPa. The spacing of post-tensioning stations in the Kabaigon and Pickerel River Bridges is 1,016 and 1,067 mm, respectively. There are 23 post-tensioning stations in the Kabaigon River Bridge and 43 in the Pickerel River Bridge. The decks of the two bridges have not been re-stressed since the second re-stressing.

6.3.2 Decks with Internal Post-Tensioning

While the development of the transverse post-tensioning technique was prompted by the need to rehabilitate existing deteriorated nail-connected decks, the technique can and has also been applied to new construction mainly through internal post-tensioning.

Fox Lake Bridge The first application of stress-laminated wood decks in a new structure was in the Fox Lake Bridge in the northern Ontario, Canada. This bridge, a view of which is shown in Fig. 6.9, is effectively a three-span continuous bridge with inclined legs that are monolithic with the deck; the spans are 4.11, 4.72 and

4.11 m in length. For the construction of this bridge, the laminates forming the deck and inclined legs were assembled on the ground, and were positioned, as shown in Fig. 6.10 by means of a relatively light crane. The whole structure was assembled in about 1 week by labour that was unskilled with the exception of the operator of the jacks used for applying the post-tensioning force. The 286 mm thick deck of this bridge is post-tensioned internally by 25 mm diameter Dywidag bars at a spacing of 1,250 mm. The total length of the bridge contains 11 post-tensioning stations. Further details of the structure are provided by Taylor and Walsh (1983).

The Fox Lake Bridge was tested soon after its construction, and was found to have ample strength for the very heavy logging trucks that use this bridge. The deck was stressed in August 1981 to an average interlaminate pressure of about 1.00 MPa. Within 2 weeks of the first stressing, and then again in November 1981, the deck was re-stressed to the same level. Since then the deck has not been re-stressed.

6.3.3 Prestress Losses

It is well known that a specimen of wood tends to deform permanently when large compressive loads are applied perpendicular to its grain. The magnitude of permanent deformation, which results from a combination of relaxation and creep, depends upon the duration and magnitude of loads. In an SWD, the lateral pressures applied by the post-tensioning system are high enough to cause the deck width to shrink permanently by a small amount over a period of time. Clearly, such shrinkage of the wood deck causes a loss of prestress, which has to be accounted for in the design of the post-tensioning system.

Observed Losses A measure of the state of prestress in a deck is provided by the average of the forces in all post-tensioning bars. The inter-laminate pressure corresponding to the average force is assumed to be the average pressure. This average pressure, shown in Fig. 6.11 corresponding to two sets of observations in the Pickerel River Bridge is used to quantify the loss of prestress in SWDs.

The average inter-laminate pressures in the five decks described in Sect. 6.3.1 and 6.3.2 are plotted in Fig. 6.12 against time (Bakht et al. 1994). In this figure, discrete points corresponding to observations at somewhat irregularly-spaced time intervals are joined by straight lines merely to facilitate readability. With the help of Fig. 6.12, several significant and far-reaching observations can be made regarding SWDs with steel tendons:

(a) Irrespective of the initial level of prestress, the average inter-laminate pressure eventually drops below the minimum level of 0.35 MPa required by the CHBDC (2006a) without having a detrimental effect on the performance of the deck, it being noted that the detriment manifests itself as longitudinal cracks in the asphalt surfacing.

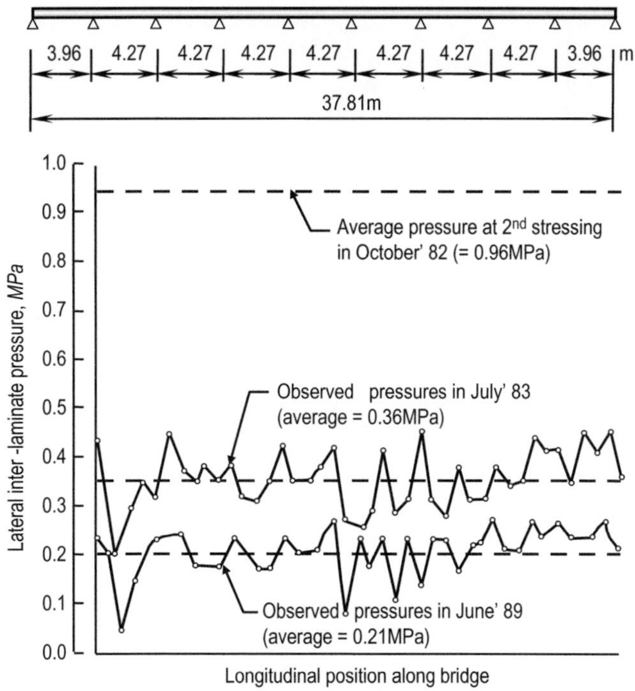

Fig. 6.11 Variations in the inter-laminate pressure along the length of the Pickerel Bridge

(b) The trend of pressure plotted against time indicates that the prestress losses generally cease to accumulate, but after a long time.
(c) The magnitude of short-term prestress loss occurring soon after stressing decreases with a decrease in the level of initial prestress.
(d) The seasonal variations in the average inter-laminate pressure are small compared to the long-term variations and the variations within the deck.

In light of the above observations, it is obvious that monitoring and maintaining prestress levels in SWDs incorporating steel bars should become regular bridge maintenance activities. To appreciate the magnitude of deck shrinkage leading to prestress losses discussed earlier, the deck of a particular bridge, the Kabaigon River Bridge, is considered; its deck is about 10 m wide and other details are as noted earlier. In order to apply an average inter-laminate pressure of 1.15 MPa, each of the 16 mm bars has to be tensioned to 148 kN under which force a bar elongates by about 39 mm. A drop in the pressure of 0.74 MPa, which occurred soon after second stressing, implies that during this time, the force in each bar dropped by 95 kN and the deck shrank by about 25 mm. It is obvious that the steel bars, which were used in all the early SWDs, have too high an axial stiffness. A preferable prestressing system is one in which the post-tensioning components are so flexible that under the same force their elongation is significantly greater than the expected shrinkage of the deck. If it is ensured, by using a highly flexible post-tensioning system, that the prestress losses will be small, then the initial lateral pressures in the

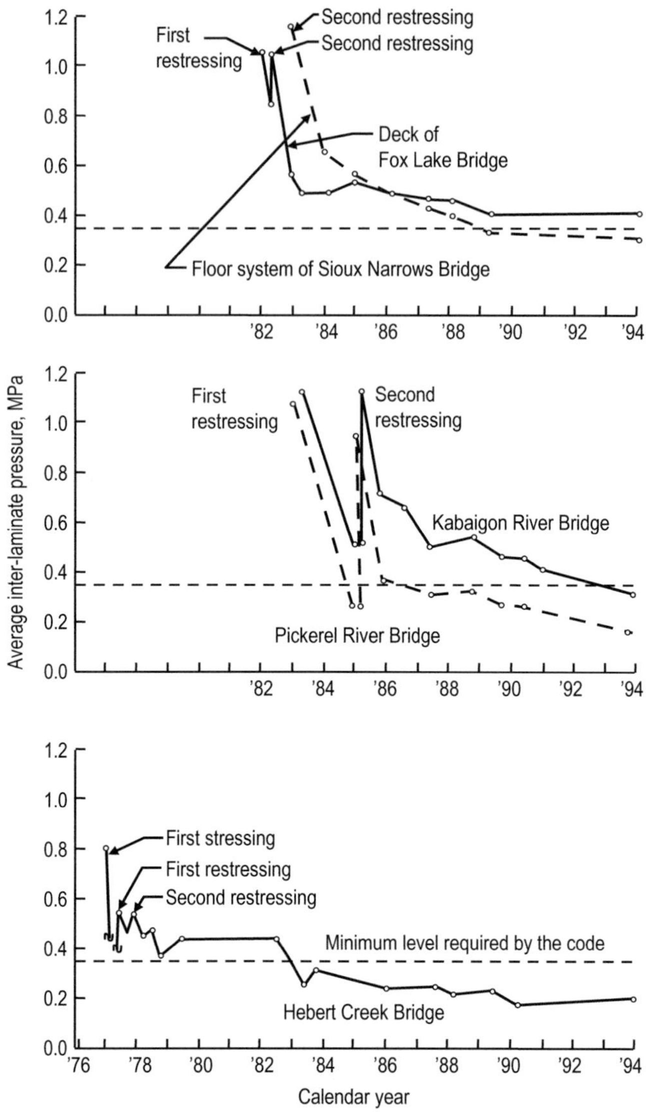

Fig. 6.12 Observed levels of prestress in five stress-laminated wood decks

deck can be reduced with the cumulative desirable effect of incurring even smaller time-dependent shrinkage of the deck and consequently smaller prestress losses.

The authors believe that the minimum level of prestress can be safely lowered to 0.20 MPa, and that the SWD can be stressed according to the following two steps instead of the three required by the OHBDC (1992a, b):

(a) Stress the deck to an equivalent inter-laminate pressure of about 1.0 MPa irrespective of the species of wood.

(b) About 24 h after the first stressing, de-stress the deck to an equivalent intermediate pressure of 0.35 MPa.

It is noted that the authors and some of their research colleagues have developed a post-tensioning system for the SWD, which comprises cables of aramid fibres; the tensile strength and modulus of elasticity of which are respectively about 1.75 and 0.41 times those of high strength steel. The low stiffness of the aramid fibre cables, which may be as low 1/6th of the steel bars, has been found to reduce the prestress losses very significantly. Mufti et al. (1993) have provided an account of the effort to develop anchorages for the aramid fibre cables to be used in conjunction with SWDs. These cables are also discussed in Sect. 6.5. Further discussion on the mechanics of prestress losses in SWDs is provided by Bakht and Jaeger (1994, 1996).

6.4 Steel: Wood Composite Bridges

Timber decks on steel girder bridges usually have their laminations laid perpendicular to the girder axes. Consequently, the relevant modulus of elasticity of wood when acting compositely with the girders is a very small transverse modulus of elasticity. Mainly for this reason, no attempt seems to have been made in the past to make wood decks composite with the girders.

Taking advantage of the much larger modulus of elasticity of wood in the longitudinal direction, a wood-steel composite bridge has been developed by Bakht and Tharmabala (1987), in which the laminates of the wood deck run along the girders. The deck is transversely post-tensioned, and the composite action between the deck and steel girders is achieved through concrete bulkheads which are formed as follows. (a) After assembling the prestressed deck on the girders, large holes are drilled through the deck at selected locations directly above the girder flanges; (b) shear studs are then installed through the holes on the girder flanges; and (c) the holes are filled with expansive concrete forming the bulkheads, which transfer the horizontal shear between the deck and the girders quite effectively.

The SWD of the steel-wood composite bridge is supported by transverse diaphragms as well as by the longitudinal girders. The analysis of the deck under concentrated loads becomes difficult because of the orthotropic nature of the deck. Erki and Bakht (1992) have provided simplified methods of analysis for these decks; they have also suggested that for optimum beneficial effect, the spacing of the diaphragms should be between 1.2 and 1.5 times the spacing of the girders.

The concept of the steel-wood composite bridge was first applied to the North Pagwachewan River Bridge, located in Northern Ontario, Canada. This bridge, which was completed in the fall of 1993, has a simply-supported span of 50.0 m, and as shown in Fig. 6.13, has five welded plate girders at a spacing of 2.5 m. The stress-laminated deck is 286 mm deep. A view of the deck of this bridge from below

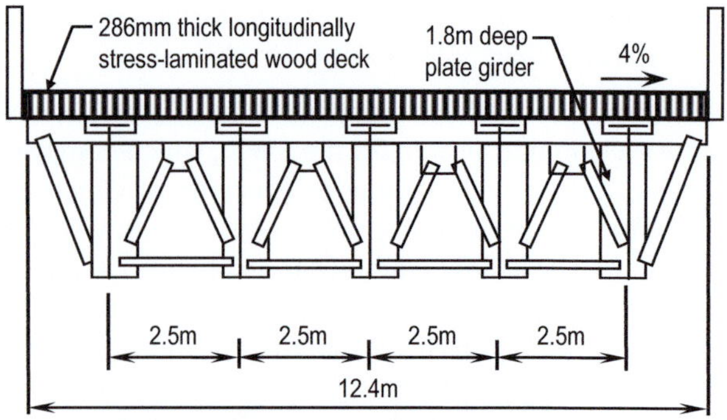

Fig. 6.13 Cross-section of a steel-wood composite bridge

Fig. 6.14 The deck of the steel-wood composite North Pagwachewan River Bridge in Ontario, Canada, viewed from below the bridge

the deck can be seen in Fig. 6.14. A proof test on this bridge has shown that the composite timber deck has improved considerably the flexural stiffness of the steel girders (Bakht and Krisciunas 1997).

Other steel-wood composite bridges built in Ontario, Canada, have been described by Krisciunas (1996).

6.5 Stressed-Log Bridges

Dimension lumber with square faces are cut from tree logs which themselves have round faces. It is believed that the densest part of the cross-section of a tree lies near its round, outer face. This densest and possibly strongest portion of the tree has to be discarded as not being suitable for structural applications. The impetus to find the structural use for uncut logs came from discarded wooden poles, which after used in Canada for telephone and electricity lines.

A concept was developed in which the discarded poles or logs are trimmed on only two parallel faces. The logs are then stacked together against their trimmed

Fig. 6.15 A prototype stressed-log bridge during construction

faces and laterally stressed. A bridge designed by this concept is called a *stressed-log bridge*. The logs are topped with a layer of concrete mixed with low-modulus synthetic fibres.

Figure 6.15 shows a photograph of a prototype stressed-log bridge, stressed with aramid fibre cables, before the casting of the concrete layer. The design and construction of the bridge are described by Bakht et al. (1996). The deck of this bridge was post tensioned according to the scheme described above; partly because of this scheme and partly due to the use of post-tensioning tendons with considerably smaller axial stiffness than that of steel tendons, the prestress losses in the prototype stressed-log bridge observed for many years were found to be negligible.

It can be appreciated that the post-tensioning techniques also permit the use of those timbers which, because of the large number of faults in individual pieces or because of having a very coarse grain structure, are not normally considered suitable for structural applications. By considerably reducing the exposed area of the wood, the prestressing technique can enhance the durability of the structure. It is foreseen that even the logs of coconut trees, which until recently had been considered unsuitable for structural purposes, can be employed to advantage in stressed-log bridges.

The stressed-log bridge with aramid fibre tendons and stainless steel anchors was found to be too expensive. By replacing the aramid fibre cables with tendons made of glass fibre reinforced polymers (GFRPs), the concept was found to be suitable for field application. The first stressed-log bridge with GFRP tendons was constructed in 1996 in Northern Ontario, Canada. As described by Bakht et al. (1997), the anchors for the GFRP tendons were constructed using off-the-shelf steel tubes. A photograph showing a longitudinal section through the anchor is presented in Fig. 6.16.

A photograph of the first application of the stressed-log bridge with GFRP tendons during construction is presented in Fig. 6.17, in which the initial prestressing of the deck through steel rods can be seen. About 24 h after the first stressing, some of the prestressing was lost. The steel bars were stressed again to

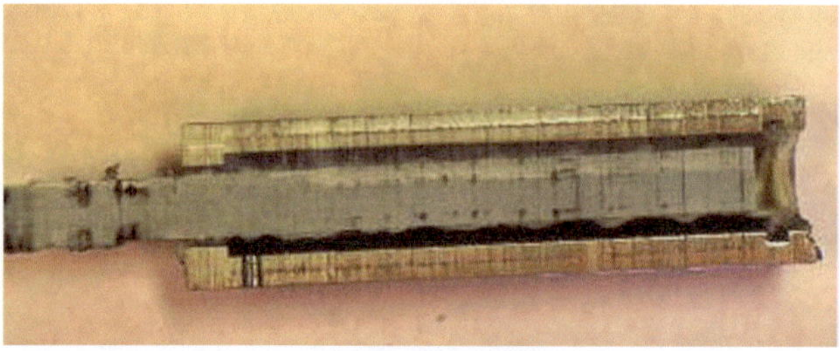

Fig. 6.16 Photograph showing longitudinal section through an anchor for a GFRP tendon

Fig. 6.17 The first field application of the concept of a stressed-log bridge with GFRP tendons

raise the inter-log stress to about 1.0 MPa. The steel tendons were then replaced by GFRP tendons.

A warning should also be given of the fact that a very large amount of energy is stored in the post-tensioning rods, the sudden release of which can turn the rods into lethal devices. When structures are designed with prestressed wood, adequate precaution should be taken to contain the rods in case of accidental breakage.

6.6 Grout-Laminated Bridges

A grout laminated wood deck (GLWD) comprises wood laminates, or logs trimmed to two vertical faces, and held together by internal grout cylinders, which are reinforced with rods of steel or GFRP. There are two methods of constructing GLWDs. In one scheme of construction, the deck is compressed laterally by means of tendons in regularly-spaced transverse holes. After stressing the deck, the holes are filled with a grout. The prestressing forces are removed from the rods after the grout has set, thus putting the grout cylinders in compression. For ease of reference, the deck resulting from this method of construction is referred to as the post-tensioned GLWD.

Fig. 6.18 Test on a grout laminated deck

In the other method of construction, the deck is compressed by an external prestressing system; and the transverse holes containing the rods are filled with grout. After the grout has set, the external prestressing system is removed inducing a tensile force in the grout cylinders most of which develop small cracks. The deck obtained by this form of construction is referred to as the reinforced GLWD.

Tests on full-scale models have confirmed that a GLWD with two rows of grout cylinders in compression has better load distribution characteristics than a GLWD with only one row of grout cylinders in tension (Mufti et al. 2004). One of the models with one row of grout cylinders can be seen in Fig. 6.18.

Although the concept of a GLWD has yet to be applied in field, it has great potential.

6.7 Stressed Wood Decks with FRP Tendons

The CHBDC (2006a) permits the use tendons made of glass or aramid fibre reinforced polymers (FRPs) in both stress laminated wood decks and stressed-log bridges, collectively called stressed wood decks. As noted in Chap. 8, both aramid and glass fibres have very high tensile strength, but a fairly low modulus of elasticity. The CHBDC design provisions for stressed wood decks with FRP tendons are similar to those for decks with steel tendons, except for the stressing procedure.

The CHBDC (2006a) requires that the initial post-tensioning in the FRP tendons should be such as to bring the interface pressure between laminates or logs to approximately 0.8 MPa, regardless of the species of wood. The prestressing forces are then required to be reduced 12–24 h after initial post-tensioning to an average interface pressure of 0.35–0.44 MPa, at which level the stresses in aramid and glass FRP tendons should not exceed 0.35 and 0.25 times the respective tensile strengths of the FRPs. It is expected that wood decks stressed with low-modulus FRPs will not suffer substantial prestress losses.

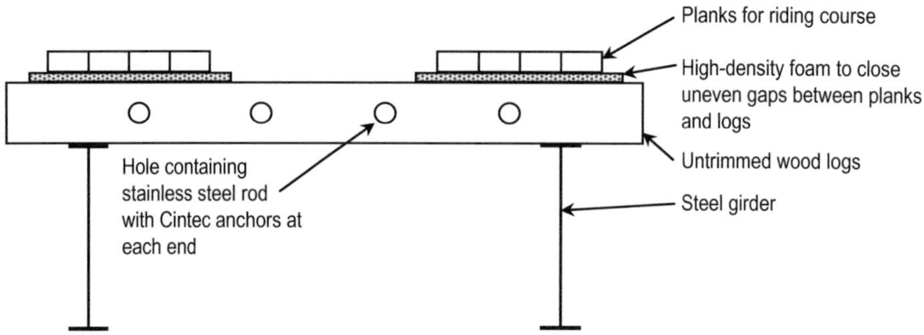

Fig. 6.19 Cross-section of a steel girder bridge with anchored log panels

6.8 Anchored Log Decks

Research was recently conducted at the university of Manitoba, Canada, to develop an alternative to stressed-log decks and grout-laminated wood decks that is cost-effective and deployable for use on temporary winter road bridges of Manitoba, Canada. The need for developing a new system arose not only for developing preassembled deck panels, but also from the high cost of trimming logs containing creosote treatment. It was also found that it was not easy to contain the grout in the holes of grout-laminated decks.

The new deck system, which is planned to be used for temporary single-lane bridges comprising two steel girders, is called the anchored log deck. The anatomy of a pre-assembled panel, about 2.5 m wide and 2.0 m long, is illustrated in Fig. 6.19.

The panels are connected to the girders with steel clip plates, and are not composite with the girders. A panel of an anchored log deck consists of untrimmed logs, usually ten in number with each log having an average diameter of 250 mm. The logs are inter-connected with stainless steel rods passing through four holes at the mid-height of the logs, and are anchored at their ends by means the proprietary Cintec Anchors, encased in a fabric socks that prevent the leakage of grout from out of the holes. The riding surface over each wheel path is provided by four 250 × 50 mm planks, placed side-by-side. The uneven gap between the planks and logs is filled by a high-density foam. Further details of the anchored log deck and static and fatigue test on its full-scale model are provided by Murison (2014).

A photograph of a full-scale model of the anchored log deck is shown in Fig. 6.20, in which the high-density foam can be seen oozing under the planks. The planks are bolted to the panels and have their ends staggered at the ends of the panels in such a way that when the panels are put together, the planks provided a smooth riding surface.

Fig. 6.20 Photograph showing the full-scale model of the anchored log deck prior to testing

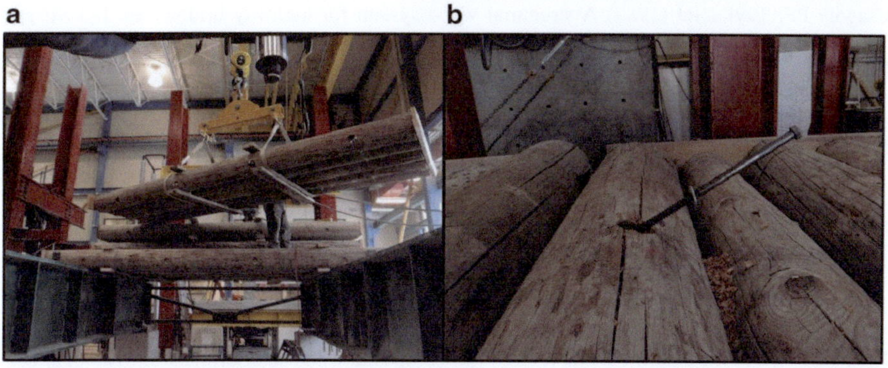

Fig. 6.21 Photographs showing the construction of the full-scale model of the anchored log deck: (**a**) panels being placed on girders, (**b**) inclined lag bolts joining two adjacent panels

The placing of the anchored log panels can be seen in the photograph in Fig. 6.21a. As shown in Fig. 6.21b, the adjacent panels were inter-connected by inclined lag bolts.

While the tests reported by Murison (2014) showed that the full-scale model of the anchored log deck had adequate capacity under static and fatigue loads, it was found that the distribution of loads between adjacent panels could be improved by a more effective joining system, which should be easy to dismantle and re-assemble.

References

AASHTO (1992) Standard specifications for highway bridges, 13th edn. American Association of State Highway and Transportation Officials, Washington, DC

Bakht B, Jaeger LG (1991) Load sharing factors in timber bridge design. Can J Civ Eng 18(2):312–319

Bakht B, Jaeger LG (1994) Revisiting prestress losses in stress-laminated wood decks. In: Proceedings of the CSCE annual conference, Winnipeg

Bakht B, Jaeger LG (1996) On the use of springs in SWDs. Can J Civ Eng 23(4):982–985

Bakht B, Krisciunas R (1997) Testing of a steel-wood composite prototype bridge. Struct Eng Int 1 (97):35–41

Bakht B, Tharmabala T (1987) Steel-wood composite bridges and their static load response. Can J Civ Eng 14(2):163–170

Bakht B, Jaeger LG, Klubal J (1994) Prestress losses in stress-laminated wood decks. In: Proceedings of the 4th international conference on short and medium span bridges, Halifax

Bakht B, Maheu J, Bolshakova T (1996) Stressed log bridges. Can J Civ Eng 23(2):490–501

Bakht B, Lam C, Bolshakova T (1997) The first stressed log bridge. In: Proceedings, US-Canada-Europe workshop on bridge engineering, Zurich, pp 155–162

CHBDC (2006a) Canadian highway bridge design code, CAN/CSA-S6-06. Canadian Standards Association, Toronto

CHBDC (2006b) Canadian highway bridge design code, Commentary on CAN/CSA-S6-06. Canadian Standards Association, Toronto

Csagoly PF, Taylor RJ (1980) A structural wood system for highway bridges. In: IABSE proceedings international association for bridge and structural engineering, Zurich, pp 35–80

Das S (2010) Shear strengths of timber beams with end splits. A thesis submitted to the Faculty of Graduate Studies of The University of Manitoba in partial fulfillment of the requirements of the degree of Master of Science, Department of Civil Engineering, University of Manitoba

Erki M-A, Bakht B (1992) Analysis of the decking of steel-wood composite bridges. In: Proceedings of the annual conference of the Canadian Society for Civil Engineering, Quebec City, pp 81–90

Gangarao HVS, Latheef I (1990) System innovation and experimental evaluation of stress-timber bridges. Transportation research record 1291, Transportation Research Board, Washington, DC

Jaeger LG, Bakht B (1990) Effect of butt joints on the flexural stiffness of laminated timber bridges. Can J Civ Eng 17(5):859–864

Krisciunas K (1996) Ontario's experience with composite wood/steel bridges. In: National conference on wood transportation structures, Forest Products Lab, vol 94, p 313

Madsen B (1992) Structural behaviour of timber. Timber Engineering Ltd, North Vancouver

Mitchell A (2014) Trees of North America. Collins & Browm, London

Mufti AA, Bakht B, Maheu J (1993) An example of the use of CAD/CAM in structures research. In: Proceedings, annual conference of the Canadian Society for Civil Engineering, Fredericton, New Brunswick

Mufti AA, Bakht B, Svecova D, Limaye V (2004) Failure tests on full-scale models of grout laminated wood decks. Can J Civ Eng 31(1):133–145

Murison E (2014) Development of a timber deck for the winter roads of Manitoba. Masters thesis, Department of Civil Engineering, University of Manitoba

NLGA (2014) Standard rules for grading Canadian lumber. National Lumber Grading Authority, Surrey

OHBDC (1992a) Ontario highway bridge design code, 3rd edn. Ministry of Transportation of Ontario, Downsview

OHBDC (1992b) Commentary: Ontario highway bridge design code. Ministry of Transportation of Ontario, Downsview

Oliva MG, Dimokis A (1988) Behaviour of stress-laminated timber highway bridge. J Struct Eng, ASCE 114(8)

OPSS – 907 (1992) Ontario provincial standard specifications. Construction specifications for structural wood system, Toronto

Sarisley EF Jr, Accorsi ML (1990) Prestress level in stress-laminated timber bridge. J Struct Eng, ASCE 116(11)

Taylor RJ, Walsh H (1983) A prototype prestressed wood bridge. Structural research report SRR-83-7, Ministry of Transportation of Ontario, Downsview

Chapter 7
Soil-Steel Bridges

7.1 Introduction

A *soil-steel bridge* is a structure made with a shell of manufactured curved corrugated steel plates and an envelope of *engineered soil*, which is well-compacted backfill composed mainly of well-graded granular soil. A photograph of a soil-steel bridge serving as a grade-separation structure is presented in Fig. 7.1. These structures are, however, also used to convey water, in which case they are appropriately referred to as culverts. A soil-steel bridge with a twin conduit serving as a culvert is shown in Fig. 7.2.

For spans of up to 25 m, soil-steel bridges are generally more economical than their conventional counterparts. In North America, soil-steel bridges are typically about 30 % cheaper than the conventional bridges such as concrete slab bridges and slab-on-girder bridges.

Despite soil-steel bridges being in existence for more than a century, their terminology has not been standardized. When the Ontario Highway Bridge Design Code (OHBDC) was first introduced in 1979, it was decided to refer to these structures as *soil-steel structures*. Consistent with a textbook on the subject (Abdel-Sayed et al. 1993), the Ontario term is used except that the word *structures* is replaced by *bridges*. It is noted that the term structure was preferred over *bridge* because not all these structures are bridges. For example, soil-steel structures have also been used for avalanche protection. Some of the commonly used terms used in conjunction with soil-steel bridges are defined in the following.

Arching is the effect produced by the transfer of vertical pressure between adjoining soil masses above and adjacent to the conduit.

Bedding is the prepared portion of the engineered soil on which the conduit invert is placed.

Compaction is the process of soil densification, at specified moisture content, by the application of pressure through kneading, tamping, rodding, or vibratory action of mechanical or manual equipment.

B. Bakht, A. Mufti, *Bridges*, DOI 10.1007/978-3-319-17843-1_7

Fig. 7.1 A soil-steel bridge serving as a grade separation structure

Fig. 7.2 A twin-conduit soil-steel bridge serving as a culvert

Conduit is the term used to refer to the bridge opening in a soil-steel bridge rather than to the metallic shell as is done often.

Conduit wall is the metallic shell of the soil-steel bridge, which in its fully assembled form is also referred to as the *pipe*.

Crown is the highest point or the transverse section of the conduit.

Deep corrugations are the structural plate corrugations with a pitch between 380 and 400 mm and a rise between 140 and 150 mm.

Depth of cover is the vertical distance between the top of the roadway above the conduit and the crown.

Engineered soil is the selected soil of known properties placed around the conduit in a prescribed manner.

Foundation is the term used for the ground on which a soil-steel structure is constructed.

Haunch is the portion of the conduit wall between the springline and the top of the bedding or footings if present.

Invert is the portion of the conduit wall contained between the haunches.

Longitudinal direction in the context of the conduit refers to the direction of the conduit axis; it is noted that *longitudinal direction* in the other chapters of the book refers to the direction of flow of traffic on the bridge.

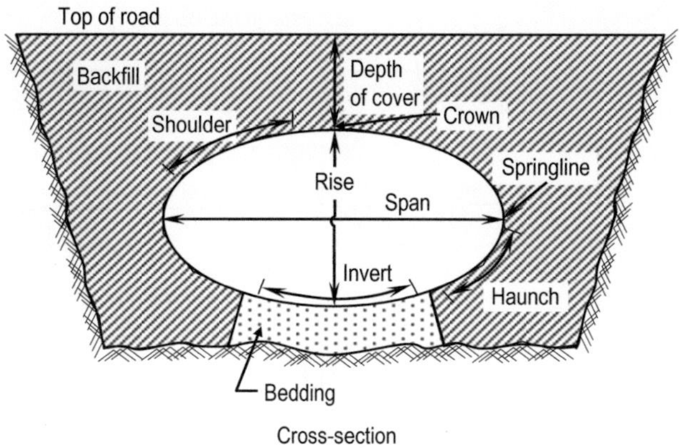

Fig. 7.3 Illustration of the terminology pertaining to soil-steel bridges

Rise is the maximum vertical clearance inside a conduit at a given transverse section.

Shallow corrugations are the structural plate corrugations with a pitch between 150 and 230 mm and a rise between 50 and 65 mm.

Shoulder is the portion of the conduit wall between the crown and springline.

Span is the maximum horizontal clearance inside a conduit at a given transverse section.

Springline is the locus of the horizontal extremities of transverse sections of the conduit.

Structural backfill is the envelope of engineered soil, including the bedding placed around the conduit in a controlled manner.

Transverse direction is the direction perpendicular to the conduit axis.

Some of the terms defined above are illustrated in Fig. 7.3.

The conduits of soil-steel bridges come in a variety of shapes, which are illustrated in Fig. 7.4; this figure also defines D_h and D_v, which are used for the design of the structures.

Soil-steel bridges can be constructed easily but not without strict adherence to well-established procedures. These structures are also fairly easy to design. Until some time ago, the corrugations of steel were *shallow*, a term defined above. With shallow corrugations, the largest span of a soil-steel bridge was about 18 m. With the advent of *deep corrugations*, introduced by the Canadian industry, the spans of soil-steel bridges can be as large as 24 m.

The purpose of this chapter is to introduce the subject of soil-steel bridges and to present briefly the design and construction procedures. A more exhaustive account of the soil-steel structures with shallow corrugations can be found in the textbook by Abdel-Sayed et al. (1993).

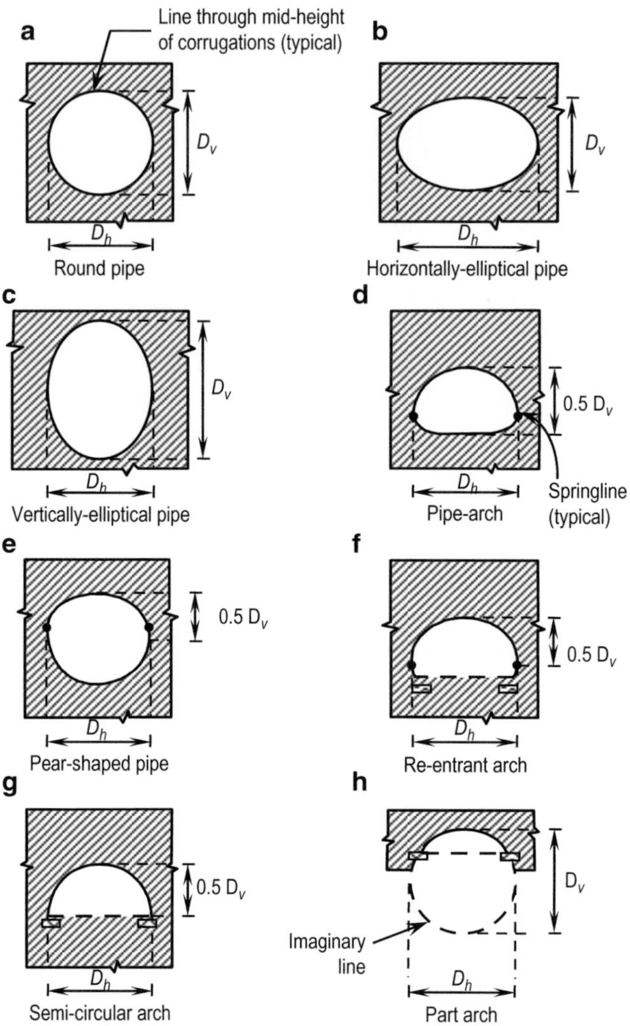

Fig. 7.4 Various conduit shapes used in soil-steel bridges: (**a**) round pipe, (**b**) horizontally elliptical pipe, (**c**) vertically elliptical pipe, (**d**) pipe arch, (**e**) pear-shaped pipe, (**f**) re-entrant arch, (**g**) semi-circular arch, (**h**) part arch

7.2 Mechanics of Behaviour

Consider two small tin cans, the type used for packing food, with their lids removed. Place one on the ground on its side and ask a child to stand on it. As expected, the tin can will be deformed to failure quite easily. Now take the other tin can and place it on the ground as before but this time pack some sand around and above it so that the tin can forms a conduit through the sand. Now, let an adult stand

Fig. 7.5 A soil-steel bridge under construction

on it. It will be found that the tin can is able to withstand the weight of even the adult without suffering noticeable deformations.

The tin can encased in sand is a scaled-down model of a soil-steel bridge which is composed of curved corrugated steel plates bolted together and surrounded with an envelope of carefully compacted granular backfill. The tin can model describes eloquently the manner in which the metallic shell sustains the applied loading. The soil envelope transforms the applied loading and its own gravity force into radial pressures that act on the shell.

On its own, the metallic shell of a soil-steel bridge is so flexible in bending that, in order to maintain its cross-sectional shape during construction, it sometimes has to be braced by ties and props. In fact, the shell is so weak in flexure that if it were scaled down by principles of structural modelling to the size of a tin can, it would be too flimsy for manual handling. The bare metallic shell of a soil-steel bridge during construction can be seen in Fig. 7.5.

Mufti et al. (1989) have studied quantitatively the mechanics of behaviour of soil-steel bridges; some of their findings are summarized in this section.

7.2.1 *Infinitely Long Tube in Half-Space*

For the finite element analysis of soil-steel bridges, it is usual to assume that the metallic shell is infinitely long, and that it is buried in a space, or half-space that extends to infinity on one side of the axis, so that one transverse slice of the structure, which is shown conceptually in Fig. 7.6, is similar in behaviour to any other transverse slice. The advantage of this assumption is that the very complex three-dimensional nature of the actual structure can be investigated with a relatively simple two-dimensional plane-strain idealization, in which deformations perpendicular to the plane of the idealized structure are assumed to be zero. The simplification afforded by the two-dimensional idealization is not without disadvantages, although it provides a useful tool to study the behaviour of soil-steel bridges. The behaviour of soil-steel structures with the help of this idealization is discussed first, its disadvantages are discussed later.

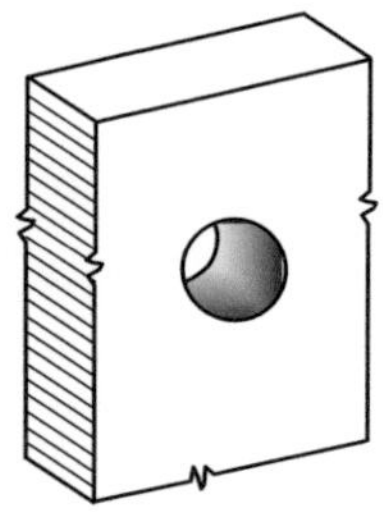

Fig. 7.6 Transverse slice of a soil-steel bridge

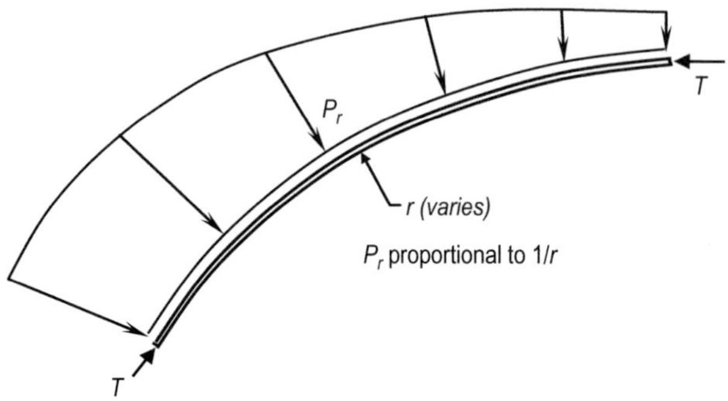

Fig. 7.7 Illustration of the condition of no moment in a curved beam subjected to axial force and radial pressure

Bending Effects It can be shown that if a segment of an idealised shell of varying radius of curvature r is subjected only to radial pressures, which vary inversely with r, as shown in Fig. 7.7, then the shell remains free from any moments, i.e. it is subjected to only axial thrust.

In an actual structure, the radial pressures on the metallic shell do vary approximately, although not precisely, in inverse proportion to the radius of curvature. Because of the radial pressures not being exactly proportional to $1/r$ and because of load effects imposed during the construction process, the shell of an actual soil-steel structure is subjected to some bending moments; however, these are limited by the plastic moment capacity of the shell. The corrugated plates with shallow corrugations are quite weak in flexure, because of which these plates can develop plastic hinges under fairly low moments. The hinges are usually formed at the crown and at shoulders as shown in Fig. 7.8. For loads applied subsequent to the formation of the plastic hinges, the corrugated metal plate shell behaves like a ring with a few hinges. Such a ring, even when subjected to radial pressures that are not proportional to the inverse of the radius of curvature, does not permit the formation of substantial bending moments. It is emphasized that plates with deep corrugations have very high flexural rigidity, because of which moments in these plates cannot be ignored.

In soil-steel bridges made with shallow corrugated plates, it is usual to ignore bending moments and design the metallic shell for thrust only. It should be noted, however, that the moments can be ignored only if the conduit walls are flexurally quite weak.

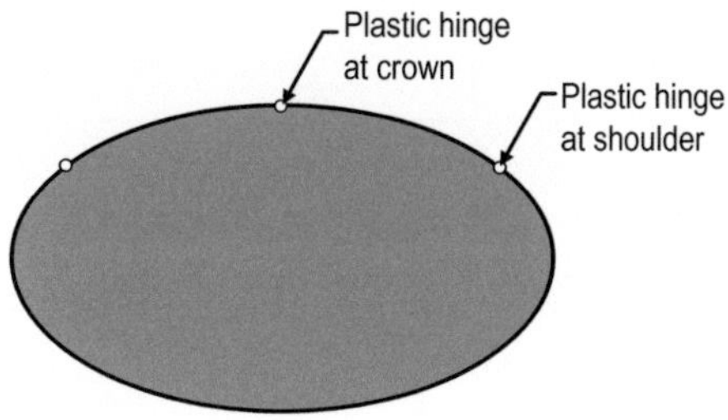

Fig. 7.8 Location of plastic hinges around the conduit wall that may be formed during construction

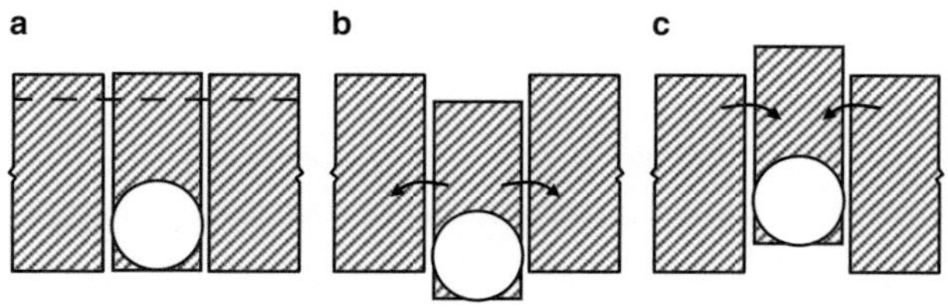

Fig. 7.9 Various forms of arching: (**a**) no arching, (**b**) positive arching, (**c**) negative arching

Although the moments in the metallic shell with shallow corrugated plates may not be taken into account in the design process, their effect on the integrity of the structure cannot always be ignored. Excessive bending deformations in the metallic shell result mainly from the inability of the backfill to sustain the radial pressures, which develop around the conduit due to the interaction of the soil and the shell. As noted earlier, the development of substantial amounts of bending moments is inhibited by the relatively small flexural rigidity of the metallic shell and by the formation of plastic hinges at strategic locations. However, the absence of substantial amounts of bending moments does not eliminate large bending deformations of the plate, which are detrimental to the integrity of the structure.

Arching Consider a plane-strain slice of a soil-steel bridge in which points along every horizontal line have the same deflections due to the gravity load of the soil. In such a structure, no transference of the load takes place from the column of soil immediately above the conduit to the adjacent columns of soil. Clearly, such a condition can take place only if the load-deformation characteristics of the body of soil displaced by the conduit are the same as those of the conduit itself. The condition, in which loads from one column of soil are not transferred to another, is shown schematically in Fig. 7.9a; it can be regarded as the no-arching condition.

If the column of soil containing the conduit were to deflect more than the adjacent columns as shown in Fig. 7.9b, then it would drag down the two adjacent columns of soil along with it and in so doing would pass on some of its own load to these columns. The effect of this load transference is to relieve the conduit wall of some of the load that it would have sustained in the absence of the load transference. This condition in a soil-steel bridge is referred to as the positive arching condition.

The condition of negative arching is clearly that in which the conduit wall is called upon to sustain more load than it would have sustained if there were no load

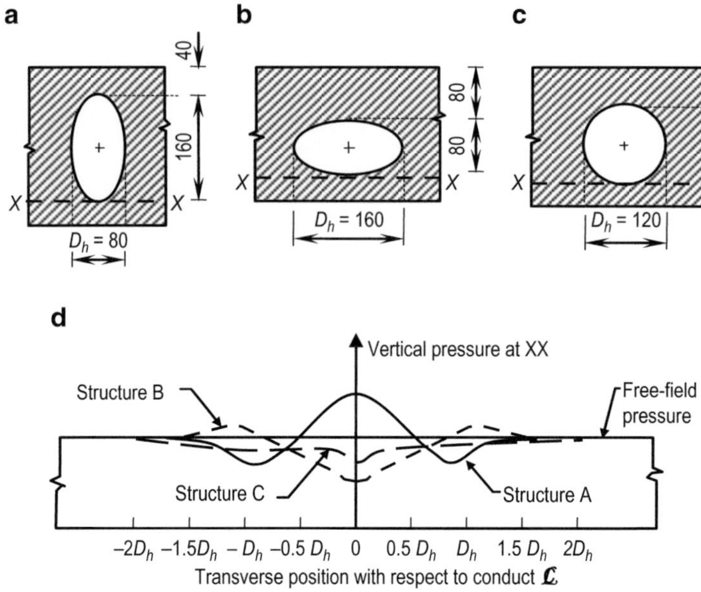

Fig. 7.10 Vertical pressures under the conduits: (**a**) vertical ellipse, (**b**) horizontal ellipse, (**c**) round pipe, (**d**) vertical pressures at invert

transference from the column of soil immediately above the conduit to the adjacent ones. This condition, as shown in Fig. 7.9c, can occur when the adjacent columns deflect more than the middle column causing some of their loads to be transferred on to the conduit wall.

The positive and negative arching conditions should be regarded only as qualitative measures; their quantification is made impractical by the fact that different load effects are influenced differently by arching. For example, the vertical soil pressure over the crown may be reduced differently by arching than the vertical soil pressure near the springlines; moreover, the thrust, which does not remain constant along the circumference, may be affected differently than the vertical soil pressure either at the crown or at the springline.

Finite Element Analyses To gain insight into their behaviour three soil-steel bridges with conduits of different shapes were analysed by Mufti et al. (1989). The critical findings of this study were the patterns of vertical soil pressures under the pipes. It was found that, as shown in Fig. 7.10d, for the structure with vertically-elliptical conduit, this pressure below the invert is larger than the free-field pressure.

In the case of structures with round conduits and horizontally-elliptical conduits, the vertical pressure directly below the invert is much smaller than the free-field pressure as shown in Fig. 7.10d. As discussed later in the Subsection, these patterns of the distribution of vertical stress under the conduit are significant with respect to the long-term performance of the structure.

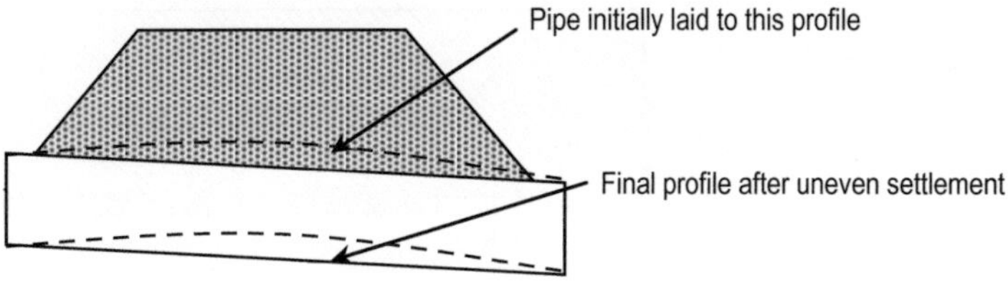

Fig. 7.11 Non-uniform settlement of the pipe in longitudinal direction

7.2.2 Third Dimension Effect

The two-dimensional idealization discussed above is not strictly valid because in practice, the pipe is subjected to nearly uniform soil weight only in its middle lengths. The effects of the load variation in the third dimension, i.e. along the length of the conduit, are discussed in this following.

Recognizing that the foundation below the middle portion of the pipe, because of being subjected to deeper fills, is likely to settle more than the foundation below its outer ends, the pipe is usually laid with an exaggerated camber in the middle as shown in Fig. 7.11. The camber is so adjusted that after the uneven settlement, the pipe eventually lies at the required gradient.

It is usual not to change the size of the plates along the length of the conduit in spite of recognizing that its middle length is subjected to heavier loads than the outer lengths. Mainly because of this practice, it is considered safe to treat the whole length of the pipe just like its middle portion. Another reason for ignoring the effect of uneven loading along the pipe length is that the pipe with its annular corrugations is perceived to have little overall flexural rigidity in its longitudinal direction. This perception is quite valid for the pipe considered on its own in which case the pipe can deform longitudinally like the bellows of an accordion, with minimal bending moments. However, when this flexible pipe is embedded in soil, the free longitudinal movement of the corrugation rings is restrained by the soil; in this case, the pipe no longer remains as flexible as it is when considered in isolation. It can be appreciated that when the longitudinal flexural rigidity of the embedded pipe is not negligible, the effect of uneven loading along its length cannot be ignored.

Longitudinal Arching If the pipe deflects unevenly along its length, then it can be foreseen that some of the load above it will be transferred from the middle portions of the pipe to the outer portions, thereby relieving the conduit wall of the middle portion of some of its load effects. This transference of load in the longitudinal direction of the conduit can be regarded as a consequence of longitudinal arching, as distinct from the arching discussed earlier in relation to the plane-strain idealization of the structure; for convenience, this latter arching can be referred to as transverse arching. It is noted that because of difficulties in analysing the structure in three dimensions, little work has been done to date to study the effect of

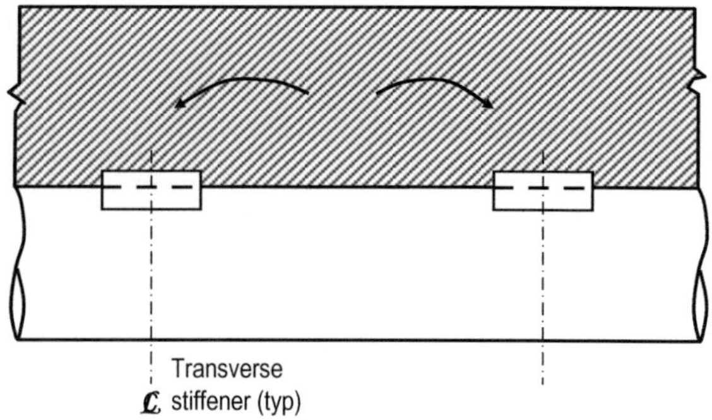

Fig. 7.12 Longitudinal arching between transverse stiffeners

longitudinal arching. The study by Girges (1993) is among the first extensive research work in this respect.

Another consequence of longitudinal arching is that when the top portions of the conduit walls are stiffened by frequently spaced transverse stiffeners, the soil load may be transferred to these stiffeners in the manner shown in Fig. 7.12. The effect of longitudinal arching between the stiffeners is clearly to relieve the portion of the conduit wall between the stiffeners of some soil weight and to transfer it to the stiffeners.

Effect of Foundation Settlement The material which should ideally be used for the backfill around the conduit is well-compacted granular material, which for all practical purposes has no time-dependent properties. The foundation of the structure, on the other hand, may not be of the same quality as the backfill. If the foundation is composed of predominantly cohesive soils, its settlements may consist of those which occur immediately after the construction of the structure and the long-term settlements, which in some cases may continue to increase for a long time. Both these settlements influence both longitudinal arching and transverse arching.

As discussed earlier and shown schematically in Fig. 7.11, the foundation under the conduit can settle unevenly in the longitudinal direction of the pipe. In addition, there is likely to be uneven settlement in the transverse direction as well. The vertical soil pressures below the invert level, discussed with respect to 2-D analyses in Sect. 7.2.1, can be used conveniently to study uneven settlement of the foundation in the transverse direction.

In the case of the structure with a vertically-elliptical conduit, the vertical soil pressures under most of the conduit are much higher than the pressures at the same level in adjacent portions of the foundation. It is obvious that in this case the foundation under the conduit will settle more than in the adjacent portions, and will create a positive arching condition.

Unlike the structure with vertically-elliptical conduits, structures with round or horizontally elliptical conduits have much smaller vertical pressures under the conduit than the corresponding pressures away from the conduit. In this case, the

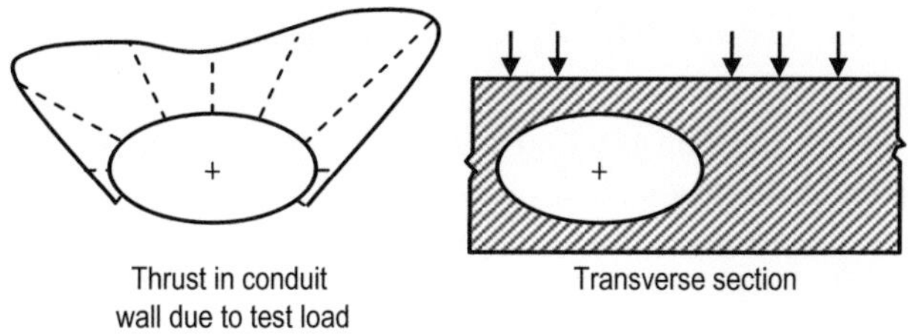

Fig. 7.13 Live load thrust around a pipe

long-term settlement of the foundation will create the situation shown schematically in Fig. 7.9d which generates negative arching.

It can be seen readily that the vertically elliptical conduit, while generating a negative transverse arching condition in the time-independent backfill, may lead to positive transverse arching condition due to time-dependent settlement of the foundation. Indeed, the reverse is true for the horizontally elliptical conduit which may generate positive arching in the backfill considered in isolation but which may also be responsible for some negative arching if the structure rests upon a foundation with significant time-dependent deformation characteristics.

Distribution of Live Loads Through field tests on three soil-steel structures with shallow depths of cover, Bakht (1981) has shown that the distribution of concentrated wheel loads of commercial vehicles in the conduit walls of soil-steel bridges is quite complex. Thrusts around the conduit wall under a test vehicle, calculated from measured strains in the conduit wall, are presented in Fig. 7.13. It can be seen that these thrusts are highly non-uniform around the conduit wall with the upper segments being subjected to higher thrusts than the portions that are remote from the applied load. Abdel-Sayed and Bakht (1982) have used analytical and experimental data to formulate the criterion of the OHBDC for load dispersion through the backfill; the same criteria are also used by the Canadian Highway Bridge Design Code (CHBDC 2006).

Dynamic Amplification of Live Load Effects In a bridge, the dynamic amplification of live load effects is caused mainly by the interaction of the dynamic systems of the vehicle and the structure. This interaction, which exists even if the riding surface of the bridge is smooth, results in a strain-time curve that is not smooth. Bakht (1981) has observed that the observed strain-time curve corresponding to a smooth riding surface in a soil-steel bridge was virtually without irregularities; this suggests that because of the considerable damping characteristics of the backfill there is little dynamic amplification of the axial strains.

An irregularity in the riding surfaces of a bridge causes a vehicle to bounce up and down thus inducing irregularities in the strain-time curves, which can be transformed into equivalent dynamic load allowance (DLA) or impact factor. Irregularities were created through 13 mm thick wooden planks placed on the riding surfaces of the soil-steel bridges tested by Bakht (1981). It was found that these irregularities led to DLA varying between 0.17 and 0.18 for single vehicles. On the basis of these test results, it was confirmed that the values of DLA specified by the CHBDC (2006) and described in Sect. 7.5 are on the safe side.

7.3 Geotechnical Considerations

Before the construction of a soil-steel bridge, the feasibility of its construction should be established by a geotechnical investigation unless prior knowledge of local subsurface conditions indicates that the approach fills and cuts will remain stable during and after construction.

The engineered backfill for soil-steel bridges should be composed of well-graded granular soils, the properties of which do not change with time. The various soils suitable for the engineered backfill are classified in Table 7.1, according to ASTM D2487 (2006).

The secant modulus of soil stiffness, E_S, depends upon the type of soil and the degree of its compaction. The values of this parameter, as specified by the CHBDC (2006), are listed in Table 7.2 for different soil groups and various Standard Proctor densities, which are defined by ASTM D698 (2007).

The values of E_S listed in Table 7.2 are plotted against Standard Proctor densities in Fig. 7.14 for both soil groups. The curves in this figure can be used to interpolate the values of E_S for intermediate values of Standard Proctor densities.

Table 7.1 Soil classification

Soil group no.	Grain size	Soil types included
I	Coarse	Well graded gravel or sandy gravel; Poorly graded gravel or sandy gravel; Well graded sand or gravelly sand; and Poorly graded sand or gravelly sand
II	Medium	Clayey gravel or clayey-sandy gravel; Clayey sand or clayey gravelly sand; and Silty sand or silty gravelly sand

Table 7.2 Values of E_S for various soils

Soil group no.	Standard proctor density, %	Secant modulus of soil E_S MPa
I	85	6
	90	12
	95	24
	100	30
II	85	3
	90	6
	95	12
	100	15

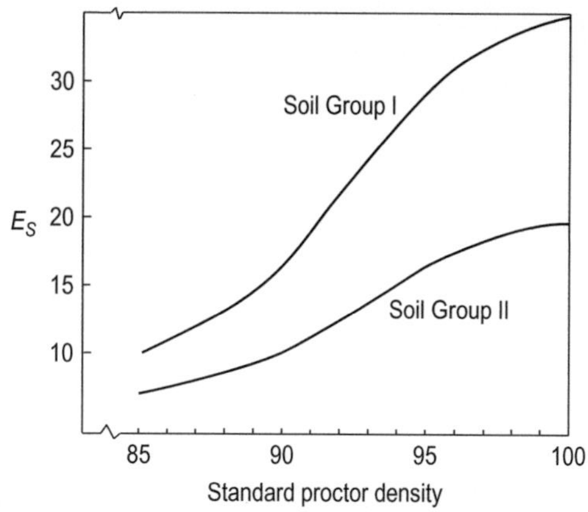

Fig. 7.14 E_S plotted against standard Proctor densities

7.4 Shallow and Deep Corrugations

Until some time ago, soil-steel bridges were usually made with plates having a 152 × 51 mm corrugation profile; this corrugation, now defined as *shallow corrugation*, is illustrated in Fig. 7.15, and its various properties are listed in Table 7.3 for thicknesses available in metric units.

The *deep corrugations* were recently introduced by the Canadian industry to use steel more efficiently. One particular deep corrugated profile, known by its trade name as Super●Cor®, is illustrated in Fig. 7.16, and its various properties corresponding to available thicknesses are listed in Table 7.4.

From Tables 7.3 and 7.4, it can be seen that the 7.0 mm thick shallow corrugated plate has the cross-sectional area, A, of about 8.7 mm^2/mm, while the deep corrugated plate with nearly the same thickness has a cross-sectional area of about 9.8 mm^2/mm, representing an increase in volume or weight of only about 13 %. The moment of inertia, I_s, of the deep corrugated plate (24,164 mm^4/mm), however, is about 9 times that of the shallow corrugated plate (2675 mm^4/mm). The radius of gyration, r, the property of the plate governing the buckling strength, of the deep corrugated plate is about 2.8 times that of the shallow corrugated plate. Thus, it can be appreciated that the deeper corrugations are very efficient in enhancing the critical properties of the corrugated plates.

The higher flexural rigidities of deep corrugated plates do add a slight complexity in the design process. The flexural rigidity of a 7.1 mm thick deep corrugated plate, i.e. the product of the modulus of elasticity of steel E and I_s, is nearly the same as that of a 140 mm thick concrete pipe. Similar to concrete pipes, the bending moments in deep corrugated plates of soil-steel bridges have to be accounted for explicitly. It is recalled that, as discussed earlier, moments in conduit walls made of shallow corrugated plates are neglected in the design of the completed structure.

Deep corrugated steel plates are also used in a ridge-over-ridge pattern; this pattern can be seen in Fig. 7.17 during a lab test on the double plates. The two plates can be made nearly fully composite by pouring concrete into the voids between the

Fig. 7.15 Profile for a shallow corrugated plate

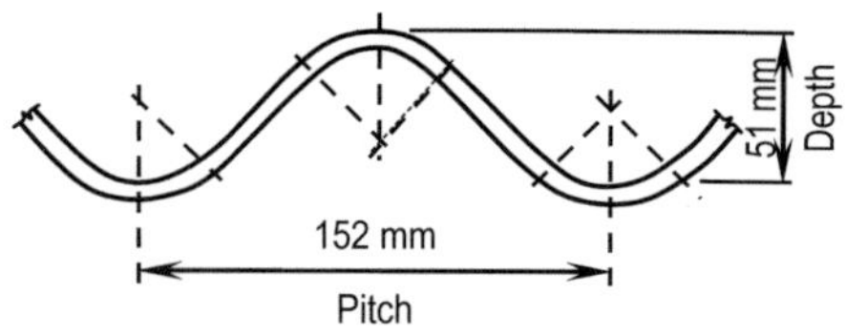

Table 7.3 Structural properties of 152 × 51 corrugated plates

Nominal plate thickness (un-coated), mm	3.0	4.0	5.0	6.0	7.0
A, area of cross section per unit length, mm^2/mm	3.522	4.828	6.149	7.461	8.712
I_s, second moment of cross-sectional area, mm^4/mm	1057	1458	1867	2278	2675
r, radius of gyration, mm	17.33	17.38	17.43	17.48	17.52

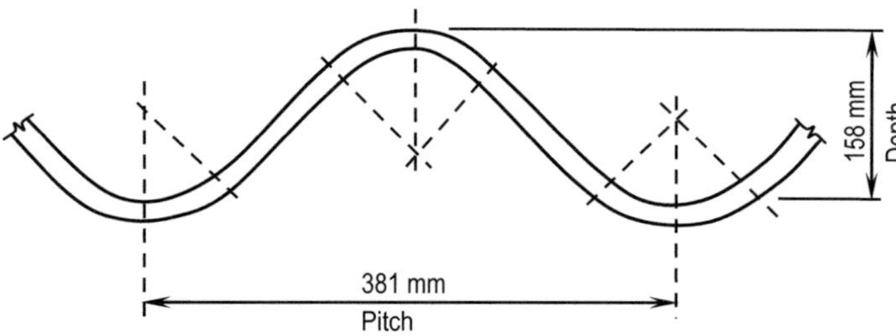

Fig. 7.16 Corrugation profile for a deep corrugated plate

Table 7.4 Structural properties of Super●Cor® plates

Nominal plate thickness (un-coated), mm	3.5	4.2	4.8	5.5	6.3	7.1	8.1
A, area of cross section per unit length, mm^2/mm	4.784	5.846	6.536	7.628	8.716	9.807	11.06
I_s, second moment of cross-sectional area, mm^4/mm	11,710	14,332	16.037	18,740	21,441	24,125	27,259
r, radius of gyration, mm	49.48	49.52	49.54	49.57	49.60	49.64	49.65

plates, it being noted that the connection between the concrete and the plates is provided by shear connectors installed on the plates.

The flexural rigidity of two 7.1 mm thick fully composite deep corrugated plates is nearly equivalent to that of a 200 mm thick concrete pipe.

Fig. 7.17 Deep corrugated steel plates in ridge-over-ridge pattern

7.5 General Design Provisions

The design provisions given in this section are sufficient for structures with shallow corrugations. However, for structures with deep corrugations, the additional design provisions given in Sect. 7.6 also apply. It is noted that the design provisions in this and subsequent sections are adopted from those of the CHBDC (2006). Bakht (2007) traces the evolution of these design provisions over 30 years.

7.5.1 Design Criteria

The CHBDC (2006) requires the consideration of both the ultimate and serviceability limit states (ULS and SLS) for the design of the conduit walls of soil-steel structures. The various limit states that are required to be considered for soil-steel bridges with shallow and deep corrugations are listed in Table 7.5 along with the corresponding material resistance factors.

For compression strength at the ULS, the conduit wall and longitudinal seams should satisfy the following condition.

$$\phi_t R_n \geq T_f \tag{7.1}$$

where, T_f, the axial thrust due to dead and live loads, is obtained from the following equation.

$$T_f = \alpha_D + \alpha_L(1 + DLA)T_L \tag{7.2}$$

The notation used in Table 7.5 and Eqs. (7.1) and (7.2) is defined in the following.

ϕ_h = the resistance factor for the plastic hinge for the completed structure
ϕ_{hc} = the resistance factor for the plastic hinge during construction
ϕ_j = the resistance factor for the failure of longitudinal seams
ϕ_t = the resistance factor for the compressive strength of the conduit wall

Table 7.5 Limit states and material resistance factors

Corrugation	Limit state	Component of resistance	Material resistance factor
Shallow	ULS	Compression strength	$\phi_t = 0.80$
	ULS	Plastic hinge during construction	$\phi_{hc} = 0.90$
	ULS	Strength of longitudinal seams	$\phi_j = 0.70$
	SLS	Deformation during construction	Not applicable
Deep	ULS	Compression strength	$\phi_t = 0.80$
	ULS	Plastic hinge	$\phi_h = 0.85$
	ULS	Plastic hinge during construction	$\phi_{hc} = 0.90$
	ULS	Strength of longitudinal seams	$\phi_j = 0.70$
	SLS	Deformation during construction	Not applicable

Table 7.6 Unit weights of different materials

Material	Unit weight, kN/m^3
Bituminous wearing course	23.5
Granular soil	22.0
Crushed rock	22.0
Fine-grained sandy soil	20.0
Glacial tile	22.0
Rockfill	21.0

α_D = the load factor for dead loads, being 1.25 for soil backfill corresponding to the nominal unit material weights being as listed in Table 7.6

α_L = the load factor for live loads, being 1.70.

DLA = the dynamic load allowance which is specified to be 0.4 for zero depth of cover, decreasing linearly to 0.1 for a depth of cover 2.0 m; for depth of cover larger than 2.0 m, DLA is specified to be 0.1.

T_D = the thrust in the conduit wall due to dead loads.

T_L = the thrust in the conduit wall due to live loads.

R_N = the nominal capacity of the conduit wall or the longitudinal seam to withstand axial thrust.

It is important to note that the value of α_L noted above (=1.70) is applicable to the CHBDC design truck, which corresponds to legally permissible upper limits of vehicle weights; this factor should be adjusted when it is applied to other design live loads.

Design codes usually do not specify construction methods as part of the design criteria. The CHBDC, however, has made an exception is this respect and has specified construction methods, site supervision and construction control for soil-steel bridges. This exception is made because the method of construction is crucial to both the short- and long-term integrity of the soil-steel bridges. Recommended procedures of construction are given in Sect. 7.8.

For soil-steel bridges with shallow or deep corrugations, the combined effects of bending moments and axial thrust arising from unfactored dead load and specified construction equipment should satisfy the following condition at all stages of construction.

$$\left[\frac{P}{P_{pf}}\right]^2 + \left|\frac{M}{M_{pf}}\right| \leq 1.0 \tag{7.3}$$

where

P_{pf} = factored compressive strength of the corrugated metal section without buckling.

M_{pf} = factored plastic moment capacity of a corrugated metal section.

$P = T_D + T_C$, in which T_D, axial thrust due to unfactored dead load, is obtained from Eq. (7.7), presented later; T_C, the axial thrust due to unfactored construction loads is obtained by using the same technique which is described in Sect. 7.5.3 for calculating T_L; for depths of cover smaller than the required minimum depth of cover, P is assumed to be zero.

$M = M_1$ (moment due to backfill to the crown level) + M_B (moment due to backfill above the crown) + M_C (moment due to construction loads), where:

$$\left.\begin{aligned} M_1 &= k_{M1} R_B \gamma D_h^3 \\ M_B &= -k_{M2} R_B \gamma D_h^2 H_c \\ M_C &= k_{M3} R_L D_h L_c \end{aligned}\right\} \tag{7.4}$$

in which:

$$\left.\begin{aligned} k_{M1} &= 0.0046 - 0.0010\ \log_{10}(N_F) for\ NF \leq 5000 \\ &= 0.0009\ for\ N_F > \ 5000 \\ k_{M2} &= 0.018 - 0.004\ \log_{10}(N_F)\ for\ NF \leq 5000 \\ &= 0.0032\ for\ N_F > \ 5000 \\ k_{M3} &= 0.120 - 0.018\ \log_{10}(N_F)\ for\ NF \leq \ 100,000 \\ &= 0.030\ for\ N_F > 100,000 \\ R_B &= 0.67 + 0.87[(D_v/2D_h) - 0.2] \\ &\quad for\ 0.2 \leq (D_v/2D_h) \leq 0.35 \\ R_B &= 0.80 + 1.33[(D_v/2D_h) - 0.35] \\ &\quad for\ 0.35 \leq 0.50 \\ R_B &= D_v/D_h \\ &\quad for\ (D_v/2D_h) > 0.50 \\ R_L &= [0.265 - 0.053\ \log_{10}(N_F)]/(H_c/D_h)^{0.75} \leq \ 1.0 \\ L_c &= A_c/k_4 \end{aligned}\right\} \tag{7.5}$$

k_4 is obtained from Table 7.7 and N_F is obtained from the following equation.

$$\begin{aligned} N_F &= E_s(1000 D_h)^3/EI \\ &(D_h\ is\ in\ m) \end{aligned} \tag{7.6}$$

Table 7.7 Values of k_4 for calculating equivalent live loads

Depth of cover, m	k_4, m Two wheels per axle	Four wheels per axle	Eight wheels per axle
0.3	1.3	1.5	2.6
0.6	1.6	2.0	2.8
0.9	2.1	2.7	3.2
1.5	3.7	3.8	4.1
2.1	4.4	4.4	4.5
3.0	4.9	4.9	4.9
4.6	6.7	6.7	6.7
6.1	8.5	8.5	8.5
9.1	12.2	12.2	12.2

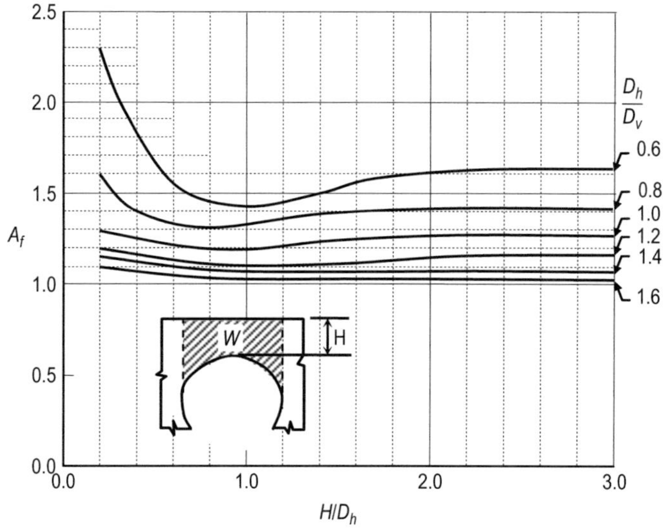

Fig. 7.18 Chart for A_f

7.5.2 Dead Load Thrust

The CHBDC (2006) requires that the dead load thrust T_D is calculated from:

$$T_D = 0.5(1.0 - 0.1C_S)A_f W \tag{7.7}$$

where

A_f = a coefficient whose values are obtained from Fig. 7.18 according to the ratios H/D_h and D_h/D_v with H being the depth of the soil cover above the crown and D_h and D_v being as defined in Fig. 7.4.

W = the nominal dead weight of the column above the conduit as defined in Fig. 7.18.
C_S = the axial stiffness parameter defined as follows:

$$C_S = \frac{E_S D_V}{EA} \tag{7.8}$$

in which

E_s = secant modulus of soil whose value can be obtained from Table 7.2, corresponding to the soil classification defined in Table 7.1.
E = modulus of elasticity of conduit wall material, which can be assumed to be 2.0×10^6 MPa.
D_V = dimension relating to the cross-section of the conduit wall as defined in Fig. 7.4.
A = cross-sectioned area of the conduit wall/unit length, which can be obtained from Tables 7.3 and 7.4 for shallow and deep corrugations, respectively.

7.5.3 *Live Load Thrust*

The live load thrust T_L, which in reality has non-uniform values around the conduit, as discussed earlier, is assumed to have the same value throughout the conduit wall for design purposes. The CHBDC requires that its value be the smaller of the values obtained from the following two equations.

$$T_L = 0.5 D_h \sigma_L m_f \tag{7.9}$$

$$T_L = 0.5 l_t \sigma_L m_f \tag{7.10}$$

where

(a) l_t is the distance between the outermost axles of the design or construction vehicle (including the tire footprints) placed in accordance with (c) (i) plus $2H$;
(b) m_f is the modification factor for multi-presence of vehicles in more than one lane as discussed in Chap. 1; and
(c) the load case yielding the maximum value of $\sigma_L m_f$ governs, and σ_L is obtained as follows:

  (i) within the span length, position as many axles of the design truck or construction vehicle at the road surface as would give the maximum total load;
  (ii) distribute the rectangular wheel loads through the fill down to the crown level at a slope of one vertically and one horizontally in the transverse direction of the conduit and two vertically to one horizontally in the longitudinal direction; and

(iii) obtain the uniformly distributed pressure σ_L by assuming that the total wheel loads considered in item (i) are uniformly distributed over the rectangular area that encloses the individual areas obtained in item (ii).

For σ_L and l_t to be realistic, the configuration of the design loading must be similar to actual vehicles. It can be demonstrated readily that a design loading formulated only on the basis of the equivalence of bending moments and shear forces in simply supported beams can fail to induce the same load effects in a soil-steel bridge as those induced by actual heavy vehicles.

7.5.4 Conduit Wall Strength in Compression

The factored nominal capacity, $\phi_\tau R_N$, of the conduit wall to sustain axial thrust is given by:

$$\phi_t R_n = A f_b \tag{7.11}$$

where

f_b = the compressive failure stress of the conduit wall, which can be obtained by the simplified method of the CHBDC (2006) described in the following.

For calculating f_b, the conduit wall is divided into lower and upper segments, separated from each other by two symmetrical radial lines with their centre at the centre of curvature of the arc at crown, and with an angle θ_0 (in radians), from the vertical calculated as follows.

$$\theta_0 = 1.6 + 0.21 \log \left[\frac{EI}{E_m R^3}\right] \tag{7.12}$$

In the upper segments, the conduit wall moves away from the soil, whereas in the lower segments, the wall moves towards the soil, because of which its compressive strength is higher than that of the wall in the upper segment. The compressive strength, f_b, is calculated as follows:

(a) for $R \le R_e$

$$f_b = \phi_t F_m \left\{ F_y - \frac{F_y^2}{12E}\left(\frac{KR}{r}\right)^2 \frac{1}{\rho} \right\} \tag{7.13}$$

(b) for $R > R_e$

$$f_b = \frac{3\phi_t \rho F_m E}{\left(\frac{KR}{r}\right)^2} \tag{7.14}$$

where

(i) ϕ_t = material resistance factor obtained from Table 7.5;
(ii) F_m = 1.0 for structures for single conduits; and
(iii)

$$F_m = \left(0.85 + \frac{0.3S}{D_h}\right) \leq 1.0 \text{ for structures with multiple conduits} \tag{7.15}$$

where S is the least transverse clear spacing between adjacent conduits and D_h corresponds to the largest conduit. As noted by Abdel-Sayed et al. (1992), F_m is derived from consideration of the reduction of the lateral support to the conduit wall due to the proximity of another conduit.

(iv)

$$R_e = \frac{r}{K}\left\{\frac{6E\rho}{F_y}\right\}^{0.5} \tag{7.16}$$

(v)

$$\rho = \left(\frac{H}{R_c}\right)^{0.5} \leq 1.0 \tag{7.17}$$

(vi)

$$K = \lambda\left\{\frac{EI_S}{E_m R^3}\right\}^{0.25} \tag{7.18}$$

(vii) E_m for the lower segments is the same as E_S; but for the upper segments, it is calculated as follows.

$$E_m = E_S\left\{1 - \left(\frac{R_C}{R_C + H + H'}\right)\right\} \tag{7.19}$$

When the conduit wall is supported by a combination of backfills having two different properties, E_m should be based on the lower value of E_S for the two materials; and

(viii) λ for the segments of the conduit wall of structures except part-arches with simple radius of curvature and a rise-to-span ratio of less than 0.4 is calculated as follows:

$$\lambda = 1.22\left\{1.0 + 1.6\left(\frac{EI_S}{E_m R_c^3}\right)^{0.25}\right\} \tag{7.20}$$

For all other cases, λ is assumed to be 1.22.

Some of the notation used in Eqs. (7.11), (7.12), (7.13), (7.14), (7.15), (7.16), (7.17), (7.18), and (7.19) has already been defined; others are defined in the following.

K = a factor representing the relative stiffness of the conduit wall with respect to the stiffness of the adjacent soil mass.
ρ = a reduction factor for buckling stress in the conduit wall.
R_c = radius of curvature of the conduit at crown.
R = radius of curvature of the conduit at the reference point under consideration.
F_y = yield stress of steel.
H' = half the vertical distance between crown and springline.

7.5.5 Longitudinal Seam Strength

The bolted longitudinal seams of structures with both shallow and deep corrugations are required by the CHBDC (2006) to be designed to sustain axial thrust. This code permits only two bolting arrangements for longitudinal seams in plates with shallow corrugations; these arrangements are shown in Fig. 7.19. Nominal strength corresponding to the two permissible bolting arrangements for seams in shallow corrugated plates can be obtained from Table 7.8 which corresponds to plates jointed with 20 mm diameter bolts and which is based on test results. The plate thickness referred to in this table is the thinner of the two mating plates at the seam.

The strengths of longitudinal seams of deep corrugated plates, as specified by ASTM (2003) are noted in Table 7.9. Additional information on strengths of longitudinal seams of both shallow and deep corrugations is provided by Lee et al. (2007).

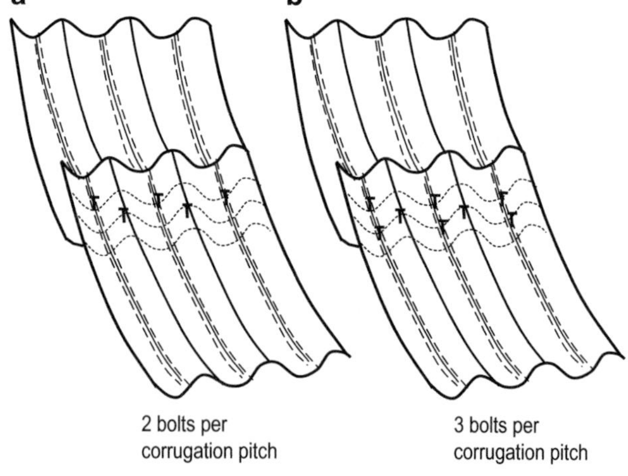

Fig. 7.19 Two bolting arrangements for plates with shallow corrugations

Table 7.8 Nominal ultimate strength of longitudinal seams of shallow corrugated steel plates (ASTM 2003)

Nominal plate thickness, mm	3	4	5	6	7
Strength in kN/m for bolting arrangement of Fig. 7.19a	750	1100	1400	1800	2050
Strength in kN/m for bolting arrangement of Fig. 7.19b	–	–	–	–	2550

Table 7.9 Nominal ultimate strength of longitudinal seams of deep corrugated steel plates (ASTM 2003)

Nominal plate thickness, mm	3.4	4.2	5.5	6.2	7.0
Strength in kN/m	1482	1811	2361	2699	3037

Based on the work of Mikhailovsky et al. (1992), it is recommended that the bolting arrangement of Fig. 7.19a be used in such a way that the bolts closer to the visible edge of the seam lie in valleys. This preferred arrangement prevents the bolt hole from cracking if the conduit wall is subjected to excessive bending deformations.

7.6 Design with Deep Corrugations

In addition to the design requirements given in Sect. 7.5, the conduit wall of the completed soil-steel structure with deep corrugations should also satisfy the following condition for the combined effect of bending and axial thrust.

$$\left[\frac{T_f}{P_{pf}}\right]^2 + \left|\frac{M_f}{M_{pf}}\right| \leq 1.0 \tag{7.21}$$

where T_f is obtained from Eq. (7.2) and P_{pf}, M_f, and M_{pf} are obtained as follows.

$$P_{pf} = \phi_h A F_y \tag{7.22}$$

$$M_f = |\alpha_D M_1 + \alpha_D M_D| + \alpha_L M_L(1 + DLA) \tag{7.23}$$

where

$$M_1 = k_{M1} R_B \gamma D_h^2 \tag{7.24}$$

$$M_D = -k_{M2} R_B D_h^2 H_e \tag{7.25}$$

in which

H_e = smaller of H and D_h

$$M_L = \frac{k_{M3} R_U D_h A_L}{k_4} \tag{7.26}$$

where k_{MI}, k_{MI}, k_{MI}, and R_B are obtained from Eq. (7.5), A_L is the weight of the second axle of the CHBDC CL-W Truck, k_4 is obtained from Table 7.7 and R_U is obtained from:

$$R_U = \frac{0.265 - 0.053 \log_{10} N_F}{(H/D_h)^{0.75}} \tag{7.27}$$

where N_f is obtained from Eq. (7.6).

The factored plastic moment capacity, M_{pf}, is obtained from:

$$M_{pf} = \phi_h M_P \tag{7.28}$$

7.7 Other Design Criteria

Design criteria of the CHBDC (2006) other than those discussed in Sects. 7.5 and 7.6 are given in this section.

7.7.1 Minimum Depth of Cover

The CHBDC (2006) requires that the minimum depth of cover over a conduit in metres in a soil-steel structure with shallow corrugations should be the largest of 0.6 m, $(D_h/6)(D_h/D_v)^{0.5}$, and 0.4 $(D_h/D_v)^2$. This requirement safeguards against the failure of soil cover under eccentric vehicle loads.

For soil-steel bridges with deep corrugations, the minimum depth of cover is required to be the smaller of 1.5 m and the minimum depth of cover required for the structure with deep corrugations but having the same conduit size.

7.7.2 Deformations During Construction

For all conduit shapes, the upward or downward crown deflection during construction is required to be limited to 2 % rise. If struts and ties are used during the assembly of the pipe or during backfilling, they should be removed before they restrict the downward movement of the pipe.

7.7.3 Extent of Engineered Backfill

The *engineered backfill*, specifications for which are given in Sect. 7.8, should extend transversely on each side of the conduit to at least the smaller of 5.0 m and one-half of the conduit span, and vertically up to the minimum depth of cover specified in Sect. 7.7.1.

7.7.4 Differences in Radii of Curvature and Plate Thickness

The radius of curvature, R, of the conduit at any location should not be less than 0.2 R_c, and the ratio of the radii of curvature of mating plates should not be more than 8.

Further, the difference in the thickness of plates meeting at a longitudinal seam should not exceed 1 mm if the thicker plate has a thickness between 3.1 and 3.5 mm. Where each plate thickness exceeds 3.5 mm, there is no restriction in the difference of plate thicknesses at a longitudinal seam.

7.7.5 Footings

The earlier design codes (OHBDC, 1979, 1983, 1993; CHBDC 2000) were not explicit in requiring that footings of soil-steel structures with arch shapes be designed for the horizontal force that develops at the footings of the structure. The CHBDC (2006) now explicitly requires that consideration should be given to resisting the horizontal reactions that develop in footings because of soil pressures on the conduit wall.

The CHBDC (2006) does not specify how to calculate the horizontal reaction in the footings of arch structures. However, measurements of earth pressures by Bakht in 1978 (unpublished), and those by Vaslestad et al. (2007) taken over a period of 21 years suggest that the lateral earth pressure on the conduit wall of a soil-steel bridge with shallow corrugations follow the pattern illustrated in Fig. 7.20; in this figure, it can be seen that lateral earth pressure in the upper region of the conduit conforms to 'Rankin' earth pressure that is employed for designing retaining walls. However, in the lower regions, the lateral earth pressure drops to zero at the invert. Abdel-Sayed et al. (1993) note that this drop in pressure is due to the placing of backfill in shallow layers and also due to the flexibility of the conduit wall. As illustrated in Fig. 7.20, the total force due to horizontal earth pressure on the pipe is borne by the footings at the bottom and the conduit wall at the top. In the latter case, the reaction becomes the thrust in the conduit wall.

In the absence of rigorous analyses supported by field observations, it is recommended that in the footings of arch structures with shallow corrugations, the horizontal reaction at the footings be assumed to be 65 % of the reaction that

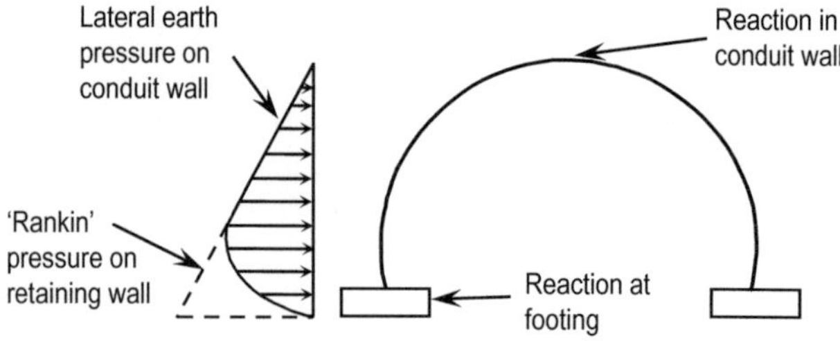

Fig. 7.20 Horizontal reaction at footing of an arch structure

would be obtained by assuming that full active (Rankin) pressure acts on the pipe. For the relatively rigid conduit walls made with deep corrugations, this ratio is recommended to be 75 %.

7.8 Construction

It is essential to realize that construction procedures are the most important factor responsible for the structural integrity of a soil-steel bridge. The CHBDC (2006) specifies that appropriate construction procedures and controls should be noted on the construction drawings. These procedures and controls are given briefly in this sub-section. For a more exhaustive treatment of the subject, the reader is referred to the textbook by Abdel-Sayed et al. (1993).

7.8.1 Foundation

The *foundation* of a soil-steel bridge is the natural ground on which the structure, including the backfill, is erected. When this foundation has markedly non-uniform settlement properties within the extent of the conduit, appropriate measures, including the removal of unsuitable materials, should be taken to avoid a detrimental effect on the structure. Engineering judgement is recommended to be exercised for establishing the extent of soil removal from the foundation and for in-situ soil improvement.

For the specific case of the foundation of a pipe-arch, the following is recommended:

(a) When the foundation comprises dense to very dense cohesionless material or stiff to hard cohesive material, no treatment is required.
(b) For soft to firm cohesive foundations, trench reinforcement should be provided according to the scheme of Fig. 7.21.
(c) For loose to compact cohesionless foundations, trench reinforcement should be provided either according to Fig. 7.21 or by in-situ compaction.

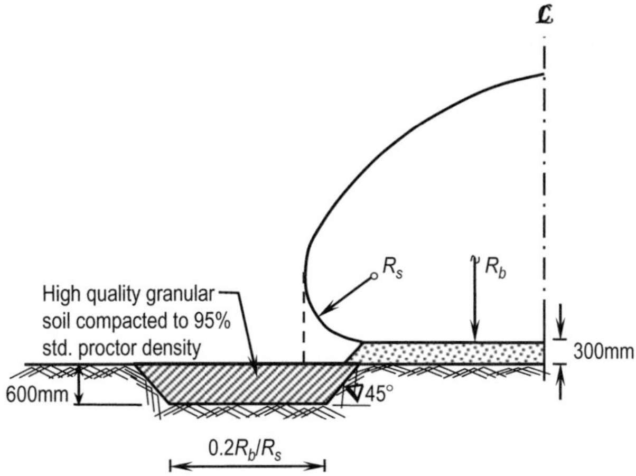

Fig. 7.21 Reinforcement of the foundation under the haunches of pipe-arches

The need for reinforcing the foundations of pipe-arches arises from the very high radial pressures under the haunches of these pipes. As noted by Bakht and Agarwal (1988), a foundation which can settle significantly under these radial pressures will cause the conduit wall to undergo excessive bending deformations.

7.8.2 *Bedding*

The bedding on which the lower segment of the pipe rests, should consist of stone-free granular material. It should be pre-shaped in both the longitudinal and transverse directions to accommodate the conduit invert. The top 200 mm layer of bedding should be left un-compacted so that the ridges and valleys of the bottom segment of the pipe maintain uniform contact with the bedding material.

7.8.3 *Assembly and Erection*

The assembling and erection of the conduit wall should be such that no permanent set results in any portion of the wall. Bolts at longitudinal seams should be arranged in accordance with one of the two arrangements shown in Fig. 7.19. When the arrangement of Fig. 7.19a is used, the bolts in the row closer to a visible edge of the mating plate should be in the valleys and those in the other row should be on the ridges. The initial torque on the bolts should be between 200 and 340 N.m. The following minimum percentages of bolts should be tested after completion of erection of the conduit walls and before backfilling:

(a) circumferential seams – 5 % of bolts in each circumferential seam; and
(b) longitudinal seams – 5 % of bolts in each longitudinal seam in each plate.

The test bolts should be selected in a random manner and the installation should be considered acceptable if the above torque requirements are met in at least 90 % of the bolts tested. If struts are used to support the conduit wall during backfilling, they should be removed before they start restricting the free downward movement of the crown.

7.8.4 Engineered Backfill

The material for the engineered backfill placed in the immediate vicinity of the pipe should be boulder-free and should be either Group I or II soils as specified in Table 7.1 with compaction corresponding to the modulus of soil stiffness used in the design. The fill should be placed and compacted in layers not exceeding 300 mm of compacted thickness, with each layer compacted to the required density prior to the addition of the next layer. The difference in levels of backfills on the two sides of a conduit in the transverse direction should not exceed 600 mm.

The degree of compaction for the engineered backfill should not be less than 85 % of the Standard Proctor density. The extent of the engineered backfill should be as specified in Sect. 7.7.3. Within a distance of 300 mm of the pipe, the backfill should be free of stones exceeding 80 mm in size. Heavy compaction equipment should be avoided within 1.0 m of the pipe. It is recommended that light compaction equipment be used in the proximity of the pipe. The compaction equipment should travel along the length of the pipe as can be seen in Fig. 7.22.

7.8.5 Headwalls and Appurtenances

Soil-steel bridges are highly susceptible to damage by hydraulic effect. It is, therefore, essential that when the structures are designed for hydraulic service they should be provided with headwalls and cut-off appurtenances. When a conduit wall at one of its ends is cut at a plane inclined to the vertical, the continuity of the ring is no longer maintained in the bevel, because of which the bevelled ends of the pipe should be designed as earth-retaining structures. The CHBDC (2006) does not encourage skew bevelled ends, in which the conduit wall at one end of the bevel becomes a large free-standing retaining wall, which can nevertheless be designed as a mechanically stabilised retaining wall by tying it to deadman anchors embedded in the fill behind the wall (Essery and Williams 2007). The larger end of the skew bevel of a structure, designed as a mechanically stabilized wall with the stabilization provided by the patented wire mesh, can be seen in Fig. 7.23.

Fig. 7.22 Light compaction equipment in the vicinity of the pipe

Fig. 7.23 Conduit wall at a skew bevel stabilised by wire mesh

7.8.6 Site Supervision and Control

Construction drawings should require that the engineer designated by the owner as being responsible for inspection or supervision is experienced in the design and construction of soil-steel bridges. The inspection and supervision of construction should be provided on the following basis.

(a) For structures with spans between 3 and 6 m, the work should be inspected by an engineer or the representative of the engineer at the completion of foundation; bedding; assembly of pipe; and placement of backfill under the haunches, up to the springline, up to the crown, and up to the level of minimum specified cover.

(b) For structures with spans greater than 6 m but less than or equal to 8 m, the inspection should be as in (a) above, and in addition, daily inspection under an engineer's supervision should be made during the backfilling operations until minimum specified cover is attained.

(c) For structures with spans greater than 8 m or those structures in which special features are used, continuous inspection and supervision by an engineer should be provided.

7.9 Special Features

In its simplest form, a soil-steel bridge contains no other structural elements than the compacted backfill and the metallic shell made only out of lapped, corrugated steel plates. This simple structure, even with plates of the largest available thickness (being 7.0 mm) cannot usually have spans larger than about 9 m with shallow corrugated plates, and about 12 m with deep corrugated plates. Spans larger than the above-cited limits are possible only if, by some means, the load effects in the conduit wall are reduced or its load carrying capacity is increased; these means are referred to as *special features* for soil-steel bridges. Some of the special features are discussed in this section. For details of other special features, reference may be made to the textbook by Abdel-Sayed et al. (1993).

Depending upon the dominant manner in which they enhance the load-carrying capacity of a soil-steel bridge, the various special features can be grouped into the following three categories.

(a) Features which reduce load effects in the conduit wall;
(b) features which increase the strength of the conduit wall by reinforcing it; and
(c) features which increase the strength of the conduit wall by stiffening the soil and thus enhancing the stiffness of the radial support to the conduit wall

The various special features are described in the following according to the three categories noted above.

7.9.1 Reduction of Load Effects

As discussed in Sect. 7.2.1, positive arching occurs in soil-steel bridges when the column of soil directly above the conduit deflects downward with respect to the adjacent columns of soil. The load effects in the conduit wall due to dead loads can be reduced by inducing positive arching. Load effects due to live loads in the conduit wall can be reduced by stiffening the medium above the conduit so that the concentrated loads applied at the top of the fill disperse to greater areas at the crown level thereby inducing smaller load effects in the pipe.

Relieving Slabs A relieving slab is a horizontal, or nearly horizontal, reinforced concrete slab located in the backfill of a soil-steel bridge above its conduit. Such slabs are placed at or below the embankment level. The cross-section of a soil-steel bridge with a relieving slab at the embankment level is shown in Fig. 7.24.

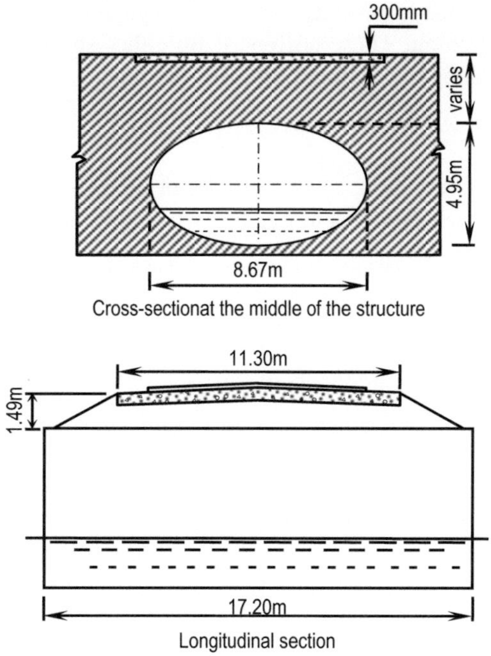

Fig. 7.24 Details of a soil-steel bridge with a relieving slab

Relieving slabs are particularly useful in structures with relatively long spans and shallow soil covers above the crown. In such structures, they serve two functions, i.e., that of reinforcing the soil cover above the conduit against shear failure, and of reducing live load effects in the conduit walls.

The live-load effects in the conduit walls are reduced by the relieving slab because it permits a much greater dispersion of the concentrated loads through the soil below, thereby reducing the resulting radial pressure on the pipe. Field testing of the structure with relieving slab shown in Fig. 7.24 and of a similar structure without the relieving slab, has shown that the presence of the relieving slab can reduce the live-load effects in the conduit wall by up to 50 % (Bakht 1985).

7.9.2 Reinforcing the Conduit Wall

The load carrying capacity of the metallic shell of a soil-steel bridge can be enhanced by attaching appendages to it. Two commonly used appendages are described in the following.

Transverse Stiffeners The conduit walls of soil-steel bridges having relatively long spans are often stiffened by circumferential stiffeners applied to the top portion of the pipe; these stiffeners are referred to as *transverse stiffeners*. The transverse stiffeners may consist of corrugated steel plates of narrow widths, having the same radius of curvature as the top segments of the pipe, and placed in a ridge-over-ridge fashion. These stiffeners are either spaced at regular intervals, or are continuous along the circumference of the pipe.

As an alternative to corrugated plate stiffeners, there are transverse stiffeners consisting of curved, rolled components. These latter stiffeners are used with pipes of very large spans. Both kinds of stiffeners have pre-drilled holes, but are attached to the pipe through holes, which are made at the site with a flame torch.

Longitudinal Stiffeners A large number of soil-steel bridges are in existence in which the pipes have been stiffened by longitudinal reinforced concrete beams located at each of the two shoulders; these beams are also known as *thrust beams*.

The proponents of thrust beams claim that:

(a) the vertical faces of these beams permit a better degree of compaction of the backfill in their vicinity;
(b) these beams promote a better distribution of the effects of live loads in the longitudinal direction;
(c) these beams isolate the top segment of the pipe thereby rationalizing the mathematical model that is commonly used in conjunction with its design.

One notable example of a soil-steel bridge with thrust beams is that of the Cheese Factory Bridge in Ontario, Canada, which has a part-arch type of construction with a record span of 18.0 m. This structure can be seen in Fig. 7.25 at the construction stage when the fill is being compacted just above the crown. The thrust beams, encasing the lower ends of the transverse stiffeners, were cast when the backfill was raised to near the thrust beams.

Vaslestad et al. (2007) have presented measured earth pressure data on a horizontally elliptical soil-steel bridge having a span and rise of 10.78 and 7.13 m, respectively. The structure has a 4.2 m deep soil cover above the crown and is installed with concrete thrust beams. At one transverse section, the earth pressure was measured at the crown, middle of the vertical face of a thrust beam, springline, haunch and invert. If the lateral pressure pattern was as shown in Fig. 7.20, the lateral soil pressure at the middle of the thrust beam would have

Fig. 7.25 The Cheese Factory Bridge in the final stages of construction

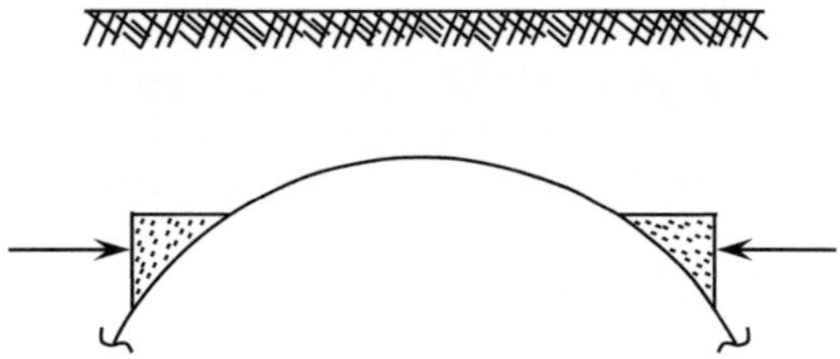

Fig. 7.26 Thrust beams providing horizontal support to metallic arch

been nearly half the corresponding pressure at the springline. The observed lateral pressure at the middle of the thrust beams was, however, nearly the same as that at the springline. This observation suggests that, as illustrated in Fig. 7.26, the thrust beams act as 'ties' to the metal arch contained within them.

7.9.3 Reinforcing the Backfill

It is imperative for the integrity of a soil-steel bridge that its backfill around the pipe continue to provide adequate support to the pipe during the lifetime of the bridge. Customarily, adequate support to the pipe can be ensured by selecting a well-graded granular material for the backfill and compacting it to a dense and uniform medium. Realizing that such an ideal medium is sometimes difficult to attain, a few techniques have been developed to enhance, or maintain, an adequate stiffness of the backfill. Two such techniques are discussed in the following.

Concreting Under Haunches The radial soil pressures under the haunches of pipe-arches are particularly high because of the relatively small radius of curvature of the conduit wall at these locations. In order for the soil to sustain the high radial pressures without yielding significantly, it is necessary that the backfill under the haunches be more densely compacted than elsewhere. Difficulty of access, however, makes it difficult to compact the backfill in these critical zones.

Bakht and Agarwal (1988) have shown that under the haunches of pipe-arches, the conventional compacted backfill can be replaced with advantage by low-strength and high-slump concrete. Before this technique was applied, many experts in the field of soil-steel bridges were apprehensive of the floating up of the pipe during concreting and of the undesirable stresses that may be induced in the conduit wall at the junction of the concrete and soil backfill.

The potential problem of the uplifting of the pipe during concreting was overcome by placing concrete in two layers and staggered longitudinal segments. Fears of damage to the conduit wall by the hard-point effect were also laid to rest by the fact that the structure does not show any sign of distress even after 18 years of service.

Controlled Low Strength Material A well-compacted backfill composed of granular materials is essential for the structural integrity of a soil-steel bridge. Even if the material selected for the backfill is good, minor and sometimes inadvertent departures from good compaction practice can manifest themselves into noticeable distress in the structure. Errors in the compaction procedure can be eliminated entirely by replacing the compacted engineered backfill with controlled low strength material (CLSM). A summary of their proposal is presented herein.

The CLSM, a flowable mixture of granular soil, cement, fly ash and water can be placed with minimal effort and supervision. By gravity alone, it can flow to the most difficult-to-reach nooks and corners of the structure. A case is made herein for using CLSM in soil-steel bridges. As pointed out by several authors (Brewer 1990; Brewer and Hurd 1991; Clem and Hook 1992), several specifications are used to achieve this mixture. Some of the proportions specified by different authorities are given in Table 7.10.

Cement proportions vary from 1.4 % to 2.8 % and those of fly ash from 0 to about 11 %. The water-cement ratio, ranging between 5 and 10, is extremely high compared to that used in concrete. When the water-cement ratio is so high, it has virtually no effect on the strength of the mixture, which is controlled instead by the content of cement and fly ash.

The CLSM sets within about 24 h to about one-tenth of the 28-day compressive strength which can vary between 0.7 and 5.0 MPa. It is interesting to note that in some CLSM applications, the emphasis is on keeping the strength low enough to permit easy re-excavation.

Several aspects of CLSM make it particularly suited for use as backfill in soil-steel bridges; these aspects are discussed in the following.

The CLSM having a very high slump, being 160–200 mm, is so flowable that by gravity alone it can reach even those nooks and crannies, which cannot be accessed easily for compacting the backfill by conventional methods. To achieve greater flowability, the granular soil should have rounded rather than angular particles. In desert regions, the sand particles are usually well-rounded, making it difficult to

Table 7.10 Mix proportions for CLSM

	Weight in kg/m^3 (percentage by weight) specified by				
Material	Brewer (1990)	American Conc. Pvmt. Assoc.	Iowa Dept. of Hwys.	S. Carolina DOT	Ohio DOT
Soil	1543 (72.2)	1661 (72.0)	1543 (72.5)	1691 (77.9)	1727 (78.4)
Water	356 (16.7)	297 (12.9)	347 (16.5)	272 (12.5)	297 (13.5)
Cement	59 (2.8)	100 (4.3)	59 (2.8)	30 (1.4)	30 (1.4)
Fly ash	178 (8.3)	250 (10.8)	178 (8.2)	178 (8.2)	148 (6.7)

achieve good compaction. This shortcoming of desert sands can be turned into an advantage by using them in CLSM and thereby achieving enhanced flowability.

There are two distinct aspects of the stiffness of the backfill that are brought to bear on the integrity of the metallic shell of a soil-steel bridge. One aspect relates to the arching action, which controls the load effects in the shell such as the thrust. The other aspect mainly concerns the radial support provided by the soil to the metallic shell. This support enables the shell to sustain high thrusts. As discussed later, the influence of the stiffness of the CLSM backfill is minimal on the load effects in the shell; however, its influence on the capacity of the metallic shell to sustain compressive axial loads is of paramount importance.

The measure of the stiffness of the backfill, which affects the deformations and buckling capacity of a buried pipe, is a parameter called the modulus of soil reaction, *E'*. As shown by several researchers (Hartley and Duncan 1987), the value of this parameter depends not only upon the engineering properties of the soil, but also upon the depth of embedment of the reference station. It can be demonstrated that a qualitative comparison of *E'* values of different backfills having time-independent characteristics can be made simply by comparing the values of their respective moduli of elasticity.

Brewer (1990) has shown that a maximum compressive strength of 0.7 MPa can be achieved for CLSM mixed according to the proportions given in Table 7.10 under his name. For this mix design, Brewer has presented the experimental values of the modulus of elasticity. These values correspond to a mean of 6.5 MPa and a standard deviation of 3.3 MPa. Despite the wide scatter of these values, it seems reasonable to use the mean value of the modulus of elasticity as the representative one. This is because the interaction of the soil and the shell extends over a very wide area, as a result of which incipient buckling of a portion of the shell in the vicinity of a zone of softer backfill is likely to be followed by internal re-arching thereby averting the failure by transferring loads to the stiffer backfill. The effect of this internal redistribution is equivalent to 'smearing' the stiffness, in which case the mean stiffness can be used as the effective one.

A modulus of elasticity of 6.5 MPa is similar to that of medium grain backfills compacted to between 85 % and 95 % Standard Proctor density. It is thus concluded that the CLSM mix described above is similar in its stiffness characteristics to fills commonly used in soil-steel bridges. Higher stiffness can be achieved by rearranging the mix proportions. It is emphasized that to enable a rational design of the metallic shell, representative values of *E'* for the intended CLSM should be obtained. The technique used by Hartley and Duncan (1987) may be found useful in this context.

Soil-steel bridges carrying water through the conduit must have adequate inlet and outlet protection against damage by hydraulic forces. Part of this damage is related to the loss of fine particles caused by the water flowing through the fill behind the conduit wall. The CLSM, permeable like the granular backfill, is less susceptible to the loss of fine particles because of its particles being held together by a chemical, rather than frictional, bond. It is expected that a soil-steel bridge with

CLSM backfill is likely to be more resistant to hydraulic damage than its counterpart, the conventional backfill.

The CLSM is similar in many aspects to the conventional backfill. However, in certain important aspects, it is so significantly different as to require special considerations for use in soil-steel bridges. Some of these special considerations are discussed in the following.

It can be appreciated that when a flowable mixture is poured around a pipe, the buoyancy force exerted by the fluid can be high enough to float the pipe and dislodge it from its bedding. Bakht and Agarwal (1988) have observed that the pipe floated up initially by up to 6 mm, when high slump concrete was poured around the haunches of a pipe-arch in two layers and staggered longitudinal segments. When CLSM is placed around a pipe in a single pour, the uplift can be even higher unless the pipe is pre-loaded preferably inside the conduit. The pre-loading may consist of dumped soil or even heavy equipment that should be left in the pipe for at least 12 h after the first pour. As discussed later, it is necessary to cast the CLSM in at least three lifts poured at least 24 h apart.

The backfill around the metallic shell performs two distinct roles. In one role, it is responsible for a very large portion of the load that the shell is called upon to sustain, and in the other it provides the necessary support to the shell to enable it to sustain the induced thrust. In the case of conventional compacted backfill, a layer of loose soil simply adds weight to the shell; it is able to perform the latter role only after it has been compacted and overlaid by subsequent layers. Since the backfill is compacted in thin layers, the load-sustenance aspect of the backfill does not lag far behind a load-inducing one. In this case, it is assumed, quite justifiably, that the backfill performs the two roles simultaneously.

The above, however, is not true for the CLSM, which loads the shell as soon as it is poured, but takes many hours to set and to develop significant stiffness to support the conduit wall. It is suggested that there are at least three critical stages governing the CLSM pouring sequence in soil-steel bridges. The first lift should be poured up to the springline, which is the level where the pipe cross-section is at its widest. The backfill for this lift lies partly under the pipe thus exerting buoyancy forces. Until the fill in this lift has attained sufficient stiffness, the pipe should be kept weighted down by the temporary dead load.

The second lift should be poured up to the level of the crown. Until the backfill reaches the crown, the load effects induced in the conduit wall by the dead load are not very high. However, the fill placed above the crown begins to induce high load effects. It is for this reason that the CLSM fill above the crown should be poured only after the backfill in the lower lifts has gained sufficient stiffness to assist the metallic shell to sustain the additional thrust.

From the standpoint of economy it may be desirable to limit the next i.e. the third, pour up to the minimum depth of cover required by the design code. Beyond the level corresponding to the minimum depth of cover, any kind of backfill is permitted by some authorities. In some cases, it may be found more convenient and economical, however, to continue to use CLSM even for the backfill at higher levels.

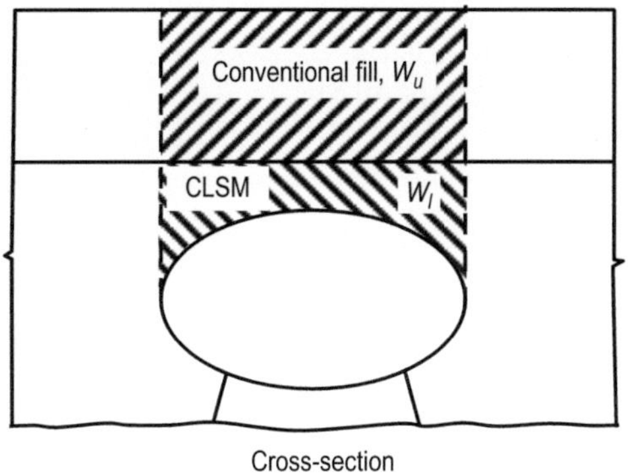

Fig. 7.27 Definition of terms used in conjunction with Eq. (7.28)

In soil-steel bridges with conventional compacted fills, the dead load effects in the metallic shell are obtained by taking account of the inherent stiffness of the backfill that leads to the arching action mentioned above. As discussed earlier, a lift of fluid CLSM induces load effects in the metallic shell long before it can affect the load distribution; consequently, it cannot participate in the arching action related to load effects induced by its own weight. It can be appreciated that the current methods of calculating dead load effects, e.g. those noted earlier, are not suitable for soil-steel bridges with CLSM. For these bridges, new methods should be developed by taking account of the relevant construction stages.

Abdel-Sayed et al. (Abdel-Sayed et al. 1993) have suggested that the dead-load thrust, T_D in the conduit wall for the case under consideration can be obtained by adapting Eq. (7.2) as follows, provided that the foundation of the structure is relatively unyielding:

$$T_D = 0.5W_l + 0.5(1.0 - 0.1C_s)A_f W_u \tag{7.29}$$

in which, as shown in Fig. 7.27, W_l is the weight per unit length of the CLSM fill directly above the conduit and W_u is the weight per unit length of the rest of the fill directly above the conduit; the other notation is as defined in conjunction with Eq. (7.2).

7.10 Examples of Recent Structures

All examples of soil-steel structures given in this chapter so far are those of Canadian structures made with plates having shallow corrugations. During the past decades or so, soil-steel bridges are also being used extensively in Australia, Korea and several European countries (Mattsson and Sundquist 2007). Three examples of soil-steel structures made with deep corrugations are given in the following.

Fig. 7.28 The Stockton soil-steel bridge in UK (Photo courtesy of Atlantic Industries Ltd., Canada)

7.10.1 A Soil-Steel Bridge in the UK

The Stockton soil-steel bridge (Fig. 7.28) was constructed in 2004 as a railway underpass. The structure, made with plates having deep corrugations, has two conduits each with a span of 15.75 m and with length of 35 m. The parent 7 mm thick plate is stiffened with 7 mm thick and 762 mm wide plates at a centre-to-centre spacing of 1524 mm. The stiffening plates are placed on the parent plate in ridge-over-ridge pattern, which pattern can be seen in Fig. 7.17. The voids between the parent and stiffening plates are filled with concrete. Composite action between the two plates is ensured by 12 mm dia. and 80 mm long shear studs welded to both of the plates in the voids at a spacing of 813 mm. The metallic arches with a reentrant angle 8.36° rest on concrete spread footings. The 1.8 m gap between the two conduits is partly filled with CLSM.

It is significant to note that the Stockton structure was designed by the CHBDC draft provisions, which are presented in this chapter.

7.10.2 An Animal Overpass in Poland

Poland, experiencing fast growth in its economy, has a resulting fast growth in the building of its roads. The authorities in Poland have become aware that the fast growth of its highways has had an adverse effect on the wildlife on the both sides of these highways (Wysokowsky et al. 2007). To minimize adverse impact on wildlife, several animal underpasses and overpasses are being built across new highways. One aesthetically pleasing example of an overpass for large animals is the four-conduit structure across A2 Motorway in Poland. The structure, described by Bednarek and Czerpak (2007), can be seen in Fig. 7.29.

Each of the two middle conduits has a span of 17.7 m and a rise of 5.5 m. The metallic shell in these conduits comprises 7 mm thick Super●Cor® parent plates stiffened continuously over their length and circumference by 7 mm thick plates in a ridge-over-ridge pattern.

Fig. 7.29 An animal overpass over A2 Motorway in Poland (Photo, courtesy of ViaCon Group, Poland)

Each of the two side spans has a span and rise of 9.36 and 8.1 m, respectively. The metallic shell in these two conduits is made with 7 mm thick plates having a 200 × 55 corrugation profile.

As noted by Bednarek and Czerepak (2007), the top of the overpass is provided with wooden fences so that the animals on the overpass are not distracted by the sight and the noise of the traffic.

7.10.3 A Bridge for a Mining Road in Alberta, Canada

The soil-steel structure under consideration is on the Cheviot Mine Haul Road, crossing the Whitehorse Creek in Northern Alberta, Canada; it is a semi-circular arch structure with a span and rise of 24 and nearly 12 m, respectively. The shell comprises a 7.11 mm thick Super●Cor® steel plate stiffened continuously with a similar plate in a ridge-over-ridge pattern with voids between the two plates installed with short shear connectors and filled with 30 MPa concrete.

The metallic shell of the structure can be seen in Fig. 7.30 during the early stages of construction. It is emphasized that to date, the Whitehorse Creek soil-steel structure holds the record for having the largest span for a soil-steel structure. The conduit of the structure is 30 m long.

The top of the backfill above the conduit slopes in the transverse direction of the conduit, with a slope of about 6 % going down towards the north. The backfill of the structure comprises mechanically-stabilized soil, known by the trade name of 'Atlantic Wire Wall'. The extent of the Atlantic Wire Wall at one end of the structure is shown in Fig. 7.31. The fill above the stabilized soil is regular well-compacted granular soil.

The structure routinely carries mining equipment that is about 15 times as heavy as the heaviest highway vehicles.

Fig. 7.30 The metallic shell of the Whitehorse Creek soil-steel structure during early stages of assembly (Photo courtesy of Atlantic Industries Ltd., Canada)

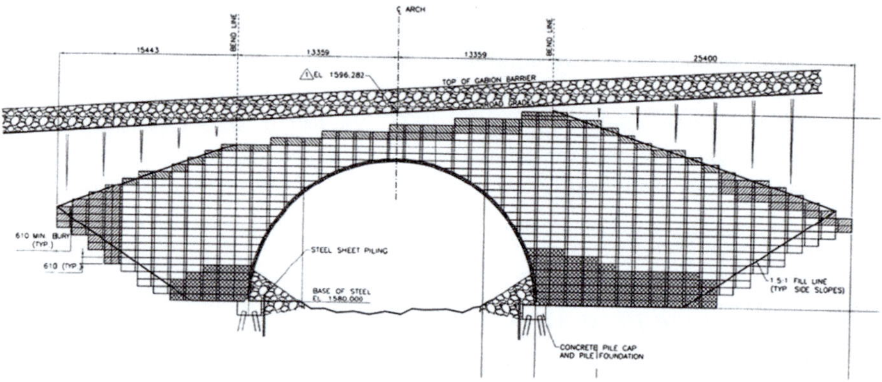

Fig. 7.31 Extent of stabilized fill at one end of the structure (Drawing courtesy of Atlantic Industries Ltd., Canada)

References

Abdel-Sayed G, Bakht B (1982) Analysis of live load effects in soil-steel bridges, Transportation research record no. 878. Transportation Research Board, Washington, DC

Abdel-Sayed G, Bakht B, Selig ET (1992) Soil-steel structure design – Third Edition of OHBDC. Can J Civ Eng 19(4):545–550

Abdel-Sayed G, Bakht B, Jaeger LG (1993) Soil-steel bridges. McGraw-Hill, New York

ASTM A796/A796M-06 (2003) Standard practice for structural design of corrugated steel pipe, pipe-arches and arches for storm and sanitary sewers and other buried applications. American Society for Testing Materials (ASTM) International, West Conshohocken, Pennsylvania

ASTM D2487 (2006) Standard practice for classification of soils for engineering purposes (Unified soil classification system). American Society for Testing Materials (ASTM) International, West Conshohocken, Pennsylvania

ASTM D698-00ae1 (2007) Standard test methods for laboratory compaction characteristics of soil using standard effort (12,400 ft-lbf/ft3 (600 kN-m/m3)). American Society for Testing Materials (ASTM) International, West Conshohocken, Pennsylvania

Bakht B (1981) Soil-steel structure response to live loads. ASCE J Geotech Eng 107(6):779–798

Bakht B (1985) Live load response of a soil-steel structure with a relieving slab, Transportation research record no. 1008. Transportation Research Board, Washington, DC

Bakht B (2007) Evolution of the design methods for soil-metal structures in Canada. Archives of Institute of Civil Engineers, Poland. Buried Flex Steel Struct 1:7–22

Bakht B, Agrawal AC (1988) On distress in pipe-arches. Can J Civ Eng 15(4):589–595

Bednarek B, Czerpak A (2007) Animal crossing built over A2 motorway in Poland. Archives of Institute of Civil Engineers, Poland. Buried Flex Steel Struct 1:45–51

Brewer WE (1990) The design and construction of culverts using controlled low strength material-controlled density fill (CLSM-CDF) backfill. Structural performance of flexible pipes. Bolkema, Rotterdam

Brewer WE, Hurd JO (1991) Economic considerations when using controlled low strength material (CLSM-CDF) as backfill. Paper no. 91-0309, Transportation Research Board Annual Meeting, Transportation research record no. 1315, Washington, DC, pp 28–37

CHBDC (2000) CAN/CSA-S6-00, Canadian highway bridge design code. Canadian Standards Association, Toronto

CHBDC (2006) CAN/CSA-S6-06, Canadian highway bridge design code. Canadian Standards Association, Toronto

Clem DA, Hook W (1992) How ready mixed concrete producer views flowable fill. Transportation Research Board Annual Meeting, Washington, DC

Essery D, Williams K (2007) Buried flexible steel structures with wire mesh reinforcement for cut plates. Archives of Institute of Civil Engineers, Poland. Buried Flex Steel Struct 1:65–79

Girges YF (1993) Three-dimensional analysis of composite soil-steel structures. Ph.D. thesis, Department of Civil Engineering, University of Windsor, Windsor

Hartley JD, Duncan JM (1987) *E'* and its variation with depth. ASCE J Transp Eng 113(5):538–553

Lee JK, Ho Choi D, Yang Yoon T (2007) Seam strength corrugated plate with high strength steel. Archives of Institute of Civil Engineers, Poland. Buried Flex Steel Struct 1:129–143

Mattsson L, Sundquist H (2007) The real service life and repair methods of steel pipe culverts in Sweden. Archives of Institute of Civil Engineers, Poland. Buried Flex Steel Struct 1:185–193

Mikhailovsky L, Kennedy DJL, Lee RWS (1992) Flexural behaviour of bolted joints of corrugated steel plates. Can J Civ Eng 19(5):896–905

Mufti AA, Bakht B, Jaeger LG (1989) Mechanics of behaviour of soil-steel structures. In: Proceedings of the annual conference of the Canadian Society for Civil Engineering, vol 1A, pp 130-150

OHBDC (1979, 1983, 1992) Ontario highway bridge design code. Ministry of Transportation of Ontario, Downsview

Vaslestad J, Kunecki B, Johansen TH (2007) Twenty one years of earth pressure measurements on buried flexible steel structure. Archives of Institute of Civil Engineers, Poland. Buried Flex Steel Struct 1:233–244

Wysokowsky A, Staszczuk A, Bednarek B (2007) Decrease of negative impact of transport infrastructure investments on natural migration of the wildlife. Archives of Institute of Civil Engineers, Poland. Buried Flex Steel Struct 1:287–295

Chapter 8
Fibre Reinforced Bridges

8.1 Introduction

8.1.1 General

In 1989, the authors of this book and their friend L.G. Jaeger (1926–2013) wrote a brief article inquiring if the time had come in Canada for the use of the advanced composite materials, now known as fibre reinforced polymers (FRPs), in civil structures (Mufti et al. 1989). The premise for the apparently affirmative answer to their query lay in the promise of the high durability of FRPs in concrete in corrosive environments, in which steel-reinforced concrete deteriorates rapidly and is thus not sustainable. Since the late 1980s, the considerable research into the use of FRPs in civil structures has led to centres of excellence, and several series of international conferences with voluminous proceedings recording a very large body of research into the subject.

The article by Mufti et al. (1989) was prompted by a provocative and futuristic paper by Professor Urs Meier, who had proposed a suspension bridge with a span of 8.4 km across the Strait of Gibraltar. Cables of even the strongest steel cannot sustain even their own dead load for such a large span. Professor Meier, then head of the Swiss Federal Laboratories for Materials Testing and Research (EMPA), had proposed the cables to be made of carbon fibre reinforced polymer. Intrigued by the paper, the authors and Leslie Jaeger visited the Swiss laboratories administered by Professor Meier at that time, and were soon converted by the extraordinary experimental evidence.

B. Bakht, A. Mufti, *Bridges*, DOI 10.1007/978-3-319-17843-1_8

During the past three decades, Urs Meier has made significant contributions to the application of FRPs in civil engineering structures. His contributions have been recognized internationally. In 2007, he was awarded a founding fellowship of the International Society for Structural Health Monitoring of Intelligent Infrastructure (ISHMII). Some time ago, the Royal Military College of Canada awarded him the honoris causa doctorate degree, a rare honour. The authors are very proud of their friendship with him.

Popular literature used to contain references to an ideal building material: a super-strong highly durable polymer, reported to have been used in aircraft and space structures. Is there such a building material? If there is, why not avoid the perpetual maintenance of bridges by building them with this ideal material rather than with the all-too-readily-corroding steel?

This chapter attempts to provide an answer to this question, and goes on to note that there are new and highly durable building materials that have already been, and are being, used in bridges not only for new construction but also for strengthening existing structures; the latter subject is dealt with in Chap. 9.

Arguably, steel has been the structural engineer's material of choice for a long time. It is only with the help of this material that man has been able to span larger distances than were possible with other conventional building materials, namely concrete, wood and masonry. Notwithstanding its admirable qualities, steel has one very significant shortcoming: This is its tendency, like those of other common metals, to revert to its natural oxide state. Because of steel's tendency to corrode readily, engineers have often wished for an ideal building material, which is at least as strong as steel but is also far more durable by being not subject to corrosion.

A misconception should first be addressed. There is no such known material as a super-strong polymer with strength similar to that of steel. There are a few synthetic fibres, e.g. carbon, aramid and glass that are even stronger in tension than some steels; however, they are weak in shear, so that for handling and placement they have to be embedded in filler materials that are relatively strong in shear but nowhere near as strong in tension as the fibres themselves.

The weakness of the synthetic fibres in shear is similar to that of a hemp rope, which can be severed easily by applying cyclically reversing shearing forces by hand.

Taken together, the filler materials and synthetic fibres mentioned above comprise fibre reinforced polymers (FRPs). Carbon fibre reinforced polymer (CFRP), aramid fibre reinforced polymer (AFRP) and glass fibre reinforced polymer (GFRP) have already been used in structural applications. A great variety of FRPs exists within each of these three categories. The least expensive and most-readily available of the three FRPs is GFRP.

Low-modulus fibres, such as polypropylene and nylon, are mixed with concrete to produce fibre reinforced concrete (FRC). It is important to note that the addition of low-modulus fibre increases the tensile strength of concrete only marginally, but also provides resistance to the cracking, which would otherwise occur due to volumetric changes in concrete.

8.1.2 Definitions

The various terms used specifically in conjunction with the use of FRP and FRC in bridges are defined in the following.

Bar – a non-prestressed FRP element, with a nearly rectangular or circular cross-section, used to reinforce a structural concrete component.
Continuous Fibres – aligned fibres whose individual lengths are significantly greater than 15 times the critical fibre length.
Critical Fibre Length – the minimum length required to develop the full tensile strength of a fibre in a matrix.
Fibres – small diameter filaments of materials of relatively high strength, being aramid, carbon, glass low modulus polymer or steel.
Fibre Reinforced Polymer – a fibre reinforced composite material with a polymeric matrix and continuous fibres.
Fibre Reinforced Concrete – Fibre-reinforced composite in which the matrix is Portland cement concrete or mortar, and in which the fibres are discontinuous and randomly distributed.
Fibre Volume Fraction – the ratio of the volume of fibre to the volume of the fibre reinforced composite.
Matrix – a term used commonly for the continuous material in a fibre-reinforced composite component which contains aligned or randomly distributed fibres.
PAN-Based Carbon Fibres – carbon fibres derived from acrylic nitrile.
Pitch-Based Carbon Fibres – carbon fibres derived from petroleum or coal-based products.
Randomly Distributed Fibre Reinforcement – discontinuous fibres distributed uniformly, but randomly, in a matrix.
Rope – an assembly of bundled continuous fibres.
Sheath – a protective and effectively continuous cover for a bar, rope or tendon, bonded.
Tendon – a bar or rope used to impart prestress to structural components.

8.1.3 Abbreviations

The various abbreviations used in conjunction with the new building materials and their design methods are as follows:

AFRP Aramid fibre reinforced polymer
CFRC Carbon fibre reinforced concrete

CFRP	Carbon fibre reinforced polymer
CSCE	Canadian Society for Civil Engineering
FRC	Fibre reinforced concrete
FRP	Fibre reinforced polymer
GFRP	Glass fibre reinforced polymer
ISIS	Intelligent Sensing for Innovative Structures
NSMR	Near surface mounted reinforcement
PAN	Polyacrylonitrile
SLS	Serviceability limit state
TSC	Technical subcommittee
ULS	Ultimate limit state

8.1.4 Scope of the Chapter

This chapter provides brief information about the use of FRPs for new construction in bridges. For historical information, the reader is referred to two books edited by Mufti et al. (1991, 1992) the former referring to developments in Europe and the latter to those in Japan. ISIS Canada Manual (2001) provides useful design information.

8.2 Fibre Reinforced Polymer

The principal use of FRP in structures has been in the form of fibre reinforced polymer defined earlier. In this section, the use of fibres in the form of FRP components is discussed.

8.2.1 Structural Properties of Fibres

Stress-strain relationships of carbon, aramid and glass fibres are compared in Fig. 8.1 with that of high strength steel. It can be seen that all these fibres are stronger than steel; the modulus of elasticity, quantified by the slope of the stress-strain curves, of carbon fibres is similar to that of steel, whilst those of aramid and glass fibres are smaller. Because the moduli of elasticity of these fibres are of the same order of magnitude as that of steel, they are also referred to as *high modulus* fibres. There are other fibres which are significantly weaker and softer. Examples of these latter fibres, which are often referred to as *low modulus fibres*, are polypropylene and nylon; their stress-strain characteristics are also shown in Fig. 8.1. It can be seen in this figure that all both high- and low-modulus fibres fail suddenly, and their behaviour remains essentially elastic up to failure.

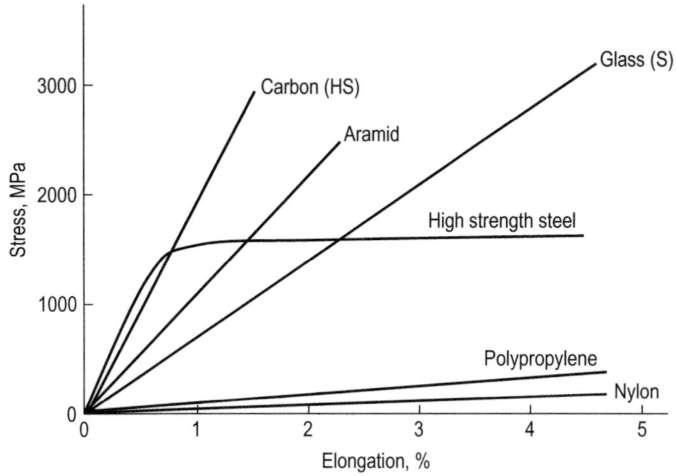

Fig. 8.1 Stress-strain relationships of various fibres compared with that of steel

Figure 8.1 should be regarded as only schematic. In fact a variety of fibres having different properties exists within each category. For example, in the general category of carbon fibre there are PAN- and PITCH-based fibres, which have been defined earlier, with further sub-divisions in each of these sub-categories. Within each specific category of fibres, there exists a statistical distribution of structural properties. Representative values of the relevant structural properties of the various fibres are listed in Table 8.1.

For comparison with the properties listed in Table 8.1, it may be noted that the unit weight, tensile strength and modulus of elasticity of high strength steel are approximately 7870 kg/m^3, 1800 MPa and 200 GPa, respectively.

8.2.2 Design Considerations

The physical properties and their methods of determination for steel and concrete are well-established and understood, so that a good designer can account for them in the design process almost instinctively. Such is not the case for FRPs, for which the current test methods, due to lack of standardization, may lead to misleading conclusions. Further, the strength of structural components with FRP is dependent upon: (a) load duration, (b) stress level, (c) load history, (d) temperature, and (e) moisture content. In 2010, the Canadian Standards Association issued a set of specifications (CSA-S807 2010) in which the FRPs are classified on the basis of their fibres, strength, stiffness and durability; this document is useful to confirm that the processes of manufacturing FRPs are standardised and could be certified.

It is important to realize that there is an interdependence of these factors, so that the effect of one factor cannot always be determined in isolation whilst holding the other factors at a constant level. As an example, it is noted that after a sustained

Table 8.1 Typical properties of various fibres

High/low modulus fibres	General category	Specific category	Unit weight kg/m^3	Tensile strength (MPa)	Modulus of elasticity (GPa)
High	Carbon	High strength carbon	1720	2800	180
		Graphite	1400	1700	250
		PAN Fibres	1400	3800	227
		PITCH-Based Fibres	1400	590	30
	Aramid	Kevlar 49	1450	2700–3500	120
		Kevlar	1450	2900	60–130
	Glass	E-Glass	2500	1500–2500	70
		S-Glass	2500	4800	86
Low	Polypropylene		1150	550–690	4
	Nylon		1150	100	3

loading of 30 years at a certain stress level in one particular GFRP under dry conditions the strength could be about, say, 95 % of the short term strength. For the same stress level and duration, the strength under wet conditions may be reduced by up to 50 %.

An approach to establishing the properties of FRP components may be to derive them through creep tests rather than the conventional stress-strain type of tests. Creep-rupture strength and creep modulus can then be derived from the logarithmic curves of properties plotted against time for different ranges of the various factors.

From the above discussion, it is obvious that the nominal strengths of FRPs are dependent upon their usage; thus, for example, the strengths developed for indoor applications of permanently imposed dead loads may not be directly relevant to bridges. It is important that the nominal strengths of FRPs intended to be used in bridge design be expressly developed for that purpose.

The Canadian Highway Bridge Design Code (CHBDC, 2006) seems to be the only national design code that contains design provisions for the design and rehabilitation of bridge components with FRPs; the provisions for rehabilitation are discussed in Chap. 9.

8.2.3 *The Most Economical FRP*

As noted earlier, three fibres are currently used for making FRPs for structural applications, these being aramid, carbon and glass. Basalt fibres are also coming to the fore, but these fibre are only a subset of glass fibres. There is little doubt that for providing the same strength, or stiffness, glass fibre reinforced polymer (GFRP) is the most economical and easily available FRP. Until recently, experts were not in full agreement about whether GFRP reinforcement in tension was stable in the alkaline environment of concrete. The opinions of these experts, who considered the GFRP to be unstable in the concrete, were generally based on lab studies, in which the effect of alkalis generated in concrete on glass fibres in the GFRP was simulated by immersing the GFRP in alkaline solutions. For example, Sen et al. (2002) reported that GFRP bars in a strong alkaline solution, and stressed to 25 % of their failure loads failed in 15–25 days, and most of the bars stressed to 15 % of the failure load failed in 42–173 days. As a part of a research project on studying the feasibility of restraining bridge deck slabs with pretensioned concrete straps, several straps, with the GFRP tendons stressed to 55 % of the 5th percentile tensile strength of their individual ultimate strength, were built in the University of Manitoba (Banthia 2003). Three of these specimens were already 365 days old when it was checked that the tendons were holding their strains. In a discussion paper, Mufti et al. (2003) wondered why these specimens had not failed. Could it be that the oxygen ions, necessary for the chemical reaction between glass and alkali are more freely available in the alkaline solution than in the hardened concrete?

To remove any doubt about the durability of GFRP in concrete exposed to natural environment, an extensive study was undertaken several years ago under the sponsorship of ISIS Canada. Nine cores were taken from each of five GFRP-reinforced concrete bridges in Canada, which were built 6–8 years before the removal of the cores. Three cores from each bridge were given to each of three teams of material scientists and experts in durability, for microscopic and chemical analyses. The findings from these analyses have confirmed that the concerns about the durability of GFRP in alkaline concrete, based on simulated laboratory studies in alkaline solutions, are unfounded (Mufti et al. 2007); a micrograph from this reference is reproduced in Fig. 8.2, in which it can be seen that the glass fibres of the GFRP are intact despite being in concrete exposed to the elements for about 8 years.

In its first edition, the Canadian Highway Bridge Design Code (CHBDC 2000) permitted GFRP only as secondary reinforcement. As a result of the durability study noted above, the second edition of the CHBDC (2006) permits the use of GFRP as primary reinforcement and prestressing tendons in concrete components. The maximum stresses in the GFRP, however, are not permitted to exceed 25 % of its ultimate strength.

8.3 Fibre Reinforced Concrete

The synthetic fibres whose use in FRP has been described in Sect. 8.2 may also be mixed randomly in chopped form in concrete. The concrete thus formed is referred to as fibre reinforced concrete (FRC). Synthetic fibres in cementitious matrices comprising cement may be used for either or both of crack control and improving the energy absorption characteristics of concrete.

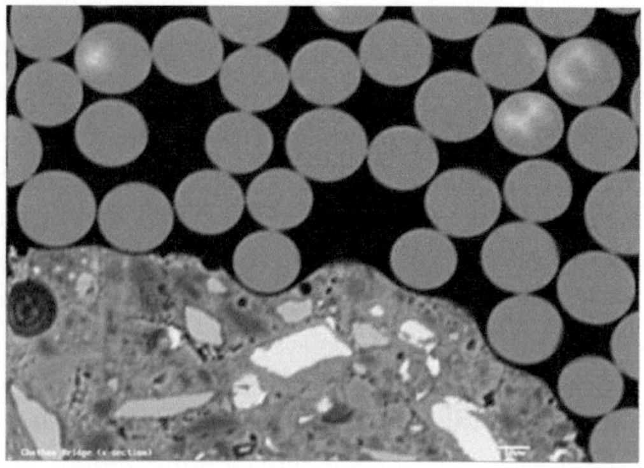

Fig. 8.2 Micrograph of a portion of cross-section of GFRP removed from an 8-year old bridge

8.3.1 FRC with Low Modulus Fibres

Low modulus fibres, such as polypropylene and nylon fibres, are used only for the control of cracks in concrete due to volumetric changes. One example of the use of FRC with low moduli fibres is the externally restrained deck slab described in Chap. 4. Banthia and Batchelor (1991) have noted that the main advantages of the polypropylene fibres are its chemical inertness, its light weight and its ability not to absorb moisture.

8.3.2 FRC with High Modulus Fibres

Of the high modulus fibres, only carbon fibre is recommended to be used in FRC. There is a lack of published data on the use of aramid fibres in FRC, and glass fibre is believed by some experts to be unstable in concrete because of its reaction with the alkalies that are produced when cement paste interacts with moisture (Sen and Issa 1992).

Carbon fibre is approved by the CHBDC (2006) for use in FRC. Apart from its high first cost there is no other disadvantage that should be a cause for concern in the use of carbon FRC in bridge applications.

Recommended Usages It is recommended that FRC with different fibres be restricted to certain specific applications, the recommended applications being those summarized in Table 8.2. This table also contains the maximum and minimum permitted values of the fibre volume fraction V_f along with the recommended values of this fraction.

Most of the fibres can be mixed with concrete with the help of a conventional mixer; however, carbon fibres with volume fractions greater than about 3.0 % require the use of special-purpose mixers.

Table 8.2 Recommended uses and fibre volume fractions of FRC

Fibre	Recommended applications	Fibre volume fraction, V_f x 100 (%)		
		Minimum	Maximum	Recommended
Carbon	1 stay-in-place formwork for deck slabs 2 repairs including patching	1.0	3.0	2.0
Steel	1 Control of thermal and shrinkage cracks in deck slabs and deck slab overlays 2 repairs including patching	0.75	1.5	1.0
Low modulus polymers	1 control of thermal and shrinkage cracks in deck slabs, deck slab overlays and barrier walls	0.5	1.5	1.0

Fibres should be mixed uniformly in the concrete. If the dispersion is not uniform, the compressive strength of the concrete will drop. Because of this, the uniformity of fibre dispersion in concrete can, in some cases, be confirmed by conducting compressive strength tests on FRC and corresponding plain concrete; a reduction in the compressive strength of FRC over the corresponding plain concrete specimens is an indication of non-uniform dispersion of the fibres. ASTM C 1399 has specified a test method for determining the effect of fibres on the post-cracking strength of FRC; this method, which has also been specified by the CHBDC (2006), is discussed in Sect. 8.5.4.

8.4 Earlier Case Histories

Until late 1980s, the use of FRPs in civil structures was non-existent in Canada. Realizing this situation, several Canadian structural engineers requested their learned society, the Canadian Society for Civil Engineering (CSCE), to establish a committee on the use of FRPs in bridges and other structures. The CSCE initiative was supported by the Federal Government of Canada in early 1990s to finance the ACMBS (Advanced Composite Materials in Bridges and Structures) Network.

The CSCE initiative eventually led to ISIS (Intelligent Sensing for Innovative Structures) Canada. This network of Centre of Excellence, founded in 1995 in 2007 incorporated 33 researchers and 256 students from 15 Canadian universities. In a relatively short period of 12 years, ISIS Canada had been responsible for about 40 demonstration and research projects incorporating FRPs and FRC. The network has brought the use of FRPs and structural health monitoring to the forefront of Canadian civil structures.

In an effort to learn first-hand about the use of FRPs in bridges, a delegation from the CSCE visited several countries in Western Europe in 1990, and another such delegation visited Japan in 1992. The authors were members of both of these delegations, with the second author of this book being the leader of both the delegations. Bridges incorporating FRPs in Europe and Japan, which were seen by the authors, are described in this section, along with the first fibre reinforced bridge in Canada.

Since the construction of the pioneering fibre reinforced bridges described in this section, a very large number of such bridges have since been built around the world.

Information about the use of FRPs and FRC in civil structures can be found in the proceedings of the several conferences. Bank and Teng (2007) have published nearly 150 photographs of construction and research projects involving the use of FRPs in civil structures.

8.4.1 Bridges in Germany

Germany took a leading role in the use of FRP in large highway bridges. The Ullenbergstrasse Bridge in Düsseldorf is the first example of such application. Opened to traffic in 1986, this two-span post-tensioned slab bridge, with spans of 21.3 and 25.6 m respectively, has been post-tensioned with GRFP tendons.

In appearance, the Ullenbergstrasse Bridge, a view of which is shown in Fig. 8.3, looks like any other post-tensioned slab bridge. However, there are several aspects of this bridge, which are of great interest to structural engineers. Two of these aspects are discussed below.

The technique for attachment of anchorages to GFRP tendons is neither as straightforward nor as well-developed as it is in the case of steel tendons. Because of this, the anchorages are sometimes attached to the tendons in the factory under controlled conditions. The tendons are already encased in the ducts before the anchorages are fixed at both ends of the carefully measured length of the tendons. An assembly of the tendons, already-encased in ducts and with anchors at both ends, can be seen in Fig. 8.4 ready to be delivered for placement at the bridge site. Similar assemblies of tendons were installed in the formwork of the Ullenbergstrasse Bridge before the concrete was poured.

The techniques for attaching anchorages to the GFRP tendons, being still in the evolution stage, are far from perfect. Some tendons are known to have slipped out of their anchorages; others broke prematurely due to nicks left by the peeling of the

Fig. 8.3 The Ullenbergstrasse Bridge in Germany

Fig. 8.4 Assembly of FRP tendons with anchors, ready to be transported to the site

Fig. 8.5 The Marienfelde Bridge in Germany

sheaths with the help of a knife. In the case of an internal prestressing system, the replacement of a broken tendon is not a practical proposition. It is interesting to note that in the Ullenbergstrasse Bridge with the internal prestressing system, a fairly large number of spare ducts are provided; these ducts were empty but could be used to house conventional prestressing tendons in the case of unforeseen damage to the experimental GFRP tendons.

The modulus of elasticity of GFRP is low, leading to a strain as high as 2 % at the stress level during initial prestressing. Such high strains clearly lead to very large elongation of the prestressing tendons, which should be accommodated by significantly long anchorages and enough space behind the anchors. The Ullenbergstrasse Bridge is provided with quite large chambers at each end. The access to each chamber is through a manhole located on a sidewalk over the bridge.

The Marienfelde Bridge is another post-tensioned concrete bridge in Germany, which incorporates GFRP tendons; this is a two-span pedestrian bridge with twin girder construction having spans of 17.6 and 23.0 m, respectively. A view of this bridge can be seen in Fig. 8.5. Unlike the previous bridge, the Marienfelde Bridge has been post-tensioned by an external prestressing system, which is located in the space between the two girders. Mainly for aesthetic reasons, this space has been covered by timber planks.

The advantage of the external prestressing system in the context of FRP tendons is that an assembly of tendons and pre-attached anchors can be replaced relatively easily in the event of distress to the original tendons.

The Marienfelde Bridge has also been provided with fairly large chambers at its ends, not only to accommodate large elongation of the tendons during initial stressing but also to facilitate the monitoring of the instruments, which have been installed on the bridge. Entrance to these chambers is provided through doors at the lower level. One such entrance can be seen in Fig. 8.5.

It is important to note that other bridges similar to the post-tensioned bridges described above have not been built in Germany, the main reason being the very high initial cost of these structures. There is little information available in the technical literature about the performance of these bridges since their construction.

8.4.2 Bridges in Japan

Despite the fact that Japan probably led the world in the manufacture of AFRPs and CFRPs in early 1990s, all of its experimental bridges incorporating FRPs, except one, have short spans. Further, most of these bridges carry either pedestrian or bicycle traffic. The caution exercised by prudent Japanese engineers clearly indicates that there was still some uncertainty regarding the long-term and in-service performance of the new materials.

The authors had a chance of seeing two bridges in Japan built with FRPs. Both were demonstration off-highway concrete bridges prestressed with AFRP produced by the Sumitomo Corporation of Japan. One bridge has a slab-on-girder type of construction in which the girders are pre-tensioned and the other is a post-tensioned single box girder bridge.

The pretensioned slab-on-girder bridge has a clear span of 11.79 m and a width of 4.60 m; it comprises three girders and a deck slab that is haunched transversely over the girders. The post-tensioned bridge adjoins the pre-tensioned bridge and spans over a greater opening of 24.1 m; it consists of a 1900 mm deep and 2800 mm wide single box girder. This second bridge is shown in Fig. 8.6, from which it can be seen that the clearance of the superstructure from the ground is very small.

It is noted that seven other FRP bridges were constructed in Japan after the bridges described above. Five of these are concrete bridges pre-tensioned with CFRP tendons. In Japan, FRP-reinforced bridges are still not being built on a routine basis. The authors suspect that similarly to the situation in Germany, the very high initial cost of these bridges is the main impediment to the general acceptance of these bridges in Japan.

Fig. 8.6 A post-tensioned slab-on-girder bridge in Japan with FRP tendons

Fig. 8.7 An FRP pedestrian bridge in the USA

8.4.3 Bridges in North America

Considerable research has been conducted in both Canada and the USA in the structural use of FRP.

One organization in the USA, namely E.T. Techtonics, markets a novel prestressed truss beam bridge in which FRP trusses are prestressed by aramid fibre cables in a double king-post or queen-post type of arrangement. These bridges, an example of which can be seen in Fig. 8.7, are mainly for pedestrian use in golf courses and parks.

In 1993, Canada made its first highway bridge incorporating FRP. A view of the bridge, which was constructed in the City of Calgary in Alberta, can be seen in Fig. 8.8. It is a two-span continuous bridge with spans of 22.8 and 19.2 m, respectively, incorporating prestressed precast concrete girders of bulb-T section. Eight of the twenty-six girders of the bridge have been prestressed by two types of CFRP tendons manufactured in Japan (Rizkalla and Tadros 1994).

The 1993 draft of the Fibre Reinforced Structures section of the CHBDC required that:

Fig. 8.8 The Beddington Trail Bridge in Canada incorporating CFRP tendons

(a) A concrete beam with FRP prestressed reinforcement should contain enough non-prestressed reinforcement to enable it to sustain the unfactored dead loads; and
(b) The design of a post-tensioning system should be such that the tendons, if damaged during or after installation, could be replaced by similar or other tendons.

Gamil Tadros, one of the most well-known bridge design engineers of Canada, has designed a large number of bridges in Canada. He was the conceptual designer of the 13 km long Confederation Bridge which is regarded as the world's longest bridge over icy sea waters. He is the designer of the demonstration bridge in Calgary, which incorporates FRP and is called the Beddington Trail Bridge. Gamil Tadros is also a member of the CHBDC technical committee that wrote a set of design provisions for fibre reinforced structures, which was first published in 2000. He took heed of the two draft provisions noted above. To comply with the latter provision, he incorporated a novel scheme in the Beddington Trail Bridge by which the structure can be post-tensioned easily through an external system. The diaphragms between those girders, which incorporate FRP tendons, have been provided with adequate reinforcement and openings to accommodate external post-tensioning systems which may be, but are unlikely to be, needed in future.

8.5 Design Provisions

The first edition of the Canadian Highway Bridge Design Code (CHBDC 2000) contained design provisions for some fibre reinforced structures; the provisions were limited to only those applications in which the Technical Subcommittee

(TSC) responsible for the provisions had confidence, or had direct access to documents substantiating its performance. In particular, the provisions of the first edition of the CHBDC, drafted mainly by 1997, were limited to: fully or partially prestressed concrete beams and slabs, non-prestressed concrete beams, slabs and deck slabs, FRC deck slabs, stressed wood decks, and barrier walls.

Similarly to the TSC for the first edition, a new TSC was formulated for the second edition of the CHBDC with membership drawn from Canada and elsewhere. In particular, experts from the USA, Japan and Sweden were included in the TSC responsible for formulating the revised design provision, which are presented in this section along with their rationale. It is noted that the design provisions for both externally and externally restrained deck slabs are not given in the section, as these provisions have already been covered in Chap. 4.

8.5.1 Durability

In addition to requiring thermosetting polymers for FRPs bars and grids, when used as primary reinforcement in concrete, and for FRP tendons, the CHBDC (2006) also requires that matrices and/or the adhesives of FRP systems should have a wet glass transition temperature, T_{gw}, of less than 20 °C plus the maximum daily mean temperature as specified elsewhere in the code.

As discussed in Sect. 8.2.3, an extensive study of GFRP bars and grids taken from in-service structures has removed doubts about the durability of GFRP in concrete. Accordingly, the CHBDC (2006) permits GFRP in concrete as the main reinforcement. However, as noted earlier, the maximum stress in GFRP at the serviceability limit state (SLS) is limited to 25 % of its ultimate strength.

8.5.2 Cover to Reinforcement

Some researchers have expressed the opinion that, because of the high transverse coefficient of thermal expansion of FRP bars, the cover to these bars should be larger than that specified in the CHBDC (2006). Extensive analyses of cores removed from GFRP-reinforced structures have confirmed that structures with even smaller covers than specified in the code show no cracks despite being in service for 6–8 years (Mufti et al. 2007). As discussed by Bakht et al. (2004), the reason for the absence of cracks above FRP bars in concrete structures might be that during the setting of the concrete, the FRP bars are ‘locked’ into concrete at a higher temperature than they are likely to experience later.

Vogel (2005) has examined a number of concrete beams prestressed with GFRP and CFRP tendons with minimum cover and subjected to the thermal gradients expected in Canada; he notes that “the flexural specimens regularly monitored during the experimental program with a handheld microscope never revealed the

Table 8.3 Resistance factors for factory-produced FRPs

Application	ϕ_{FRP}
AFRP reinforcement in concrete and NSMR	0.60
AFRP in externally bonded applications	0.50
AFRP and aramid fibre rope tendons for concrete and timber components	0.55
CFRP reinforcement in concrete	0.75
CFRP in externally bonded applications and NSMR	0.75
CFRP tendons	0.75
GFRP reinforcement in concrete	0.50
GFRP in externally bonded applications and NSMR	0.65
GFRP tendons for concrete components	0.50
GFRP tendons for timber decks	0.65

presence of cracks within the cover." Aguiniga (2003) has also reported that even shallow covers over FRP bars do not lead to cracks in structures exposed to the environment.

The minimum clear cover to FRP bars is 35 mm with a construction tolerance of ± 10 mm.

8.5.3 Resistance Factors

The resistance factors specified in the CHBDC (2006) depend upon the condition of use and the method of manufacturing the FRPs. For factory-produced FRPs, in which the variability of the properties is relatively small because of the high degree of control in the manufacturing process, the resistance factors, ϕ_{FRP}, are as listed in Table 8.3.

The resistance factors for FRPs made in the field, as for the rehabilitation of structures discussed in Chap. 9, are specified to be 0.75 times the corresponding values in Table 8.3.

8.5.4 Fibre Reinforced Concrete

The effectiveness of fibre in controlling cracks is determined by the residual strength index, R_i, which according ASTM C 1399 (2004), is determined by performing two tests on a $100 \times 100 \times 350$ mm FRC beam with a simply supported span of 300 mm. In the first test, the beam is placed on a $12 \times 100 \times 350$ mm steel plate, and subjected to two knife edge loads, placed 100 mm from each support. The load is increased gradually until the crack in the beam is detected either audibly or visually. The steel plate is then removed, and the cracked beam is retested to failure under the same loading arrangement. The ratio of the failure load of the cracked beams and the load causing the beam to crack during the first test is R_i.

The CHBDC (2006) requires that R_i should not be less that 0.1, if the barrier wall or the externally restrained deck slab contains only one mesh of reinforcement. When these components contain two meshes of reinforcement, fibres are not required in the concrete.

8.5.5 Protective Measures

The CHBDC (2006) requires that exposed tendons and FRP strengthening systems that are deemed to be susceptible to damage by UV rays or moisture be protected accordingly. Also, where the externally bonded FRPs are susceptible to impact damage from vehicles, ice, and debris, consideration should be given to protecting the FRP systems. According to NCHRP (2004), "Protective coating is applied for aesthetic appeal or protection against impact, fire, ultra-violet and chemical exposure, moisture, and vandalism. FRP systems are usually durable to weather conditions, seawater, and many acids and chemicals. Mortar finish can provide protection against impact or fire. Weather-resistant paint of the family of urethane or fluorine or epoxide can provide protection against direct sunlight."

The CHBDC (2006) forbids direct contact between CFRP and metals, as the contact between carbon fibres and metals can lead to galvanic corrosion. The contact between carbon fibres and steel could, for example, be avoided by an isolation layer of an appropriate polymer.

8.5.6 Concrete Beams and Slabs

The two deformability requirements for FRP reinforced concrete beams and slabs are described in the following.

Minimum Flexural Resistance The factored resistance, M_r, is required to be at least 50 % greater than the cracking moment M_{cr}. This requirement may be waived if M_r is at least 50 % greater than M_f. If the design for the ultimate limit state (ULS) of the section is governed by FRP rupture, then the M_r is required to be greater than 1.5 M_f, where M_f = the factored moment at a section, N × mm.

The above requirement ensures that critical sections contain sufficient flexural reinforcement so that there is adequate reserve strength after the formation of initial cracks in concrete or the rupture of FRP. The rupture of FRP is allowed because some FRP bars have a very low modulus of elasticity, as a result of which the amount of reinforcement required for T-sections would become very large if it was controlled by compression failure.

Crack Control Reinforcement When the maximum tensile strain in FRP reinforcement under full service loads exceeds 0.0015, cross-sections of the component in maximum positive and negative moment regions are required to be so proportioned that the crack-width does not exceed 0.5 mm for members subject to aggressive environments and 0.7 mm for other members, where the crack width, w_{cr}, is given by:

$$w_{cr} = 2\frac{f_{FRP}}{E_{FRP}}\frac{h_2}{h_1}k_b\sqrt{d_c{}^2 + \left(\frac{s}{2}\right)^2} \tag{8.1}$$

The value of k_b in Eq. (8.1) is required to be determined experimentally, but in the absence of test data it may be taken as 0.8 for sand-coated bar and 1.0 for deformed FRP bars. In calculating d_c, the clear cover is assumed to be not greater than 50 mm. In Eq. (8.1), f_{FRP} is the stress in the tension FRP reinforcement in MPa, h_1 and h_2 are the distances from the centroid of tension reinforcement and the extreme flexural tension surface, respectively, to the neutral axis in mm, d_c is the distance from the centroid of the tension reinforcement to the extreme tension surface of the concrete in mm, s is the spacing of shear or tensile reinforcement in mm, and the other notation is either defined earlier or discussed in the following.

As noted in ACI 440 (2002), a modified version of crack width equation by Frosch (1999) was used and the bond parameter k_b was recalibrated. The value of 2 in the equation was used for predicting the maximum crack width. A value of 1.5 can be used for the mean crack width and 1.0 for the minimum crack width. The value of k_b ranges from 0.60 to 1.72 with a mean value of 1.10 and a standard deviation of 0.3. El-Salakawy et al. (2003) recommend a value for k_b of 0.8 for sand-coated bars and 1.0 for deformed FRP bars.

It is required that maximum stress in FRP bars or grids under loads at SLS be not greater than $F_{SLS}f_{FRPu}$, where F_{SLS} is as given in Table 8.4.

Design for Shear For concrete beams reinforced with steel or FRP longitudinal reinforcement and steel or FRP stirrups, the factored shear resistance, V_r, is required to be computed from:

$$V_r = V_c + V_{st} + V_p \tag{8.2}$$

where V_c, V_{st} and V_p are factored shear resistance provided by concrete, stirrups and tendons, if present, respectively. The shear contribution V_{st} is denoted as V_s if the stirrups are of steel and by V_{FRP} if they are of FRP.

The contribution of V_c, V_s and V_p are calculated according to the standard practice prescribed in the concrete section of the CHBDC (2006), except as follows:

Table 8.4 Values of F_{SLS}

F_{SLS} for AFRP	0.35
F_{SLS} for CFRP	0.65
F_{SLS} for GFRP	0.25

(a) The following equation is used for calculating V_c

$$V_c = 2.5\beta\phi_c f_{cr} b_v d_{long}\sqrt{\frac{E_{long}}{E_S}} \tag{8.3}$$

(b) The following equation is used for calculating ε_x.

$$\varepsilon_x = \frac{\frac{M_f}{d_{long}} + V_f - V_p + 0.5N_f - \left(A_{FRP} f_{po} \;\; or \;\; A_p f_{po}\right)}{2\left[E_s A_s + \left(E_p A_p \;\; or \;\; E_{FRP} A_{FRP}\right)\right]} \leq 0.003 \tag{8.4}$$

(c) For the factored shear resistance carried by FRP shear reinforcement, V_{FRP}, the following equation is used. For components with transverse reinforcement perpendicular to the longitudinal axis, V_{FRP} is calculated from:

$$V_{FRP} = \frac{\phi_{FRP} A_v \sigma_v d_{long} cot\theta}{S} \tag{8.5}$$

When the transverse reinforcement is inclined at an angle θ to the longitudinal axis, V_{FRP} is calculated from:

$$V_{FRP} = \frac{\phi_{FRP} A_v \sigma_v d_{long}(cot\theta + cot\alpha)sin\alpha}{S} \tag{8.6}$$

where in Eqs. (8.5) and (8.6), θ is obtained by conventional methods, the resistance factor, ϕ_{FRP}, is as given in Table 8.4, and σ_v is the smaller of the values obtained from the following two equations:

$$\sigma_v = \frac{\left(0.05\frac{r}{d_s} + 0.3\right) f_{FRPbend}}{1.5} \tag{8.7}$$

$$\sigma_v = E_{vFRP}\varepsilon_v \tag{8.8}$$

in which, ε_v is obtained from:

$$\varepsilon_v = 0.0001\left(f'_c \frac{\rho_s E_{FRP}}{\rho_{vFRP} E_{vFRP}}\right)^{0.5}\left\{1 + 2\left(\frac{\sigma_N}{f'_c}\right)\right\} \leq 0.0025 \tag{8.9}$$

(d) The minimum amount of shear reinforcement, A_{vmin}, is calculated from:

$$A_{vmin} = 0.06\sqrt{f'_c}^{\frac{b_w S}{\sigma_v}} \tag{8.10}$$

where σ_v is calculated by Eq. (8.7).

It is well-known that the shear carried by concrete is smaller in FRP reinforced concrete beams than in beams reinforced with a comparable amount of steel. Tariq and Newhook (2003) have listed different equations for shear carried by concrete in FRP beams. The majority of researchers conclude that the shear carried by concrete in FRP reinforced beams is $(E_{FRP}/E_s)^n$ times the shear carried by concrete in steel reinforced beams. Usually n is taken as 1/2 or 1/3. Other researchers simply assume that the shear carried by concrete in FRP reinforced beams is half that carried by concrete in steel reinforced beams.

The equation for shear capacity, (8.2), is based on the work of Machida (1996). The equation for ε_v, (8.9), is as specified in the JSCE design recommendations (1997).

The equations for the calculation of shear capacity follow the procedure given in the concrete section of the CHBDC (2006) for concrete reinforced with steel bars.

The limit on the longitudinal strain in FRP stirrups is increased from 0.002 in the CHBDC (2000) to 0.0025 in CHBDC (2006) to reflect the finding that aggregate interlock can exist up to a strain of 0.003 (Priestly et al. 1997). The stress in FRP stirrups depends on the strength of the straight portion of a bent stirrup. For bent bars, the test method is specified in CSA S806 (2002).

The equation for minimum shear reinforcement for FRP reinforced beams is based on the work of Shehata (1999).

References

ACI 440.2R-02 (2002) Guide for the design and construction of externally bonded FRP Systems for strengthening concrete structures: 45. American Concrete Institute, Farmington Hills

Aguiniga G (2003) Characterization of design parameters for fiber reinforced polymer composite reinforced concrete systems. Ph.D. thesis, Texas A&M University, Austin

ASTM C1399-04 (2004) Test method for obtaining residual strength of fiber-reinforced concrete. West Conshohocken, Pennsylvania

Bakht B, Mufti A, Tadros G (2004) Discussion of fibre-reinforced polymer composite bars for the concrete deck slab of Wotton Bridge. Can J Civ Eng 30(3):530–531

Bank LC, Teng JG (2007) FRP photo competition 2005. International Institute for FRP in Construction, Hong Kong Polytechnic University

Banthia V (2003) Transverse confinement of steel-free deck slabs by concrete straps. Masters thesis, Department of Civil Engineering, University of Manitoba, Winnipeg

Banthia N, deV Batchelor B (1991) Chapter entitled material properties of fibre reinforced concrete in Advanced Composite Materials. Can Soc Civ Eng

CHBDC (2000) Canadian highway bridge design code, CAN/CSA-S6-00. Canadian Standards Association, Toronto

CHBDC (2006) Canadian highway bridge design code, CAN/CSA-S6-06. Canadian Standards Association, Toronto

CSA-S806 (2002) Design and construction of building components with fibre-reinforced polymers. Canadian Standard Association, Toronto

CSA-S807 (2010) Specifications for fibre reinforced polymers. Canadian Standard Association, Toronto

El-Salakawy E, Benmokrane B, Desgagné G (2003) Fibre-reinforced polymer composite bars for the concrete deck slabs of Wotton Bridge. Can J Civ Eng 30(5):861–970

Frosch RG (1999) Another look at cracking and crack control in reinforced concrete. ACI Struct J 96(3):437–442

ISIS Design Manual No. 3 (2001) Reinforcing concrete structures with FRPs. ISIS Canada Research Network. University of Manitoba, Winnipeg

JSCE (Japan Society of Civil Engineers) (1997) Recommendations for design of structures testing continuous fiber reinforcing materials. JSCE concrete series no. 23. Tokyo, Japan Society of Civil Engineers

Machida A (1996) Designing concrete structures with continuous fiber reinforcing material. In: Proceedings, first international conference on composites in infrastructure (keynote paper), Tucson

Mufti AA, Jaeger LG, Bakht B (1989) Has the time come for advanced composite materials in bridges? Can Civ Eng 6(2):9–15

Mufti AA, Erki M-A, Jaeger LG (1991) Advanced composite materials with application to bridges. Canadian Society for Civil Engineering, Montreal

Mufti AA, Erki MA, Jaeger LG (1992) Advanced composite materials in bridges and structures in Japan. Canadian Society for Civil Engineering, Montreal

Mufti AA, Onofrie M, Bakht B, Banthia V (2003) Durability of e-glass/vinylester reinforcement in alkaline solution. ACI Struct J 100(2):265

Mufti AA, Onofrei M, Benmokrane B, Banthia N, Boulfiza M, Newhook J, Bakht B, Tadros G, Brett P (2007) Durability of GFRP composite rods in field structures. Can J Civ Eng 34(3):355–366

NCHRP (2004) Report 514. Bonded repair and retrofit of concrete structures using FRP composites, National Cooperative Highway Research Program. Transportation Research Board, Washington, DC

Priestly MJN, Seible F, Calvi GM (1997) Seismic design and retrofit of bridges. Wiley, New York

Rizkalla SH, Tadros G (1994) A smart highway bridge in Canada. Concr Int 16(6)

Sen R, Issa M (1992) Feasibility of fibreglass pretensioned piles in marine environment. Report prepared for Florida Department of Transport, University of South Florida, Tampa

Sen R, Mullins G, Salem T (2002) Durability of e-glass/vinylester reinforcement in alkaline solution. ACI Struct J 99(3):369–375

Shehata EF (1999) Fibre reinforced polymer (FRP) for shear reinforcement in concrete structures. Ph.D. thesis, Department of Civil Engineering, University of Manitoba, Winnipeg

Tariq M, Newhook JP (2003) Shear testing of reinforced concrete without transverse reinforcement. In: Annual conference of the Canadian Society for Civil Engineering, Moncton, GCF-340-1 to 10

Vogel HM (2005) Thermal compatibility of bond strength of FRP reinforcement in prestressed concrete applications. M.Sc. thesis, Department of Civil Engineering, University of Manitoba, Winnipeg

Chapter 9
Rehabilitation with FRPs

9.1 Introduction

The acceptance that fibre reinforced polymers (FRPs) have so far gained for new structures is not in proportion to their potential, because their short-term costs are usually higher than those for structures with conventional materials. Designers and owners have usually justified the use of FRPs in new structures on the basis of life-cycle costing. The same is not the case for rehabilitation of existing structures with FRPs. The FRPs are highly cost-effective for rehabilitation, mainly because of their light weight and ease of bonding them to existing structures. A few of the several innovative applications of FRPs in this respect are described briefly in the following Sections, it being noted that the design provisions of all these rehabilitation techniques are covered by the Canadian Highway Bridge Design Code (CHBDC 2006). It is recalled that the CHBDC defines rehabilitation as 'modification, alteration, or improvement of the condition of a structure that is designed to correct deficiencies in order to achieve a particular design life and live load level.

9.2 Rehabilitation of Concrete Components with FRPs

The CHBDC (2006) specifies design provisions for the rehabilitation of concrete structures with FRP; these provisions, which are largely based on the works of Täljsten (1994, 2004a, b), are applicable to existing concrete structures having the specified concrete strength f_c' less than or equal to 50 MPa, and strengthened with FRP comprising externally bonded systems or near surface mounted reinforcement (NSMR). If the concrete cover is less than 20 mm, NSMR is not permitted to be used. Rehabilitation of concrete structures having f_c' more than 50 MPa requires approval by authority having jurisdiction over the structure.

B. Bakht, A. Mufti, *Bridges*, DOI 10.1007/978-3-319-17843-1_9

The behaviour of concrete elements strengthened with FRP is highly dependent on the quality of the concrete substrate. Corrosion-initiated cracks are more detrimental for bond-critical applications than for contact-critical applications. The code defines bond-critical applications as those applications of FRP that rely on bond to the substrate for load transfer; an example of this application is an FRP strip bonded to the underside of a beam to improve its flexural capacity. Similarly, the contact-critical applications of FRP rely on continuous intimate contact between the substrate and the FRP system. An example of a contact-critical application is an FRP wrap around a circular column, which depends upon the radial pressure that it exerts on the column to improve its compressive strength.

Prior to developing a rehabilitation strategy, an assessment of the existing structure or elements is required to be conducted following the requirements of the evaluation section of the CHBDC (2006). Only those structures are permitted to be strengthened that have a live load capacity factor F of 0.5 or greater. It is recalled that the evaluation section of the code defines F as follows for a structural component for the ULS.

$$F = \frac{U\phi R - \sum \alpha_D D - \sum \alpha_A A}{\alpha_L L(1 + I)} \tag{9.1}$$

where

U = the resistance adjustment factor, depending upon the category of resistance; for example, its value for axial compression of reinforced concrete components is 1.11
ϕ = the resistance factor specified in the concrete section of the code with a value of 0.75 for concrete
R = nominal unfactored resistance of the component
α_D = load factor for effects due to dead loads
D = nominal load effect due to unfactored dead load
α_A = load factor for force effects due to additional loads including wind, creep, shrinkage, etc.
A = force effects due to the additional loads
α_L = load factor force effects due to live loads
L = force effects due to nominal, i.e. unfactored live loads
I = dynamic load allowance

9.2.1 *Strengthening for Flexural Components*

FRP rehabilitation systems of the externally bonded and NSMR types may be exposed to impact or fire. To provide safety against collapse in the event that the FRP reinforcement is damaged, the structures that are to be strengthened with FRP require a live load capacity factor, F, defined above, larger than 0.5. With $F > 0.5$, the structure without rehabilitation will thus be able to carry all the dead loads and a portion of the live loads. Similar stipulations can be found in CSA-S806 (2002) and ACI 440-2R-02 (2002). The requirement that $F > 0.5$ also provides some benefits under normal service conditions; the stresses and strains in all materials including, concrete, steel and FRP, are limited and the risk of creep or yielding is avoided.

In addition to the conditions of equilibrium and compatibility of strains, the calculation for ULS is to be based on the material resistance factors for the materials of the parent component and those of the FRP given in Table 8.3 in Chap. 8, which are based on the assumptions implicit in the design of the parent component, and the following additional assumptions: (a) strain changes in the FRP strengthening systems are equal to the strain changes in the adjacent concrete; and (b) the contribution of the FRP in compression is ignored.

For an externally bonded flexural strengthening system, the maximum value of the strain in the FRP is not to exceed 0.006; this conservative requirement has been formulated to avoid a possible failure by the delamination of the FRP initiating at cracks in externally bonded flexural strengthening systems (Täljsten 2000; Teng et al. 2002).

In the FRP strengthening of concrete components, the failure modes required to be considered are: (a) crushing of the concrete in compression before rupture of the FRP or yielding of the reinforcing steel; (b) yielding of the steel followed by rupture of the FRP in tension; (c) in the case of members with internal prestressing, additional failure modes controlled by the rupture of the prestressing tendons; (d) anchorage failure; (e) peeling failure or anchorage failure of the FRP system at the cut-off point; and (f) yielding of the steel followed by concrete crushing, before rupture of the FRP in tension.

For externally bonded FRP strengthening systems, the anchorage length beyond the point where no strengthening is required is not to be less than l_a given by:

$$l_a = 0.5\sqrt{E_{FRP}t_{FRP}} \tag{9.2}$$

where t_{FRP} is the total thickness of externally bonded FRP plates or sheets in mm. In addition to the above requirement, the anchorage length should be at least 300 mm; otherwise the FRP needs to be suitably anchored.

The anchorage length is of central importance if an effective strengthening design is to be achieved. A good design will always lead to concrete failure.

9.2.2 *Strengthening of Compression Components*

When a column is strengthened with FRP, the compressive strength of the confined concrete, f'_{cc}, is determined from the following equation:

$$f'_{cc} = f'_c + 2f_{\ell FRP} \tag{9.3}$$

The confinement pressure due to FRP strengthening at the ULS, $f_{\ell FRP}$, is determined from the following equation.

$$f_{\ell FRP} = \frac{2\phi_{FRP} f_{FRPu} t_{FRP}}{D_g} \tag{9.4}$$

For columns with circular cross-sections, D_g is the diameter of the column; for columns with rectangular cross-sections having aspect ratios less than or equal to 1.5 and a smaller cross-sectional dimension not greater than 800 mm, D_g is equal to the diagonal of the cross-section.

Various formulae for determining the compressive strength of FRP-confined concrete have been assessed by Teng et al. (2002), Theriault and Neale (2000), and Bisby et al. (2005). Equations (9.3) and (9.4) have been shown to provide excellent but conservative estimates of the compressive strength.

The confinement pressure at the ULS is required to be designed to lie between $0.1f'_c$ and $0.33f'_c$. The minimum confinement pressure is specified in order to ensure the ductile behaviour of the confined section, and the maximum confinement pressure is specified in order to avoid excessive axial deformations and creep under sustained loads. The limit provided is such that the factored resistance of the FRP-confined concrete does not exceed the equivalent nominal strength of the unconfined concrete; i.e., $0.8\phi_c f'_{cc} \leq f'_c$.

9.2.3 Strengthening for Shear

The shear-strengthening scheme is to be of the type in which the fibres are oriented perpendicular or at an angle θ to the member axis. The shear reinforcement is to be anchored by suitable means in the compression zone by one of the following schemes:

- The shear-strengthening scheme is to be of the type in which the fibres are orientated perpendicular, or at an angle β, to the member axis. The shear reinforcement is to be anchored by suitable means in the compression zone by one of the following schemes: the shear reinforcement is fully wrapped around the section as shown in Fig. 9.1a.
- The anchorage to the shear reinforcement near the compression flange is provided by additional horizontal strips as shown in Fig. 9.1b.
- The anchorage is provided in the compression zone as shown in Fig. 9.1c. If none of these schemes can be provided, special provisions must be made.

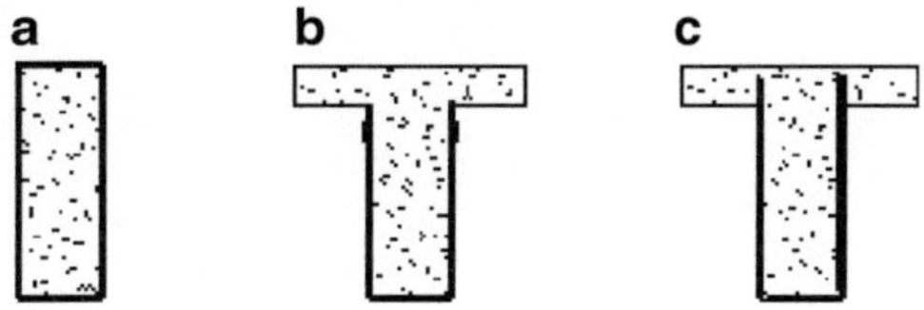

Fig. 9.1 Anchorage of externally bonded FRP shear reinforcement (**a**) fully wrapped section, (**b**) anchorage with *horizontal* strips, (**c**) anchorage in compression zone

For reinforced concrete members with rectangular or T-sections and having the FRP shear reinforcement anchored in the compression zone of the member, the factored shear resistance, V_r, is calculated from:

$$V_r = V_c + V_s + V_{FRP} \tag{9.5}$$

where V_c and V_s are calculated as for steel-reinforced sections, and V_{FRP} is obtained from the following.

$$V_{FRP} = \frac{\phi_{FRP} E_{FRP} \varepsilon_{FRPe} A_{FRP} d_{FRP} (cot\theta + cot\beta) sin\beta}{s_{FRP}} \tag{9.6}$$

where

$$A_{FRP} = 2t_{FRP} w_{FRP}. \tag{9.7}$$

For completely wrapped sections,

$$\varepsilon_{FRPe} = 0.004 \leq 0.75\varepsilon_{FRPu} \tag{9.8}$$

For other configurations, ε_{FRPe} is calculated from:

$$\varepsilon_{FRPe} = \kappa_V \varepsilon_{FRPu} \leq 0.004 \tag{9.9}$$

where for continuous U-shape configurations of the FRP reinforcement, the bond-reduction coefficient, K_V, is as follows:

$$\kappa_V = \frac{k_1 k_2 L_e}{11900\varepsilon_{FRPu}} \leq 0.75 \tag{9.10}$$

and

$$k_1 = \left(\frac{f_c'}{27}\right)^{\frac{2}{3}} \tag{9.11}$$

$$k_2 = \frac{d_{FRP} - L_e}{d_{FRP}} \quad (9.12)$$

$$L_e = \frac{23,300}{(t_{FRP}E_{FRP})^{0.58}} \quad (9.13)$$

It is noted that the value of ε_{FRP_e} is limited to 0.004 in order to maintain aggregate interlock in the evaluation of V_C.

For prestressed concrete components, V_r is the sum of V_c, V_s, V_p and V_{FRP}, where the general theory for steel reinforced concrete is used to calculate V_c, V_s and V_p, and the equations given above to calculate V_{FRP}.

9.2.4 Case Histories of Column Rehabilitation

Sheikh and Homam (2007) have described two severely deteriorated concrete columns, which were rehabilitated with GFRP. Several deteriorated columns, which are under a bridge in Toronto, Canada, can be seen in Fig. 9.2.

In 1995, using a steel formwork, one of the deteriorated columns was encased in grout of expansive cement, developed by Timusk and Sheikh (1977). A part of the formwork can be seen in Fig. 9.3a. In is important to note that no effort was made to remove either the corroded steel or concrete contaminated with chlorides from de-icing salts. About 20 h after casting the grout, the formwork was removed; and the grout layer was first wrapped in a polyethylene sheet and then with two layers of a GFRP sheet, in which most of the fibres were aligned in the circumferential direction of the column. Three days after grouting, the GFRP wrapping was instrumented with strain gauges measuring circumferential strains. The rehabilitated column can be seen in Fig. 9.3b. Sheikh and Homam (2007) report that within about 7 days after the pouring of the grout, the tensile circumferential strains in the GFRP wrapping grew to about 1500 $\mu\varepsilon$, thus effectively applying a radial prestressing pressure to the column. As noted by Erki and Agarwal (1995), the

Fig. 9.2 Deteriorated concrete columns

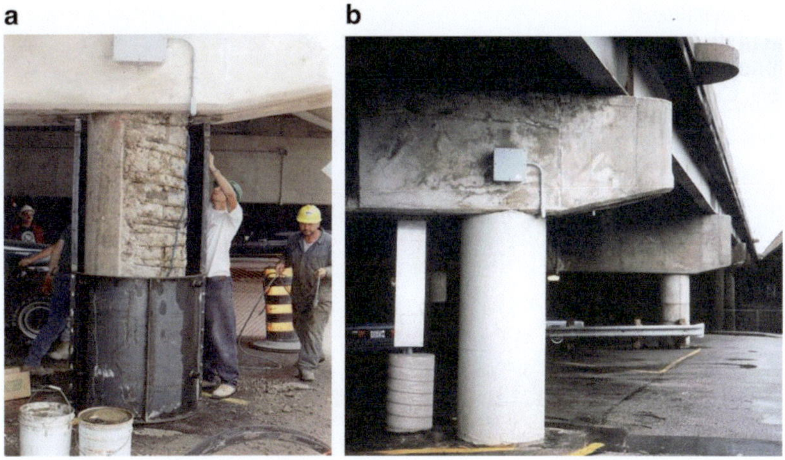

Fig. 9.3 Rehabilitation of a concrete column: (**a**) partial formwork for grout with expansive cement, (**b**) rehabilitated column

concept of using expansive grout to prestress the rehabilitated column radially was introduced by Baidar Bakht.

The strains in the rehabilitated column are being monitored periodically. It has been found that over the past 12 years, the circumferential strains in the GFRP wrapping have dropped only slightly, thus confirming that the radial pressures generated by the expansive grout exist on a long-term basis.

Several half-cells were installed in the rehabilitated column to monitor the corrosion activity of the steel reinforcement. Data collected in these half-cells, by the Ministry of Transportation of Ontario, has shown that the corrosion activity in the steel reinforcement of the rehabilitated column has decreased over 12 years of monitoring. The reduction of the corrosion activity is likely to be the result of preventing the ingress of the main elements necessary for steel corrosion, namely oxygen and water.

The case history described above confirms the effectiveness of repairing deteriorated columns by wrapping them with FRP. It is also important to note that the seismic resistance of columns, especially those with circular cross-sections, can be improved considerably by wrapping them with FRPs. One example of enhancing the seismic resistance of bridge columns is the Portage Creek Bridge in British Columbia, Canada. Wrapping the columns with GFRP sheets is described by Mufti et al. (2003); the rehabilitated columns, which are monitored by ISIS Canada in real-time, can be seen in Fig. 9.4.

Fig. 9.4 Rehabilitated columns of the Portage Creek Bridge in British Columbia, Canada

9.3 Rehabilitation of Timber Beams

It is generally believed that timber components being made of a 'low-tech' material do not require for their design the same amount of computations as components made with concrete or steel. Consistent with such belief, the design requirements for the rehabilitation of timber bridges with FRPs could not be made as complex as those for concrete components.

There were three aspects of the structural behaviour of timber that led to the formulation of clauses of CHBDC (2006) for the rehabilitation of timber beams and stringers of bridges with FRPs: (a) timber beams in flexure and shear mainly fail in tension at the location of their defects, such as knots and checks, (b) the flexural and shear strengths of timber beams of a given species are specified with respect to their grades, which relate mainly to the frequency of defects in a given piece of wood, and (c) FRPs applied to timber beams tend to 'bridge' over the defects, thus improve their grade designation. In Canada, timber is graded according to the rules of the National Lumber Grading Authority of Canada (NLGA 2014). Select structure (SS) grade timber has minimum defects, Grade 1 has medium number of defects and Grade 2 timber has the maximum defects, which could be permitted in lumber that is suitable for structural use. According to the CHBDC (2006), the cross-sections of beams and stringers have the smaller dimension of at least 114 mm, and the larger dimension more than 51 mm greater than the smaller dimension.

The CHBDC (2006) specifically refers to GFRP mainly because of economy and partly most of the tests that led to its provisions for the rehabilitation of bridges were conducted with GFRP sheets and bars. The CHBDC notes that the same provisions can also be applied to AFRP and CFRP sheets and bars. It is noted that the CHBDC requirements are mostly based on the works of Eden (2002), Hay (2004), and Hay et al. (2004).

9.3.1 General Requirements

The CHBDC (2006) requires that the drawings and specifications for the strengthening of timber stringers by FRP should address the following.

(a) Identification of the FRP strengthening system and protective coatings;
(b) Surface preparation;
(c) Shipping, storage and handling of the FRP strengthening systems;
(d) Installation details of the FRP strengthening systems;
(e) The curing conditions of the strengthening systems;
(f) The quality control of the strengthening systems;
(g) Staff qualifications;
(h) Material inspection before, during and after the completion of installation; and
(i) System maintenance requirements.

The above detailed requirements are specified because it is believed that the industry for the strengthening of timber by FRPs has not matured yet.

9.3.2 Strengthening for Flexure

Strengthening with GFRP Sheets The minimum requirements for strengthening of timber stringers for flexure with GFRP sheets are as follows.

(a) The minimum fibre volume fraction, being the ratio of the volume of the fibres to the volume of the FRP, of the GFRP in the direction of the span of the beam is 30 %;
(b) The GFRP sheet on the flexural tension face of the beam covers at least 90 % of the width of the beam and has a minimum thickness of 0.1 mm;
(c) The adhesive used for bonding the GFRP sheet to the timber beam is compatible with the preservative treatment used on the timber and the expected volumetric changes of the timber; and
(d) In the longitudinal direction of the beam, the GFRP sheets extends as close to the beam support as possible.

Strengthening with NSM GFRP Bars The minimum requirements for strengthening of timber stringers for flexure with near surface mounted (NSF) GFRP bars are as follows.

(a) The minimum volume fraction of GFRP bars is 60 %;
(b) There are at least two GFRP bars within the width of the beam;
(c) The total cross-sectional area of all GFRP bars is at least 0.002 times the cross-sectional area of the timber stringer;
(d) The depth and width of the grooves, the edge distances of the grooves and the clear distances between the grooves are as shown in Fig. 9.5;

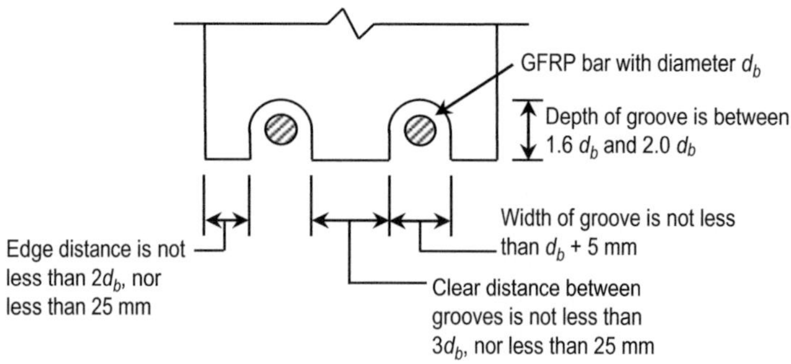

Fig. 9.5 Cross-section of a timber beam with NSM GFRP bars

(e) The grooves are cleaned with pressurised air to remove any residue before the GFRP bars are embedded in them;
(f) The adhesive used for bonding the GFRP bars is compatible with the preservative treatment of the timber and with the expected volumetric changes of the timber;
(g) In the longitudinal direction of the stringer, the GFRP bars extend as closely to the stringer supports as possible; and
(h) Each GFRP bar is held as close to the tip of the groove as possible.

Upgrading of Specified Bending Strength When the above minimum conditions are satisfied, the specified bending strength of timber can be increased by 5 % if the grade of the original stringer is Select Structural, by 20 % if the grade of the original stringer is No. 1, and by 50 % if the grade of the original stringer is No. 2.

The above specifications are formulated that by recognising that the FRPs tend to eliminate the effects of defects bridging over them (e.g., Johns and Racine 2001; Eden 2002; Hay 2004). Grade 2 timbers having the largest number of defects benefit the most by strengthening with FRPs.

9.3.3 Strengthening for Shear

The CHBDC (2006) provisions for strengthening timber stringers with GFRP, noted in the following, are applicable only when the horizontal splits in the stringers are closed mechanically before applying the strengthening scheme. As discussed by Hay et al. (2004), the mechanical closing of a horizontal split in a stringer involves the temporary closing of the gap by an external device such as grip with a hydraulic jack and keeping the gap closed permanently by a lag screw. The simple closing of the gap is expected to raise the shear strength of the timber by up to 30 %. The addition of FRP further increases the shear strength. However, in the spirit of caution, the CHBDC (2006) provisions do not include this further increase.

Strengthening with GFRP Sheets The minimum requirements for strengthening of timber stringers for shear with GFRP sheets are as follows.

(a) The minimum fibre volume fraction of the GFRP sheets along their exes is 30 % and sheets have a minimum thickness of 0.1 mm;
(b) The GFRP sheets have at least the same width as the width of the cross-section of the beam, as shown in Fig. 9.6a;
(c) As shown in Fig. 9.6a and b, the GFRP sheets are inclined to the beam axis at an angle of 45° ± 10°;
(d) The top of the inclined sheets are as close to the centreline of the beam support as possible;
(e) The adhesive used for bonding the GFRP sheet to the timber beam is compatible with the preservative treatment used on the timber and the expected volumetric changes of the timber;
(f) The lower ends of the sheets extend up to the bottom of the stringer, if no dap is present, as shown in Fig. 9.6a; and
(g) If the stringer has a dap, as shown in Fig. 9.6b, the lower ends of the sheets are wrapped around the bottom to at least half width of the beam, and the lower corners of the beam are rounded to a minimum radius of 12.5 mm.

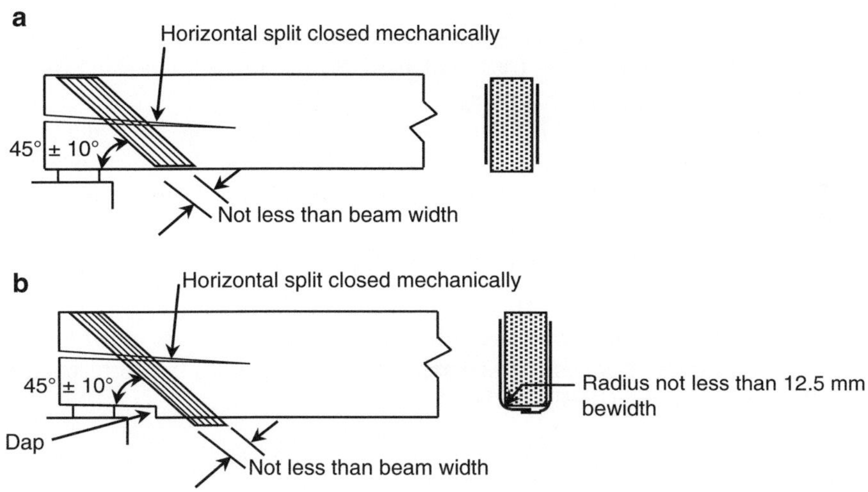

Fig. 9.6 Timber stringer with GFRP sheets for shear strengthening: (**a**) stringer without daps, (**b**) stringer with daps

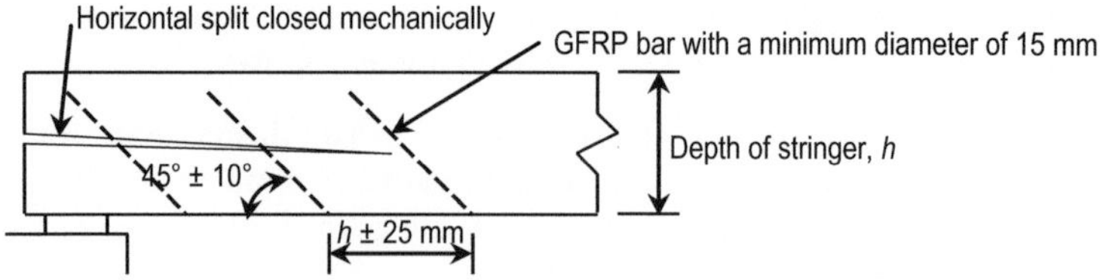

Fig. 9.7 Timber stringer with GFRP bars for shear strengthening

When the above conditions of strengthening are satisfied, the specified longitudinal shear strength of stringers can be assumed to be two times the corresponding strength specified in the CHBDC (2006).

Strengthening with GFRP Embedded Bars The minimum requirements for strengthening of timber stringers for shear with GFRP embedded bars are as follows.

(a) The minimum fibre volume fraction of the GFRP bars is 60 %;
(b) As shown in Fig. 9.7, there are at least three GFRP bars at each end of the stringer;
(c) The diameter of each bar is at least 15 mm, and the minimum diameter of a hole containing the bar is bar diameter +3 mm;
(d) The spacing of the bars along the stringer axis is the depth of the stringer $h \pm 25$ mm;
(e) The adhesive used for bonding the GFRP sheet to the timber beam is compatible with the preservative treatment used on the timber and the expected volumetric changes of the timber;
(f) As shown in Fig. 9.7, the GFRP bars are inclined to the stringer axis at an angle of $45° \pm 10°$;
(g) The tops of the GFRP bars are between 10 and 25 mm from the top of the stringer;
(h) When daps are present, the bottom of the bar nearest to a support is 100 ± 10 mm.

When the above conditions of strengthening are satisfied, the specified longitudinal shear strength of stringers can be assumed to be 2.2 times the corresponding strength specified in the CHBDC (2006).

References

ACI 440.2R-02 (2002) Guide for the design and construction of externally bonded FRP systems for strengthening concrete structures: 45. American Concrete Institute, Farmington Hills

Bisby LA, Dent AJS, Green M (2005) Comparison of confinement models for fiber reinforced polymer-wrapped concrete. ACI J 102(1):62–67

CHBDC (2006) Canadian highway bridge design code, CAN/CSA-S6-06. Canadian Standards Association, Toronto

CSA-S806 (2002) Design and construction of building components with fibre reinforced polymers. Canadian Standards Association, Toronto

Eden R (2002) Strengthening of timber bridge stringers using GFRP. Master of Science in Civil Engineering thesis, University of Manitoba, Winnipeg

Erki M-A, Agarwal AC (1995) Strengthening of reinforced concrete axial members using fibre composite materials: a survey. In: Proceedings, annual conference of Canadian Society for Civil Engineering, vol II. Ottawa, pp 565–574

Hay S (2004) Shear strengthening of timber stringers using GFRP. Master of Science in Civil Engineering thesis, University of Manitoba, Winnipeg

Hay S, Svecova D, Bakht B (2004) The effectiveness of diagonal versus vertical GFRP sheets for shear strengthening of timber. In: Conference on innovative materials and technologies for construction and restoration, University of Lecce, Lecce

Johns K, Racine P (2001) Composite reinforcement of timber in bending. In: Proceedings of the 29 the annual conference of the Canadian Society for Civil Engineering, Victoria, pp 1–6

Mufti AA, Neale KW, Rahman S, Huffman S (2003) GFRP seismic strengthening and structural health monitoring of Portage Creek Bridge concrete columns. In: Proceedings for the fib2003 symposium – concrete structures in seismic regions, Athens

NLGA (2014) Standard rules for grading Canadian lumber. National Lumber Grading Authority, Surrey

Sheikh SA, Homam SM (2007) Long-term performance of GFRP repaired bridge columns. In: Third international conference on structural health monitoring of intelligent infrastructure, Vancouver (proceedings on CD)

Täljsten B (1994) Plate bonding: strengthening of existing concrete structures with epoxy bonded plates of steel or fibre reinforced plastics. Doctoral thesis. Department of Civil Engineering, Luela University of Technology, Luela

Täljsten B (2000) FRP strengthening of existing concrete structures: design guidelines. Luela University of Technology, Luela

Täljsten B (2004a) Design guideline for CFRP strengthening of concrete structures. IABMAS, Kyoto

Täljsten B (2004b) FRP strengthening of existing concrete structures – design guidelines, 3rd edn. Luleå University of Technology, Division of Structural Engineering, Luleå. ISBN 91-89580-03-6: 230

Teng JG, Chen JF, Smith ST, Lam L (2002) FRP-strengthened RC structures. John Wiley & Sons, Ltd, West Sussex

Theriault M, Neale K (2000) Design equations for axially loaded reinforced concrete columns with fibre reinforced polymer wraps. Can J Civ Eng 27(6):1011–1020

Timusk J, Sheikh S (1977) Expansive cement jacks. ACI J Struct Eng 74(2):80–85

Chapter 10
Structural Health Monitoring

10.1 Introduction

Structural health monitoring (SHM) is the integration of a sensory system, a data acquisition system, a data processing and archiving system, a communication system, a damage detection system and a modeling system to acquire knowledge about the integrity and load-worthiness of in-service structures on either a temporary or continuous basis.

The objective of an SHM system is to monitor the behaviour of a structure accurately and efficiently, so as to assess its performance under various service loads, to detect damage or deterioration, and to determine the health or condition of a structure. The ISIS Canada design guideline on SHM (Mufti 2001) divides SHM into four categories: (a) static field testing; (b) dynamic field testing; (c) periodic monitoring; and (d) continuous monitoring. Unlike other organizations specializing in SHM, ISIS Canada includes both the static and dynamic testing of bridges as part of SHM. Although field testing of bridges is not a new activity, it has not yet been used extensively as a tool for managing structures. In the past, civil engineers have ascertained the integrity of their structures mainly by means of physical inspection, and occasionally by nondestructive evaluation (NDE). When undertaken with the help of electronic sensors and data acquisition systems, NDE becomes a part of modern SHM. For bridges, NDE usually takes the form of a bridge test. The main benefit of a static test on a bridge is the utilization of its latent strength without compromising safety. Clearly, many other advantages accrue from this single benefit, for example avoiding service interruptions, eliminating unnecessary weight restrictions on vehicles, and optimal utilization of resources.

A new term 'civionics' was coined recently to denote structural health monitoring of civil structures with help of electronic sensors (Mufti 2003). A discussion on civionics is provided in Sect. 10.2, whereas the field testing of truss bridges and slab-on-girder bridges are described in Sects. 10.3 and 10.4.

B. Bakht, A. Mufti, *Bridges*, DOI 10.1007/978-3-319-17843-1_10

10.2 Civionics

It is argued in this section that civil engineers are usually conservative in the design of new structures and the evaluation of existing structures, and that SHM will assist in providing data that could be used to fine-tune the calibration of load and strength factors leading to more efficient and economical designs and better utilization of the strengths of existing structures.

For major changes in design, construction and evaluation to be accepted, it is necessary that innovative structures be monitored for their health so that the required data bank can be developed. To assist in achieving this goal, civil engineers in Canada are developing a new discipline, which integrates civil engineering and electrophotonics under the term ‘civionics’. Two important questions are raised with respect to the general acceptance of civionics as an integral part of structures: (a) are civil structural engineers ‘risk-averse’?, and (b) can SHM be useful in managing the risk associated with structures? Before answering these two questions, it is instructive to examine various human endeavours and the degree of ‘acceptable’ risk involved in each of them. For example, the table of the risk of death involved in various activities, as proposed by Melchers (1987), is reproduced in Table 10.1.

Notwithstanding the accuracy of the risk values given in Table 10.1, it is clear that the risk involved in the failure of completed civil structures is significantly smaller than that involved, for example, in air travel. It is recalled that civil structures include bridges, public buildings, dams, offshore platforms and power plants. The glaring differences between the risks involved in various human endeavours lead to another set of questions: (a) are the low levels of risks associated with completed civil structures fixed by society?, and (b) is the society wary of taking undue risks?

It is true that the public does not expect civil structures to ever fail. However, it is suspected that the low risk levels associated with completed civil structures are a result of a combination of both the demands of society and also the awareness of the structural engineer that, unlike manufactured machines with usually well-tested prototypes, each structure is unique and nearly always without means to monitor its

Table 10.1 Risk of death involved in various activities

Activity	Risk (/Million/year)
Alpine climbing	1,500–2,000
Swimming	120
Smoking	1,000
Air travel	24
Car travel	200
Coal mining	300
Construction	150–440
Manufacturing	4,000
Building fires	8–24
Structural failures	0.1

performance during its service life under loads and circumstances, which are more difficult to forecast than those for machines. A case can be made for including SHM in important civil structures, such as large bridges and public buildings, which could be monitored on a continuous basis, thus eliminating certain unknowns about their performance, which could and do change with time. In making a case for the SHM of important civil structures, a slight increase in the notional or theoretical, rather than the actual, risk of failure is foreseen, but a significant reduction in the capital cost that society has to incur partly to construct and largely to maintain its infrastructure.

Civil Structural Design Process In some jurisdictions, structural components still continue to be designed by the working stress method, which requires that the maximum stress due to nominal dead loads and live 'service' loads is a fraction of the maximum stress that the component can withstand. The ratio of the failure stress to the 'actual' stress was, and is, still known as the factor of safety. As long as the factor of safety was a sufficiently large number, such as 2 or 3, the design was deemed to be safe. While the working stress design has served the engineering community well for a long time, it has led to structures with non-uniform margins of safety. For example, if long- and short-span components were designed to the same design specifications, or codes, the long-span components would have considerably larger margins of safety than their short-span counterparts, because the designs of large span components are largely governed by dead loads, the magnitudes of which can be forecast with more certainty than those for live loads. On the other hand, the design of short-span components is usually governed by live loads.

The works of researchers such as Cornell and Lind in the late 1960s and early 1970s laid the foundations for modern structural design codes, which are based on the concept of structural reliability (Nowak and Collins 2000). In Canada and some parts of Europe, the probabilistic-based design method is referred to as the limit states design method. In the USA, the same method is referred to as the load and resistance factor design (LRFD) method. The first limit states design code for bridge design in Canada was the Ontario Highway Bridge Design Code introduced in 1979, followed by its two editions, and then by the Canadian Highway Bridge Design Code (2000, 2006). The LRFD method was first introduced for bridge design by the AASHTO Specifications in 1994.

It is important to note that the working stress method, which is still practiced in many parts of the USA as well as in many other countries, was used to design a very large part of the current stock of infrastructure in practically every country. The earlier working stress design methods, requiring manual calculations and based on simplifying assumptions usually led to structures with large reserves of strength. By careful use of SHM, the conservatism in the risk assessment of civil structures can be reduced, leading to economies in the cost and maintenance of these structures. The use of SHM is also expected to encourage innovative designs.

The level of safety of a structural component is determined by the probability of its strength exceeding the maximum load effect it would receive during its lifetime. Such a determination requires knowledge of the statistical distribution of both the strengths of the component and the maximum loads to which the component would be subjected. For designing new components, the statistical distribution of strengths necessarily includes very low – even zero – strengths. For a component of an in-service structure, however, such low strengths need not be included in the strength distribution. A method proposed by Bakht et al. (2002) allowed advantage to be taken of the fact that a component of an in-service bridge has already been proof tested by traffic, therefore, its strength cannot be below the corresponding maximum traffic load. By utilizing the proposed method, the evaluated live load capacity of a component can be upgraded significantly, if the strength under consideration has a large statistical variation.

As discussed in Sect. 1.3, safety margins in a structural component are measured by the probabilistic-based safety index, β.

In order to explain the safety index in simple terms, consider a multi-lane bridge with four simply supported spans and a total of 28 reinforced concrete girders, as shown in Fig. 10.1. Further, assume that the point of interest is the moment capacity of the girders at their mid-spans. It is well known that, if all the reinforced girders were tested to failure, the likely finding would be that each girder has a somewhat different moment of resistance than those of the others. A few girders are likely to have a very small moment capacity, while a few others would have a very high moment capacity. The majority of girders would perhaps have their moments of resistance within a narrow band. The histogram of the moments of resistance of the girders can be represented by a continuous distribution curve, denoted in Fig. 1.22 as R (resistance). The mean and standard deviation of R are denoted as μR and σR, respectively.

During the lifetime of the bridge, each girder is likely to be subjected to a different maximum live load moment. For example, as illustrated in Fig. 10.1, a girder in Span 1 shown in solid line, might receive its maximum live load moment from two exceptionally heavy trucks, present on the bridge at the same time. Similarly, a girder in Span 2, also shown in solid line in Fig. 10.1, might experience its maximum moment under three different heavy trucks simultaneously present on the span, and so on. The lifetime maximum moments are designated as $S1$, $S2$, $S3$, $S4$, etc. A normal distribution is assumed for these moments, denoted by S in Fig. 1.22; for simplicity, it is also assumed that these moments include the effect of both dead and live loads. The mean and standard deviation of S are denoted as μS and σS, respectively.

The overlapping area of the R and S curves, shown shaded in Fig. 1.22, is a qualitative, although not quantitative, measure of the fact that there is a probability

Fig. 10.1 Girders of a four-span bridge shown in plan

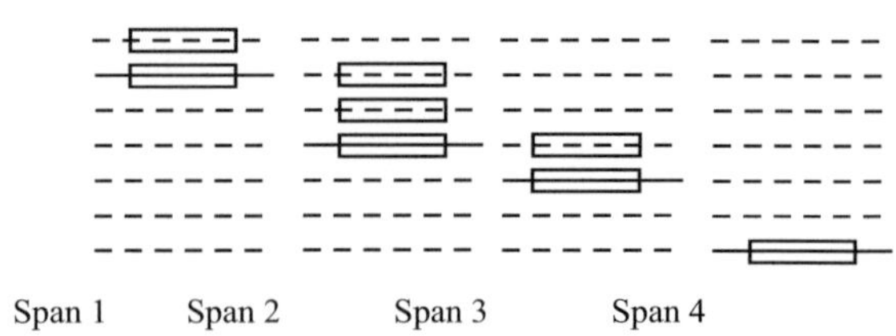

that the moment of resistance of some girders would be exceeded by the maximum applied moments. The reduction of maximum applied moments – for instance, by enforcing stricter weight regulations – would have the effect of moving the *S* curve to the left, thus reducing the area of overlap, and thereby in effect reducing the probability of failure.

A quantity g, defined by $g = R\text{-}S$, can be used to assess safety quantitatively; the frequency distribution of this quantity is shown in Fig. 1.23, in which it can be seen that a certain portion of the g curve, shown shaded, lies in the negative region. As is also shown in Fig. 1.23, the zero value of g lies a distance $\beta \times \sigma g$ to the left of μg, where σg and μg are respectively the standard deviation and mean of g. The quantity β is the quantitative measure of the probability of S exceeding R; it is referred to as the safety index. The safety index for individual components is defined by Eq. (1.11) in Chap. 1.

Similar to several other modern limit state design codes, the Canadian and American bridge design codes relate only to the notional safety index β. After determining by back calculations, that the value of β for components of existing bridges is 3.5 or greater, it was agreed amongst the experts in the field that components of new bridges should be designed for β equal to approximately 3.5. Both the Canadian and American bridge design codes are generally calibrated to $\beta = 3.5$, which corresponds to the notional, i.e. theoretical, probability of failure of a component of 1 in nearly 2000 during the lifetime of the bridge. Considering the lifetime of a bridge to be 75 years, the probability of failure corresponding to $\beta = 3.5$ translates to one failure in about 150,000/year. The fact that this probability of failure is significantly larger than the probability of failure of one in 10 million/year for completed structures (Table 10.1), underscores two very important points with respect to bridges, in particular: (a) β relates only to the theoretical failure of a component; and (b) the failure of the combination of components, being a system, has a much smaller probability of failure. Since most structural failures are caused by extreme events, such as those relating to hydraulics, β should not be taken as the real measure of the safety of a structure.

The resistance of a structure, such as a bridge, can decrease with time due to environmental and other time-dependent effects, with the result that the *R* distribution shown in Fig. 10.2 moves to the left. Similarly, the live loads that a bridge is called upon to carry can also increase with time, as has been experienced in Canada with the changes in vehicle weight regulations demanded by the economics of weight hauling by highway trucks. The increase in vehicle weights causes the *S* curve in Fig. 1.22 to move to the right. As illustrated in Fig. 10.2, the net result of the decrease of resistance and increase of loads over time is an increase of overlapping areas for the two curves.

Hence, it can be seen that the notional safety indices of the components of a bridge can decrease with time. Should such a reduction in the safety index always be a cause for concern? Fortunately, a bridge can still be used even if its safety index is smaller than 3.5, but only if there is reliable knowledge about the structure. Bakht and Jaeger (1990b) have shown that each time a bridge was tested surprising results were observed. The actual behaviour of bridges was found to be usually

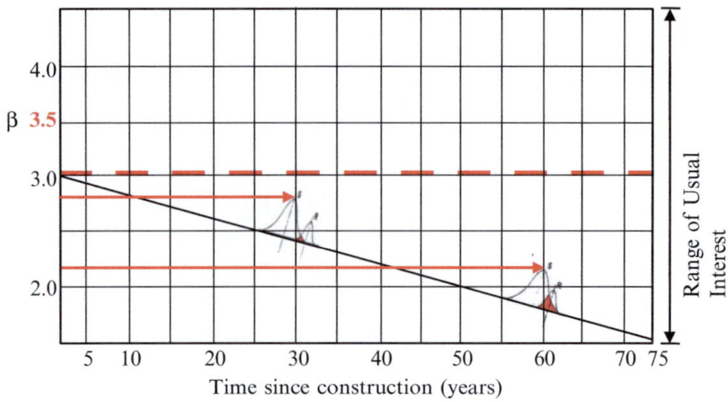

Fig. 10.2 Reduction in value of safety index with time

different than that of the mathematical model used in the original design. In most cases, the actual strengths of bridges were considerably higher than the theoretical strengths.

SHM Systems The term 'civionics' refers to the application of electronics to civil structures for the purpose of determining the state of their health. The use of sensors to monitor the response of a structure or its model to applied loads is not new, nor is bridge evaluation by field testing, which includes both diagnostic and proof testing. What is new, however, is the use of SHM through civionics. The purpose of SHM, according to Mufti (2001), is to monitor the in-situ behaviour of a structure accurately and efficiently to determine its health or condition. SHM is the integration of a system of sensors, a data acquisition system, a data processing system, an archiving system, a communications system, and a damage detection and modeling system to acquire knowledge, either on demand or on a continual basis, regarding the in-service performance of structures, as illustrated in Fig. 10.3.

In the past, civil engineers have gained knowledge about the integrity of civil structures largely by means of manual inspections, and rarely by nondestructive evaluation (NDE) and interpretation of data using conventional technologies. The structural engineering profession has relied heavily on evaluation parameters given in codes of practice that lead to conservative and often costly conclusions about the strength of existing structures. The current practice has resulted in a stock of civil structures whose health is not easy to monitor. For example, many bridges and large buildings constructed in earthquake-prone areas cannot be opened for public use immediately after a seismic event due to the time and cost involved in performing extensive safety checks. Quite often, after such extreme events the safety of these structures cannot be ascertained accurately, so it is difficult to know whether they should continue to stay in service.

In order to remain competitive in today's global economic environment, the owners of civil structures need to minimize the user costs involved with the

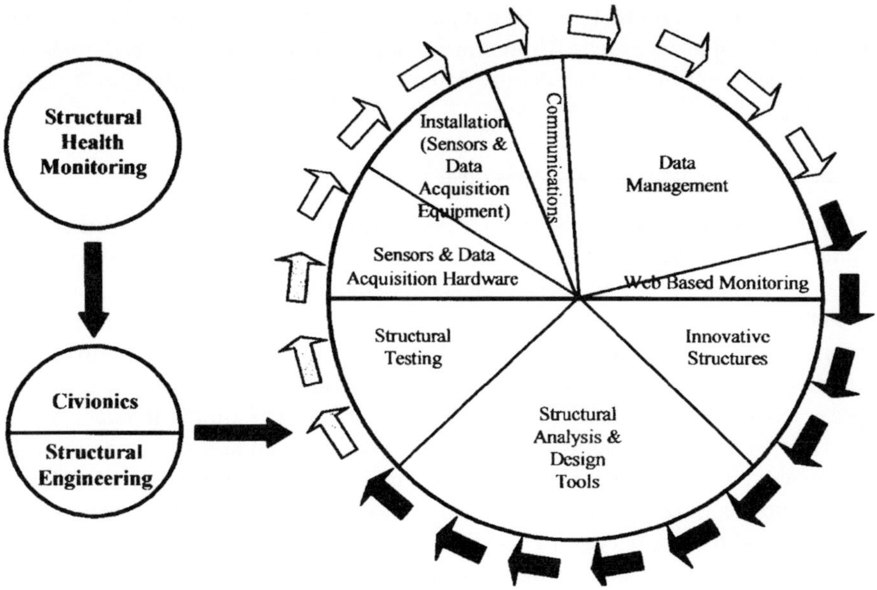

Fig. 10.3 Basic components of an SHM system

unnecessary closing of the structures and the service disruption caused by outdated and time-consuming inspections following extreme events, such as strong-motion earthquakes, hurricanes, or flash floods. In the evaluation of any structural system, it is important to be able to assess specific performance issues related to serviceability, reliability and durability. To effectively quantify the system's performance requires a means to monitor and evaluate the integrity of these large civil structures with minimal interruption of service. The SHM data, besides allowing owners to better allocate their resources towards repair, replacement or rehabilitation of the structures, will also be useful in future projects in estimating the life cycle costs of the structural system compared to the initial cost. An efficient SHM system should be autonomous and capable of continuous monitoring, inspection and damage detection of structures. Reports regarding the integrity of a structure should be automatically relayed through a local network or to a remote monitoring centre. Clearly, the development of major SHM systems will involve many disciplines including structures, materials, damage detection, sensors, data collection and intelligent processing, computers, and communication.

Although conventional NDE can be considered to be within the framework of SHM, there is a difference in terms of data interpretation between the traditional NDE and SHM. The traditional NDE techniques tend to use direct measurements at discrete time intervals to determine the physical condition of structures. For example, bridge testing could either be diagnostic or involve proof loading. For evaluation by bridge testing, historical data is generally not required. SHM techniques, using continuous monitoring, assess changes in the condition of a structure. Hence, a history of data is crucial to the technique. The direct benefits from SHM systems are very large and include: (a) monitoring and evaluating structures in real-time under service conditions; (b) reducing downtime; (c) improving safety and reliability; and reducing maintenance costs.

With the reduced downtime and improved reliability, in-service structures can be used more productively with less cost.

Civionics into Practice For SHM to become part of civil structural engineering, it should include civionics. The new discipline of civionics must be developed by civil structural engineers and electrophotonics engineers to lend validity and integrity to the process. Civionics includes the hardware and physical installation of the sensors, wires, conduits, termination and control boxes. SHM complements civionics and includes the collection and interpretation of data. A parallel for this process exists in the medical profession where a doctor requires medical instruments and technology to permit him, or her, to monitor and assess the health of the human body. In this same manner, civil engineers need civionics and SHM to monitor the health of a structure. Civionics and SHM will provide engineers with the means and the knowledge to build "smart" structures containing the necessary equipment to provide much needed information related to the health of a structure *before* things go wrong.

Realistically, it is true that consulting engineers and contractors will only invest in the development of the expertise created by graduates of the civionics discipline when they can be assured that the prospects for business are good in this field. The experience of ISIS Canada in integrating fibre optics sensors and fibre reinforced polymers into innovative structures, built across Canada, has demonstrated that such opportunities do exist.

The CHBDC (2006) has a section on the strength evaluation of existing bridges; the section is based on the concept of a target reliability index that changes with (a) system behaviour, (b) component behaviour, and (c) the level of inspection. The system behaviour relates to the effect of the failure of a component to the failure of the whole structure; the component behaviour corresponds to the ductility of its failure; and the inspection level refers to the degree of confidence in the inspection process in determining the actual condition of the bridge and its components. The effect of the three factors on the target reliability index can be explained with the help of two examples. In the first example, the component under consideration is critical to the safety of the entire structure. It can fail suddenly, such as in shear or by buckling. The component cannot be inspected, possibly because it is inaccessible. For such a component, the target reliability index β is required by the CHBDC (2000) to be 4.00. For normal traffic, the live load factor α_L corresponding to a β of 4.00 is 1.77.

For the second example, the component is such that its failure does not affect the failure of the whole structure. Also, it is subject to gradual failure with advance warning of impending failure. The inspection of the component is carried out by the evaluator and the calculations for the final evaluation account for all the information gathered during this inspection. For such a component, the target reliability index β is required by the CHBDC (2000) to be 2.50 with the live load factor αL for normal traffic being 1.35.

Depending upon the system and element behaviour and confidence in the inspection, the difference between two useable live load capacities can be as large as about 24 %. If the condition of a component of a structure were determined with the help of sensors in an SHM system, the degree of confidence in the determination will be greater than in any visual inspection, with the consequence that the evaluator of the component will be able to utilize a larger portion of its live load capacity.

The concept of the target reliability index changing with the inspection level does not exist in the design of new structures. Yet, it can be appreciated that if the designer of a structure were confident that the condition of the structure and the load that it receives would be determined continuously and accurately by an SHM system, he/she can afford to be less conservative, thus reducing the capital cost of the structure. Drawing upon the comparison between a fully instrumented aerospace structure and a civil structure without any sensors to report on its health, it can be stated confidently that the designers of civil structures are risk-averse to an extent because of the absence of information about the field performance of these structures.

The answer to the question, *"Would SHM help to change the 'risk shy' culture of civil structural engineers?"* is a conditional yes. However, the civionics discipline needs to be developed further.

10.3 Truss Bridges

Because of their economic use of materials, truss bridges were the preferred bridges until 1930s for spans larger than about 20 m. A truss is usually composed of horizontal, vertical and inclined components, which are referred to as chords. The locations where the chords are joined are referred to as the nodes of the truss, and the portion of a truss contained between two consecutive nodes is called the panel. While the connections at nodes are usually rigid in steel trusses and partially rigid in timber trusses, for simplicity of analysis and design, the nodes are assumed to be pin-connected so that the chords are consequently assumed to be in either pure compression or tension with hinge supports at their ends. In steel trusses, the tensile chords are single or double rolled steel sections; however, the chords in compression are usually built up with different rolled steel sections in steel trusses; in timber trusses, the compression chords are usually made of several timber components of rectangular cross-section placed side by side. Depending upon their own location and the location of the live loads, the inclined chords can be either in compression or tension. The compression chords are designed for buckling.

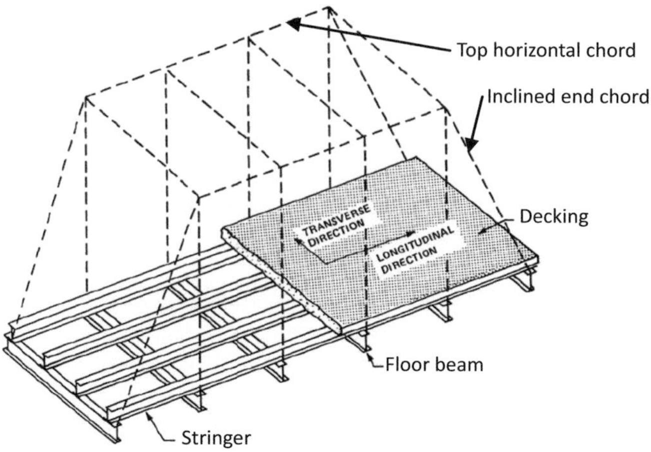

Fig. 10.4 Anatomy of a truss bridge, shown without internal inclined chords

10.3.1 General Concepts

The anatomy of a truss bridge is illustrated in Fig. 10.4, in which it can be seen that the floor system could comprise of transverse floor beams, longitudinal stringers spanning between the floor beams, and a decking that could be made of concrete, timber, metal, or in modern times of fibre reinforced polymer (FRP).

Most truss spans are simply supported, because of which the top chords of the trusses are in compression and the bottom chords in tension. Depending upon the location of the floor system with respect to the top or bottom chords of the trusses, and support to the main compression chords, the truss bridges are divided into three broad categories. The through truss bridge, illustrated in Fig. 10.4 and shown in Fig. 10.5, has its floor system near the bottom chords of the trusses. The top chords of the trusses of these bridges are laterally restrained by a system of horizontal bracing to improve their buckling capacity.

When the clearance below the trusses is not limited, it is preferable to employ deck truss bridges, in which the floor system, located near the top chords of the trusses, also provides lateral support to the horizontal compression chord. Figure 10.6 presents a view of two side-by-side deck trusses in a bridge in Toronto, Canada.

Unlike through truss bridges, pony truss, or half-through truss, bridges have no height restrictions for the vehicles. The lateral support to the horizontal compression chords of its trusses is provided by knee-braces that lie on the outer sides of the bridge, as can be seen in Fig. 10.7.

In most bridges, the trusses are statically determinate, in which the removal of a chord leads to the collapse of the truss. In some bridges, such as the one shown in Fig. 10.5, two inclined slender chords are provided in some panels in an X pattern, thus technically making the trusses statically indeterminate. It can be seen in

Fig. 10.5 A steel through truss bridge

Fig. 10.6 Partial view of two side-by-side steel deck trusses carrying a common deck system

Fig. 10.7 A pony truss, or half-through truss, bridge

Fig. 10.5 that the X-chords are very slender, and thus incapable of carrying substantial compressive loads. When a vehicle traverses the bridge, the forces in the X-chords in a panel change from tensile to compressive in one chord and from compressive to tensile in the other. The chord in compression bows out thus taking little compressive force. For all practical purposes such a truss can be regarded as statically determinate. A through truss bridge with X-type inclined chords in all but the end panels was monitored under a vehicle moving slowly. The axial forces in the two components of one of the inclined chords were calculated from observed strains at its mid-length (Bakht and Jaeger 1987). Figure 10.8 shows the strains in the instrumented components of the chord as the vehicle moves from left to right. It can be seen that the sign of the axial strains in the chord changes as the vehicle

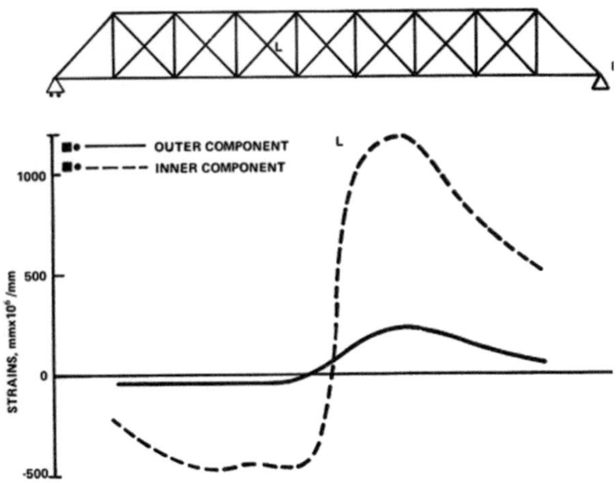

Fig. 10.8 Variation of axial strains in two components of an inclined chord in a panel as a vehicle moves from *left* to *right* (compressive strains shown as negative)

passes its location, and that the magnitude of the compressive strains in the same chord remain small as compared to the magnitude of the tensile strains; this observation confirms the above claim that an apparently indeterminate truss of the kind under discussion can be analyzed as a statically determinate truss without significant loss of accuracy.

This section deals mainly with steel trusses, because in steel chords, it is feasible to convert the observed strains into stresses by assuming that the value of the modulus of elasticity *E* of steel is 200 GPa. The computed stresses can then be used to determine the chord forces. The same facility is not available for the timber components of timber trusses, because *E* of timber is highly variable, as discussed later.

10.3.2 Buckling of Trusses

The Canadian Highway Bridge Design Code (CHBDC 2006) permits the results from a bridge test to be extrapolated to determine the live load carrying capacity of the bridge. However, one of the two conditions associated with the extrapolation is that the stability of the component is not of concern with any further increase in load. This condition is particularly relevant to truss bridges; the strengths of the trusses of these bridges are generally governed by the buckling capacity of their compression chords.

The evaluation of the load carrying capacity of truss bridges through SHM, or testing, must be complemented by rigorous buckling analysis of their trusses. As noted in the followings, the strengths of the compression chords of the trusses of all three types of bridges discussed above are governed by their buckling in or out of the plane of the trusses, and also by local buckling of the plates of their built up chords.

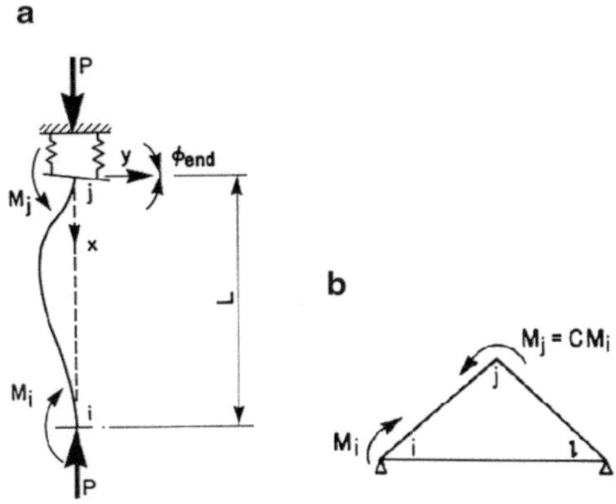

Fig. 10.9 Notation for buckling analysis: (**a**) chord in compression, (**b**) a triangular frame with rigid joints

In-Plane Buckling of Trusses As noted earlier, the basic assumption in the design and evaluation of the load carrying capacity of a truss is that all chords are connected at the nodes through pins, or hinges, so that there are no bending moments at the ends of all chords. This conservative assumption, while having served well for the design of trusses for more than 150 years, is too restrictive for the evaluation of truss bridges.

Csagoly and Bakht (1978a, b) have presented an analytical method, which shows that a steel truss can fail not only through the buckling of one its compression chords, by also through the instability of one its nodes. The notation used to differentiate between the rotational stiffness of a chord and a node is explained with the help of Fig. 10.9a, which shows a chord ij subjected to a compressive force P and a moment M_i at node i. Node j, with a finite rotational stiffness experiences moment M_j. K_i' is the total rotational stiffness at node i, K_{ij}' is the rotational stiffness of chord ij at end i, and K_{ij} is the total rotational stiffness of node i minus the rotational stiffness of chord ij at its end i. The concept of these stiffnesses can be visualized with the help of a triangular framework ijl with rigid joints, shown in Fig. 10.9b. A moment M_i is applied at node i. The total rotational stiffness, K_i', at node i is given by the following equation.

$$K_i' = K_{il}' + K_{ij}' \tag{10.1}$$

and K_{ij} is given by:

$$K_{ij} = K_i' - K_{ij}' = K_{il}' \tag{10.2}$$

To demonstrate that rotational stiffness of the nodes of a rigidly jointed framework is dependent upon the axial forces in the chords, (Csagoly and Bakht 1978a, b) have given the example of a triangular frame work ijl with a vertical force P at node i. As shown in Fig. 10.10, the increasing load causes the rotational stiffness of the compression chords to decrease, and that for the tension chords to increase,

Fig. 10.10 Stiffness values of rotational restraint in a triangular framework plotted against applied load

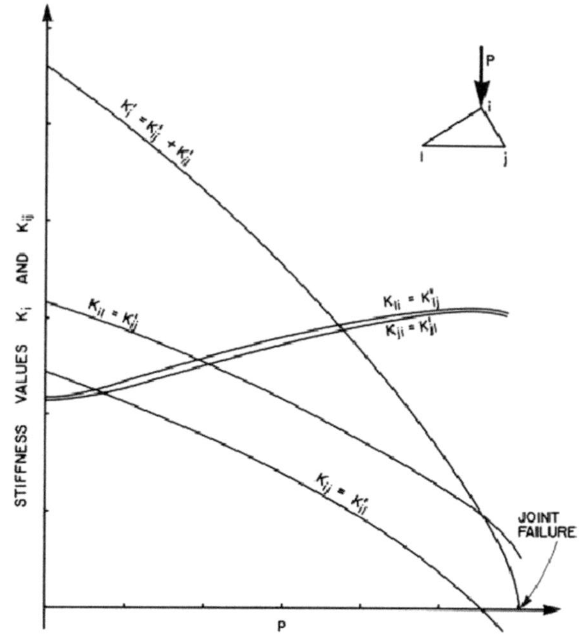

although at a slower load. The rotational rigidity at node *i*, being the sum of the rigidities of two compression chords drops more rapidly. When the nodal rigidity becomes zero, the joint becomes unstable causing the failure of the framework.

The method described above has been incorporated in a computer program called IPBUCK, the validity of which has been verified with the help two small-scale models (Csagoly and Bakht 1978a, b).

Local Buckling of Compression Chords As described by Bakht and Csagoly (1979), a steel through-truss bridge, called the North West Arm Bridge, was tested to failure; the primary purpose of the test was to validate a method of predicting the in-plane buckling behaviour of trusses, discussed later. To ensure that the bridge did not fail in tension, some of the tension chords were reinforced. Two lower nodes of the two trusses of the bridge were encased in concrete blocks, one of which can be seen in Fig. 10.11. Hydraulic jacks were placed on these concrete blocks to pull on the steel cables the lower ends of which were anchored in rocks directly below the nodes contained by concrete blocks. The pulling of the cables induced downward loads at the nodes.

Under the test loads the bridge failed prematurely due to the local buckling of the cover plate of the built-up section of one of its inclined chords. This local failure, which can be seen in Fig. 10.12a, resulted in the failure load of the bridge being about one-third less the predicted failure load.

Fig. 10.11 The North West Arm Bridge

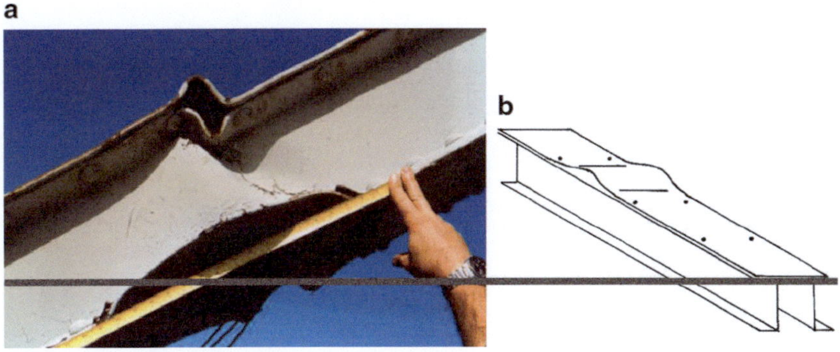

Fig. 10.12 Local failure of a compression chord of a steel truss: (**a**) photograph of failed inclined chord, (**b**) sketch showing ripple in the cover plate

It was discovered that the cause of this unexpected weakness in the strength of the compression chord was the build-up of rust between the sparsely-riveted cover plate and the channel sections. The rust between the cover plate and the channel sections caused the cover plate to ripple between the rivets, as shown in the sketch in Fig. 10.12b. The rippled cover plate, unable to sustain substantial compressive forces, reduced the buckling capacity of the built up chord.

A sound segment of the inclined chord, removed from the bridge, is shown in Fig. 10.13, in which a bulge of the cover plate can be seen between the rivets. Ordinary C-clamps were used midway between the rivets to prevent the cover plate from deforming excessively. After clamping the cover plate to the channels, the segment of the inclined chord was tested under direct compression, as shown in Fig. 10.14. It was found the clamped segment failed at about 40 % higher load than the force under which the inclined chord had failed due to local buckling in the ultimate load test.

The above discussion underscores the need for analytical investigation of the local buckling of compression of built-up chords for the evaluation of truss bridges.

Lateral Buckling of Pony Trusses The top chords of pony trusses are laterally supported by knee-braces that comprise projections of floor beams at the lower ends, and inclined members that connect the ends of the projected floor beams to the

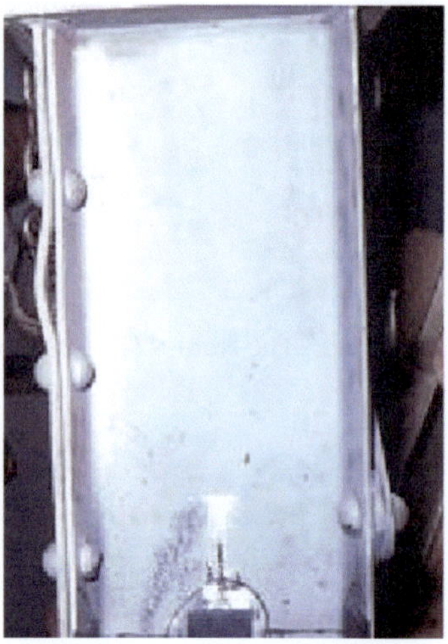

Fig. 10.13 A segment of the inclined chord showing ripple in cover plate

Fig. 10.14 A segment of the inclined chord with C-clamps

top flange. A knee-brace can be seen in Fig. 10.7. As illustrated in Fig. 10.15 vehicle load on a floor beam causes the top chords of the trusses to deflect towards the inside of the bridge, thus reducing its buckling capacity. The resistance of the top of the knee-brace to lateral movement, on the other hands, provides the spring support to the top chord, enhancing its buckling capacity.

The model used for incorporating the non-linear interaction between the detrimental and beneficial aspects of the knee-braces, addressed by Csagoly et al. (1975), is illustrated in Fig. 10.16a, b, where the former figure shows the elevation of a truss with the locations of the knee-braces, and the latter the top chord in plan with spring supports offered by the knee-braces at their top ends.

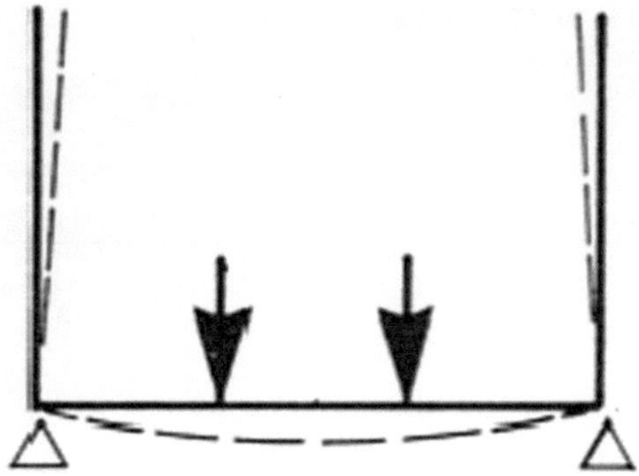

Fig. 10.15 Cross-section of a pony truss bridge at a floor beam location

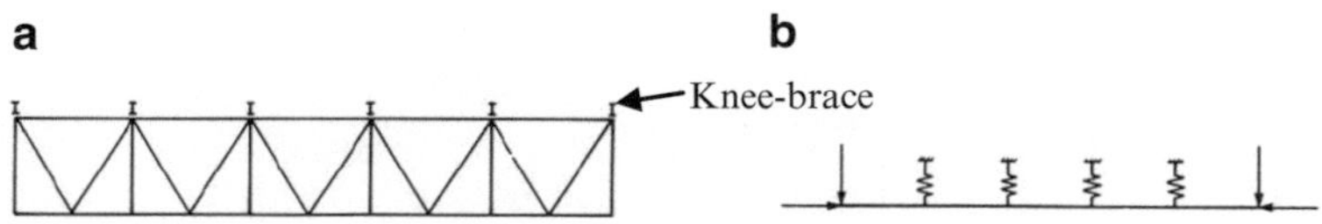

Fig. 10.16 Idealization of the top chord of a pony truss bridge; (**a**) the truss in elevation, (**b**) the top chord in plan

The method described above had been incorporated in computer program called LATBUK (Csagoly et al. 1975). Since this program, written on cards, is no longer available, the original method has been re-incorporated in a program called PTBUCKLING, written in modern FORTRAN (Mufti and Bakht 2015). The validity of the method has been verified by an ultimate load test on a pony truss bridge which was tested under a load of concrete blocks, as can be seen in Fig. 10.17. An important outcome of the ultimate load test, confirmed by the analytical method, was that the trusses of pony truss bridges are most vulnerable at the nodes joining end inclined chords with the top chords.

10.3.3 Case Histories

Field testing of truss bridges is not new. It has been practiced for centuries. An example of an early bridge test is shown in Fig. 10.18, in which a truss bound for a bridge in India was tested in England in the nineteenth century.

Simply supported truss bridges are usually simple to analyse because of the limited number of paths that a load can take, due to their low degree of structural redundancy. Tests on these bridges have shown, however, that even these bridges have certain aspects of behaviour which may surprise even experienced bridge engineers. Some significant surprises relating to truss bridges are presented in the following.

Fig. 10.17 A pony truss bridge during an ultimate load test

Fig. 10.18 Testing of a steel truss in England for a railway bridge in India in the nineteenth century (Print courtesy of Dr. Roger Dorton)

Interaction of Floor System with Bottom Chord It is well known that the breaking of the bottom chord of a statically determinate truss leads to its collapse. Figure 10.19a shows a statically determinate truss that remained intact, and continued to carry traffic, even after the breaking of its bottom chord. This dramatic example underscores the interaction between the floor system and the horizontal chords that are close to it. Figure 10.19b shows the photograph of an in-service bridge, the bottom chord of which is buckled and thus carries no tensile force. It is obvious that the tensile force at the bottom of the truss, necessary for its integrity, is sustained by the frozen bearings of the truss.

Bakht and Jaeger (1987) have described tests on two truly pin-connected steel truss bridges, which were similar in their span lengths and other dimensions. One of these through-truss bridges can be seen in Fig. 10.5. The bottom chord strains in this bridge due to a truck, plotted in Fig. 10.20a, were found to be smaller by a factor of about 15 than the strains which would have occurred if the chord had sustained all the live load force by itself. The obvious conclusion drawn from this observation is that if the bearings of the truss are functioning, the floor system must be acting with the bottom chord in sustaining the tensile forces. The observation that the floor system in pony-truss and through-truss bridges, takes a large portion of the tensile force of the truss bottom chords has been made so many times that it has virtually become a cliché. Nevertheless, a surprising feature was observed in a test on a through-truss bridge with similar dimensions, the results of which are plotted in Fig. 10.20b.

Fig. 10.19 Statically determinate trusses: (**a**) with broken bottom chord, (**b**) with buckled bottom chord

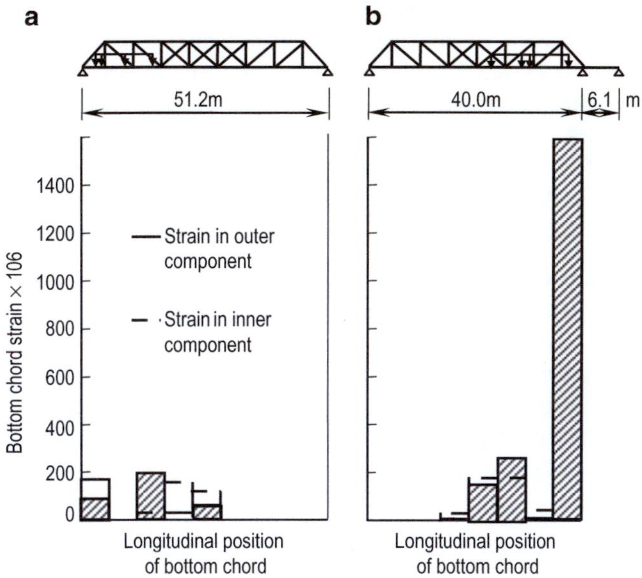

Fig. 10.20 Strains along the bottom chords of the trusses of two bridges: (**a**) bridge without approach span; (**b**) bridge with approach span

The surprising feature relates to the bottom chord strains in the panel closest to the right hand support of the latter bridge. It can be seen in Fig. 10.20b that the strains in this panel are about 15 times larger than the strains in the adjacent panel. Since the two panels have components of the same cross-section, it is obvious from simple statics that the bottom chord strains in the two panels should be very nearly the same.

The fact that the total strains in the bottom chords of the two panels are so significantly different from each other suggests that the floor system does not participate with the bottom chord in the end panel.

This unexpected behaviour is explained as follows by Bakht and Jaeger (1987). All stringers of the floor system of the bridge in Fig. 10.20b are connected to the

Fig. 10.21 A truss arch bridge

truss nodes in such a way that the bottom chord between adjacent nodes cannot deform without engaging the longitudinal stringers of the deck system. As shown in this figure, the bridge has a small approach span, which is formed by extending the stringers of the floor system beyond the pier supporting the trusses. Because of this extension, the floor system does not have a floor beam at the end node, as it does for all the other nodes. Accordingly, the bottom chord in the end panel, by deforming independently of the stringers, is called upon to sustain all the tensile force of the truss. It is obvious that the beneficial interaction between the floor system and the bottom chords of the trusses cannot always be taken for granted.

Component Interaction The beneficial effect of component interaction in truss bridges is often disregarded as being insignificant. Confirmation that the effect of interaction is not always small was provided by a test on a deck-truss type of arch bridge with a span of about 104 m, reported by Tharmabala (1986). The bridge, which can be seen in Fig. 10.21, has several transverse floor beams of the same cross-section. For investigating their strengths by a diagnostic test, the bottom flanges of the floor beams were instrumented with strain gauges attached to their respective mid-spans.

As shown in Fig. 10.22, it was found that under similar loading, the four beams closest to one of the truss supports showed considerably higher strains than the other two instrumented floor beams. The reason for this unexpected behaviour came to light after an inspection of the cross-frames under the various floor beams. The configurations of the various cross-frames are shown in Fig. 10.23. It can be seen in this figure that the cross-frames under the outer five floor beams have an X-type of bracing, which permits these floor beams to span between the trusses, as was assumed in the pre-test analysis, and as probably would have also been assumed in the design calculations. The other floor beams cannot span directly between the trusses due to the double-A type of bracing in the cross-frames under them. It was because of the integral nature of these floor beams with the cross-frames that the measured strains in the floor beams were considerably smaller than they would otherwise have been. Advantage was taken of this feature in the rehabilitation of the bridge.

Fig. 10.22 Strains in the bottom flanges of floor beams midway between trusses

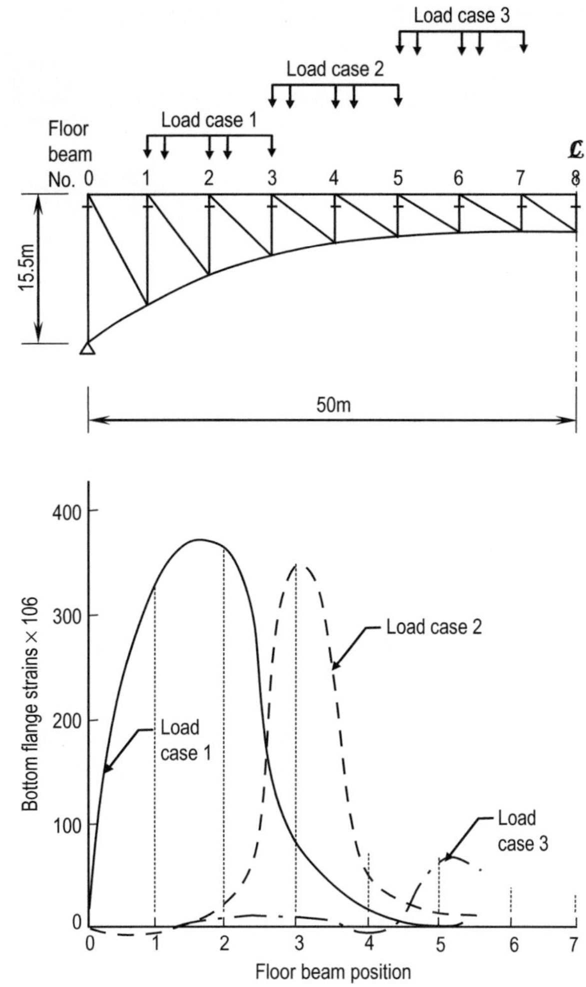

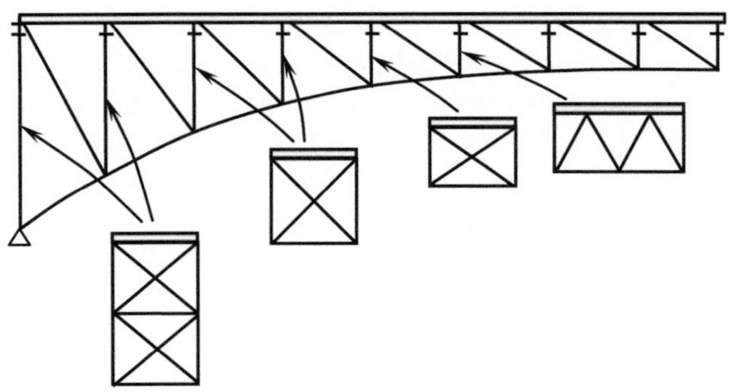

Fig. 10.23 Details of cross-frames

Fig. 10.24 A bowstring truss span

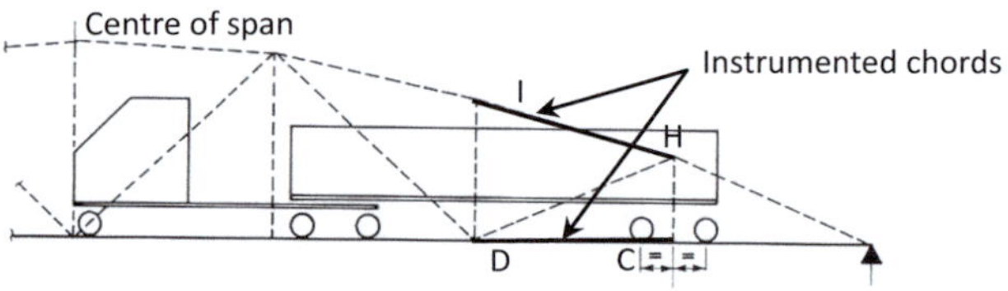

Fig. 10.25 Truck position on bowstring truss

Moments in Chords In trusses with slender chords, such as those shown in Fig. 10.5, the moments induced by rigid nodes are small and can be neglected. However, in shallow trusses with chords having substantial cross-sections, the moments induced by rigid joints are large and cannot be ignored as shown by the example of a pony truss span shown in Fig. 10.24.

The bowstring truss span of Fig. 10.24, with a length of 34.14 m, was instrumented with electrical resistance strain gauges to monitor both axial forces and moments in several of its top and bottom chords due to a 5-axle truck carrying a gross weight of 1,414 kN. As an example, Fig. 10.25 shows the instrumented bottom and top chords of one panel; the same figure also shows a position of the truck with respect to the bowstring truss span. Full results from the monitoring are presented by Bakht and Csagoly (1977).

The observed load strains in the chords were converted into stresses by multiplying them with E of steel. The stresses were then used to calculate the axial forces and moments near the two ends of the instrumented chords. The axial forces and moments in the two instrumented chords under the truck shown in Fig. 10.23 are compared in Table 10.2 with two sets of corresponding analytical results; in one set, the analysis was carried out by assuming that all nodes of the pony truss were rigid connections, and that the truss supports were simple, permitting unrestricted longitudinal movement. In the other analysis, the assumption of rigid connections was repeated, but the two supports of the truss were assumed to have full restraint against longitudinal movement.

Table 10.2 Axial forces and moments in two chords of a bowstring truss

		Axial force, kN			Moment, kN.m		
Chord	Location	From observed strains	Analytical, free bearings	Analytical, jammed bearings	From observed strains	Analytical, free bearings	Analytical, jammed bearings
Top	I	−1,023	−1,050	1,023	6.4	5.8	11.5
	H	−961	−1,050	1,023	34.4	55.4	33.3
Bottom	D	809	934	334	5.0	−0.7	7.9
	C	792	934	334	64.0	76.4	49.3

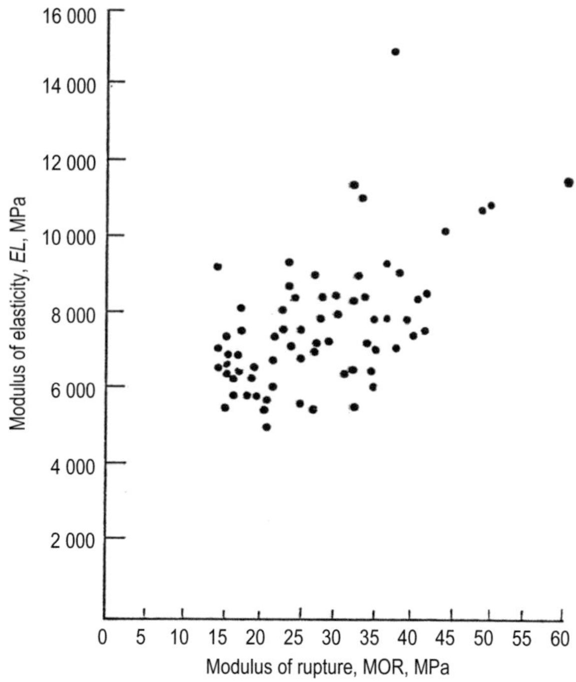

Fig. 10.26 Test data for 70 Red Pine stringers

The following observations were made by Bakht and Csagoly (1977) about the comparison between the analytical values of axial forces and moment and those computed from the observed data.

(a) The experimental axial forces in the bottom chords were between the analytical values for free and jammed bearings, but were closer to the values for free bearings.
(b) The experimental axial forces in the top chords, which are relatively insensitive to the bearing restraint, compared well with experimental values, being within $-10\ \%$ in most cases.
(c) The bending moments in the chords, induced by the rigid joints, were so large that they could not be neglected.
(d) Most experimental bending moments near the nodes were between the analytical values for free and jammed bearings, thus highlighting the important of bearing restraint on moments.

A Timber Truss Bridge Figure 10.26 shows the values of the longitudinal modulus of elasticity, E_L, and modulus of rupture (MOR) obtained by in-grade testing of 70 Red Pine beam and stringer grades (Bakht and Jaeger 1985). It can be seen in this figure that E_L of the test specimens ranges between about 5,000 and 15,000 MPa. In light of such large variability, which is typical of all timber specimens of the same species and grade, it is not feasible to determine stresses in timber from measured strains. It is for this reason that the only load effect that can

Fig. 10.27 The Sioux Narrows Bridge known as the world's longest span timber bridge in its time

be monitored realistically in a timber bridge is its deflections. While the deflections of a timber truss bridge can provide a qualitative measure of the robustness of the joints, the load capacity of such a bridge requires that the test data be complemented with extensive analysis.

The Sioux Narrows Bridge in northern Ontario, Canada, can be seen in Fig. 10.27; this bridge, with a record span of 64 m and built in 1936, was considered to be the longest span timber bridge in the world. In the middle 1970s, this bridge was scheduled to be replaced because its main timber components were perceived to have deteriorated excessively. A load test of the bridge was performed by measuring the deflections of its trusses under known truck loads. When it was found that the measured deflections were considerably smaller than the calculated deflections of trusses with deteriorated components, the bridge was kept in service for unrestricted traffic for about 30 years. After about 70 years of interrupted service, the bridge was taken down to make room for a wider bridge.

After the initial test, the old sawn timber deck of the bridge was replaced by a more efficient stress-laminated wood deck. After the refurbishment of the deck, the bridge was re-tested. During the latter test, the steel hangers of the trusses were instrumented with strain gauges, it being noted that each hanger comprised of two steel bars of rectangular cross-section. It was found the sharing of loads in two components of the hangers was highly uneven, suggesting that that the dead load forces in the components might also be highly uneven. Bakht and Jaeger (1987) have presented the following relatively simple experimental procedure to determine the ratio of the dead load forces in the two components of a hanger.

A prismatic bar with a length L between its pinned ends and subjected to a tensile force P is considered. It can be readily shown that the frequencies of vibration, f_1 and f_0, due to forces P_1 and zero, respectively, are related according to the following equation.

$$f_1^2 = f_0^2\left[1 + \left(P_1/P_E\right)\right] \tag{10.3}$$

where P_E is the Euler buckling load for the bar. The frequency f_2 corresponding to a tensile force P_2 can similarly be obtained, leading to the following equation.

$$\frac{P_1}{P_2} = \frac{f_1^2 - f_0^2}{f_2^2 - f_0^2} \tag{10.4}$$

The following equation gives the value of f_0.

$$f_0 = \frac{\pi}{2L^2}\left[\frac{EI}{m}\right]^{0.5} \tag{10.5}$$

where m is the mass per unit length of the bar and EI its flexural rigidity. If f_1 and f_2 can be determined experimentally for the two components of a tension chord, then the ratio (P_1/P_2) can be readily obtained from Eq. (15.4). In practice, f_0 is much smaller than f_1 or f_2, so that (P_1/P_2) is approximately equal to $(f_1/f_2)^2$.

The frequencies of vibration of the two components of a tension chord of a truss of the Sioux Narrows were determined by attaching a deflection transducer to the component with respect to a tripod placed on the deck, which was regarded as the fixed datum for the vibrations of the component; the monitoring arrangement is illustrated in Fig. 10.28.

The tension component was disturbed by hand to vibrate freely in the lateral direction, and in the pre-PC days the output from the deflection transducer was recorded in graphical form by a strip-chart recorder, in which the ink stylus records the dynamic lateral deflections, and the graph paper is made to advance at a

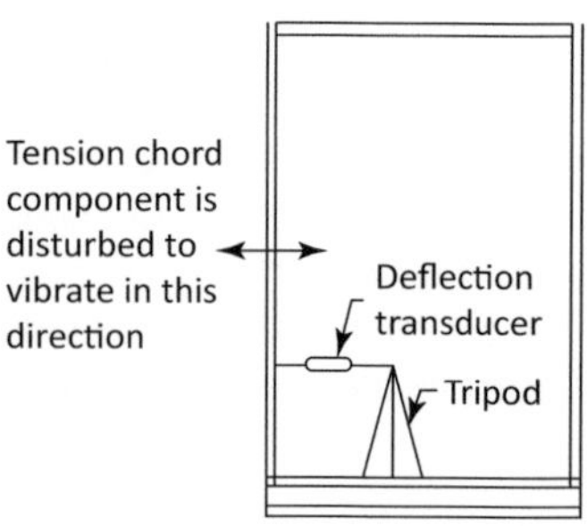

Fig. 10.28 Test arrangement for monitoring lateral frequencies of hangers in a truss

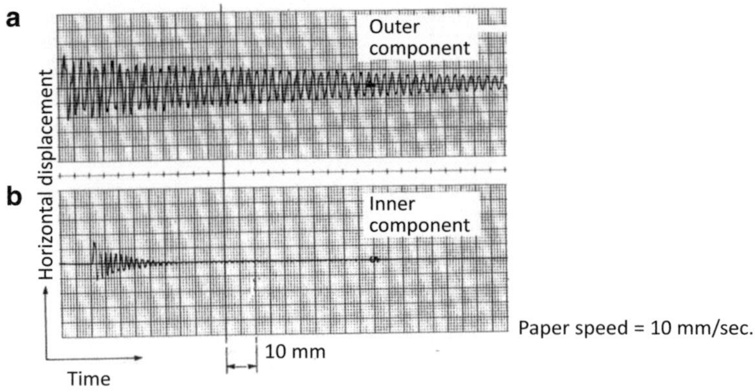

Fig. 10.29 Strip chart output: (**a**) for outer component, (**b**) for inner component

pre-determined speed. The output for the free vibrations of the outer and inner components of the hanger of the Sioux Narrows Bridge are reproduced in Fig. 10.29a, b, respectively.

From the displacement-time curves presented in Fig. 10.29a, b, it can be readily calculated that that the first frequencies of vibration of the inner and outer components of the tension chords are nearly 5.50 and 3.75 Hz, respectively. Each component was 7.93 m long and has a 63.5 × 23.8 mm cross-section. The mass m of each component was 11.81 kg/m, and its moment of inertia about the weak axis, which corresponded to the monitored frequency was 0.71×10^{-7} m^4. For a modulus of elasticity E of 200,000 MPa and assumed pin-ended boundary condition, f_0 is obtained as follows.

$$f_0 = \frac{\pi}{2 \times 7.93^2}\left[\frac{200{,}000 \times 10^6 \times 0.71 \times 10^{-7}}{11.81}\right]^{0.5} = 0.87\ Hz$$

The ratio of the tensile forces taken by the outer and inner components, P_1/P_2, is obtained from the following equation.

$$\frac{P_1}{P_2} = \frac{5.50^2 - 0.87^2}{3.75^2 - 0.87^2} = 2.22$$

The rotational restraint at the end of the long and slender component is unlikely to affect its effective length substantially. A maximum reduction of 10 % is considered, in which case the effective length of the component becomes 7.14 m, and f_0 is found to be 1.07 Hz. The value of P_1/P_2 is now calculated to be 2.25, only 1.4 % smaller than the value calculated previously. It can be seen that the vibration characteristics of multiple components of the tension chords are useful in estimating the fractions of the tension due to dead loads.

10.4 Slab-On-Girder Bridges

Amongst medium span bridges, which have spans ranging between about 20 m and 125 m, slab-on-girder bridges are most common; in these bridges, the girders are usually of steel, reinforced concrete, or prestressed concrete, and the 'deck slab' is commonly of concrete. The term 'deck slab' is used for a solid concrete slab, carrying and distributing wheel loads to the supporting members, whereas the term 'slab' is used commonly for the slabs which do not have supporting members such as girders, floor beams and stringers. The term 'girders' is used for the longitudinal supporting members, which transfer the loads to the abutments or piers of the bridge. When the spacing of the girders is greater than about 4 m, it is economical to introduce floor beams, i.e. transverse members between the girders. The floor beams are then inter-connected by longitudinal beams, called the stringers.

In modern bridges, the concrete deck slab is connected to the girders, and stringers if present, by shear connectors so that the deck slab and longitudinal components act compositely. The highway systems, however, still have older slab-on-girder bridges in which no mechanical shear connection is provided between the deck slab and the longitudinal supporting members.

As noted in several places in the book, the main consideration of an SHM system for bridges should be to optimize the number of sensors so that the most relevant information is obtained without sacrificing accuracy. To this effect, it is assumed that the stress distribution in the supporting members of the slab-on-girder bridges is predominantly uniaxial and that the stress in the perpendicular direction of the axis of the members can be safely neglected. The effect of this assumption is that strain gauges measuring strains only in one direction can be used to monitor the strains in the members, and that the stress at the instrumented location is simply equal to strain $\times E$, where E is the modulus of elasticity of the material of the supporting member. For steel, E can safely be taken as 200 GPa, and for concrete E in MPa is taken as $\left(3000\sqrt{f_c'} + 6900\right)(\gamma_c/2300)^{1.5}$ (CHBDC 2006).

10.4.1 Designing of an SHM System

The determination of the placement and number of sensors, varies from structure to structure, requiring an intimate knowledge of its perceived strengths and weaknesses, it being noted that the primary means of acquiring this knowledge are the condition survey and an analytical evaluation of the bridge. Since every bridge is unique in its behaviour, it is not possible to lay down categorical rules for the designing of an SHM system for all bridges of a given type. However, there are certain traits of behaviour that are usually present in a given type of bridge. The following example of the design of an SHM system for a seven span slab-on-girder bridge could be used as a guide.

a

b

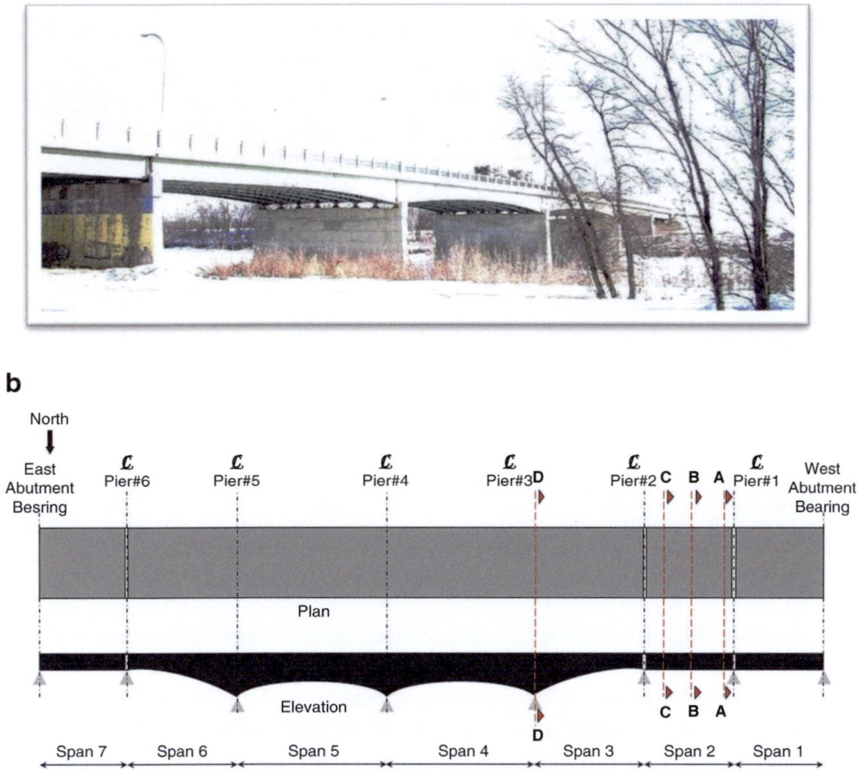

Fig. 10.30 A seven-span bridge: (**a**) photo of the bridge, (**b**) plan and elevation of the bridge

The bridge under consideration, shown in Fig. 10.30a, has seven spans, three of which are simply supported and the other four are continuous. As shown in Fig. 10.30b, starting from the east end of the bridge, the lengths of the spans are 28.8, 36.6, 48.8, 48.8, 36.6, 28.8, and 28.8 m, respectively. The deteriorated concrete deck slab of the bridge was replaced by an externally restrained deck slab with glass fibre reinforced polymer (GFRP) bars. The rehabilitated bridge was decided to be instrumented and monitored on a continuous basis. The reasons for the monitoring of the bridge are discussed in the following along with the proposed instrumentation at different sections of the bridge.

The simply supported span No. 2 of the bridge, being the second from the western end, was instrumented for weighing-in-motion of heavy vehicles. Accordingly, the girders in this span were instrumented at three transverse sections, being A-A, B-B and C-C. Section A-A was about 1 m from the centreline of girder supports at Pier 1, Section B-B was at the mid-span, and Section C-C was about 3 m away from the centreline of girder supports at Pier 2.

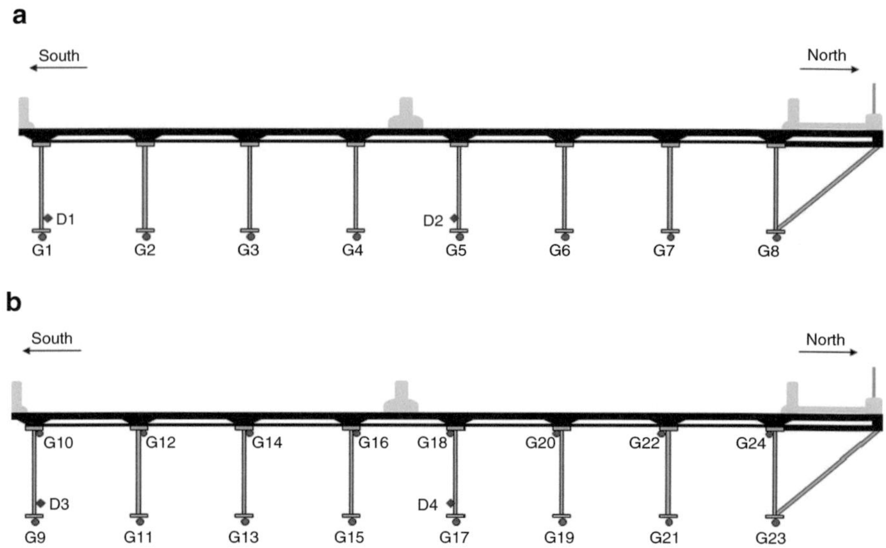

Fig. 10.31 Strain gauges on girders: (**a**) at Section A-A, (**b**) at Section B-B

As shown in Fig. 10.31a, electrical resistance strain gauges at Section A-A were installed on the underside of the bottom flanges of the girders; these gauges, labelled as G1 through G8, were primarily meant to measure the effect of bearing restraint. As noted by (Bakht and Jaeger 1990a, b) and shown later in this chapter, the bearing restraint is present even in bridges with apparently functioning neoprene bearings. Figure 10.31a also shows two dummy gauges, labelled as D1 and D2; these gauges, placed on unstressed pieces of steel attached to the girders, were meant to correct the effect of strain gauge drift and free expansion and contraction of the girders due to temperature changes.

As shown in Fig. 10.31b, electrical resistance gauges at Section B-B were placed on the underside of both the top and bottom flanges of the girders; these gauges were installed to establish the degree of composite action between the deck slab and the girders, and to determine the transverse positions of the vehicles. The transverse location of a vehicle on the bridge can be determined easily by comparing the transverse profile of longitudinal girder strains with similar transverse profiles obtained by calibration tests. The gauges at Section B-B were labelled from G9 to G24, and the dummy gauges at the same section were labelled as D3 and D4.

The gauges at Section C-C were identical to those at Section A-A, except that there were no dummy gauges at Section C-C. The gauges at Section C-C were meant to measure weights and spacing of truck axles and axle groups, it being noted that the speed of a vehicle, and hence the spacing of its axle groups can be calculated by using strain reading at all the three instrumented sections. While gauges at Section B-B can also be used to measure weights of vehicles in motion, the strain-time curves at these locations do not have well-defined peaks because of which the determination of weights of axles and closely spaced axle groups is prone to large numerical errors.

Because of concerns for the fatigue resistance of top flanges of girders over the piers, shear connectors were not provided between the girders and deck slabs in the negative moment regions. Realizing, however, that some composite action might exist in these regions, it was decided to install electrical resistance strain gauges at both the top and bottom flanges of the girders at Section D-D. The gauges had the same pattern as that shown for Section B-B in Fig. 10.31b. Another reason for the sensors at Section E-E was to gather the statistics of tensile strains in the top flanges of girders over the intermediate supports.

10.4.2 Case Histories Dealing with Boundary Conditions

While it is well known that the boundary conditions of a structural component affect significantly its load carrying capacity, there are still some misconceptions about the boundary conditions of some components in common bridges. Two examples are given in following to illustrate this point.

Bearing Restraint in Simply Supported Spans The cross-section of the single-span North Muskoka River Bridge is shown in Fig. 10.32. This bridge comprises five steel girders and a composite deck slab; its span and width are 45.7 and 14.6 m, respectively. One end of every girder rests on laminated elastomeric bearings each measuring 560 × 335 mm in plan and 64 mm in thickness. The design shear rate for each bearing is about 30 kN/mm.

A dynamic test showed the North Muskoka River Bridge to be about 20 % stiffer in flexure than could be rationalized by even a very detailed analysis in which all those components of the bridge were taken into account, which could conceivably enhance flexural rigidity. To determine the cause for the apparent discrepancy, a diagnostic static test was conducted subsequently. For this latter test, all the girders were instrumented with gauges to measure longitudinal strains at three transverse

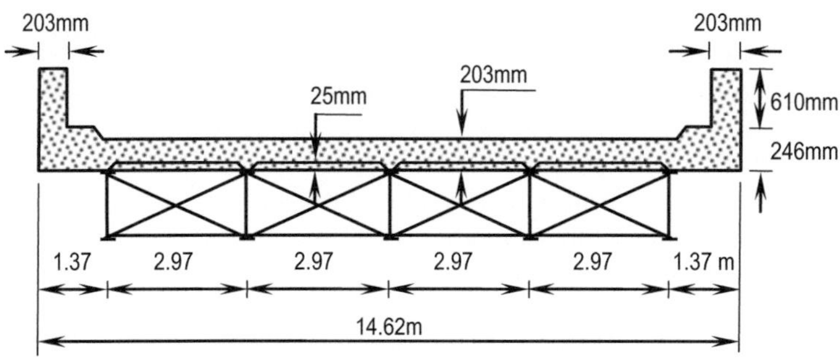

Fig. 10.32 Cross-section of the North Muskoka River Bridge (not to scale)

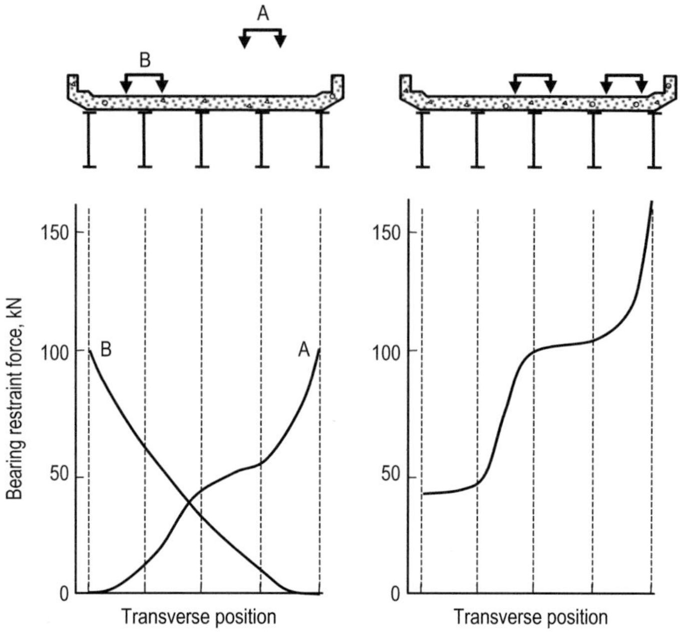

Fig. 10.33 Bearing restraint forces in the North Muskoka River Bridge

sections of the bridge, one section being near the mid-span and the other two near each abutment (Bakht and Jaeger 1990a).

Had the elastomeric bearings permitted free longitudinal movement of the girders, the live load strains in the bottom flanges near the bearings would have been tensile and very small. It was found that this was not the case. The test loads induced fairly large compressive strains in the bottom flanges near the elastomeric bearings. Bearings restraint forces computed approximately from observed strains are plotted in Fig. 10.33 for different load cases. It is interesting to note that under transversely symmetrical loads, the corresponding bearing restraint forces were not exactly the mirror image of each other, as should have been the case for an ideally symmetrical structure. Bearing restraint forces as high as 175 kN, which can be seen in Fig. 10.33, are considerably larger than a functioning elastomeric bearing would be expected to develop. Nevertheless, such large forces were really present, despite the fact that the bearings were apparently in excellent and functioning condition.

A further proof of the presence of large bearing restraint forces in the North Muskoka River Bridge was provided by comparisons of applied moments obtained from considerations of simple supports with those computed from girder strains. Figure 10.34 shows the comparison of mid-span girder moments computed from measured strains with those obtained by the familiar grillage analogy method, in which the bearing restraint forces were not accounted for. It can be readily concluded from this figure that the total moment sustained by all the girders is

Fig. 10.34 Mid-span girder moments in the North Muskoka River Bridge

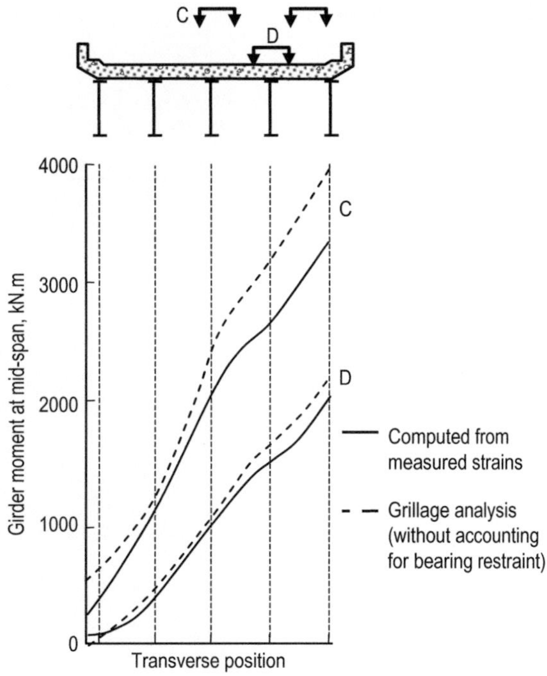

noticeably less than the corresponding applied moment obtained on the basis of simple supports; this observation confirms that the applied moments were reduced by the effect of bearing restraint.

It was found that at the time of the test, the bearing restraint in the North Muskoka River Bridge reduced the mid-span deflections due to test loads by about 12 %. This reduction is considerably smaller than the 20 % reduction observed in the previous dynamic test on the same bridge. The previous test was conducted on a relatively cool day in October and the latter on a very hot day in June. It is hypothesized that the elastomeric bearings had become stiffer in the cold temperature when the first test was conducted thereby generating higher restraint forces which consequently caused the bridge to become effectively stiffer than it was at the time of the second test.

Results of the tests on the North Muskoka River Bridge demonstrate the significant influence of the restraining effects of elastomeric bearings which may change with load level and temperature. To be able to analyse bridges with these bearings more accurately, it is essential to include their effective shear stiffness in the mechanical model.

Boundary Conditions of Columns The boundary conditions of a few columns of a 696 m long bridge with an assortment of spans are discussed in the following. The bridge was declared unsafe mainly because of buckling capacity of its 48 columns of steel I-sections; many of these columns can be seen in Fig. 10.35.

Fig. 10.35 A bridge with 48 columns

The buckling capacities of the columns of the bridge under consideration were determined analytically by using two main assumptions: (a) at its lower end each column, which was connected to the steel base plate and concrete pedestal with only two bolts, was assumed to have a hinge connection; and (b) the compressive force induced by live loads was assumed to have an eccentricity of 200 mm from the centre line of the columns. The connection of a column to its base is shown in Fig. 10.36. It is noted that the 200 mm dimension represents the distance from the centre of the column I-section to the centre of the vertical line of bolts connecting the steel beams to the columns through steel angles.

A few columns of the bridge were installed with electrical resistance strain gauges measuring vertical strains at four different horizontal sections; three gauges were placed on each flange, and one gauge on each face of the web. Two of the instrumented sections were about 150 mm away from the steel angles at the top and

Fig. 10.36 Connection of a steel column to its base

bottom of the columns, respectively. The other two sections divided the length between the top and bottom sections equally.

The observed strains were converted to stresses by multiplying them to *E* of steel (200 GPa), and from the computed stresses, were computed the axial force, the bending moment and the eccentricity of the axial loads with respect to the centrelines of the columns. The various steps in calculating the axial force and moments along the column from the observed strains are given in Fig. 10.37; from the bending moment diagram given in this figure, three observations were made by Bakht and Csagoly (1980): (a) the moment at the bottom of the column is reverse in direction as compared to the moment at the top; (b) the magnitude of the moment at the bottom is about half of that of the moment at the top; and (c) the eccentricity of the axial load with respect to the centre of the column is 74 mm. The first two observations confirm that, contrary to the assumption made in the analytical evaluation, the column is nearly fixed at its bottom. The third observation leads to much smaller bending moments due to live loads than assumed in the analytical evaluation. As noted by Bakht and Csagoly (1980), the rigid connection of the column bases to the foundations, and the much-less-than-expected live load eccentricity improved the load carrying capacity of the columns to such an extent that they were considered safe for unrestricted traffic. The bridge which was initially declared unsafe in 1973, continued to carry unrestricted traffic for about 30 years when it was replaced because of extensive corrosion damage.

An interesting observation from the monitoring described above was that, as shown in Fig. 10.38, the stresses in the column were far from being uniform. The distribution of live load stresses in the flanges was non-uniform with peaks at the centre. The stresses on the two faces of the web were not of the same magnitude. Bakht and Csagoly (1977) concluded that the difference in the stresses on the two faces of the web were probably because of local bending induced by the longitudinal stringer resting on the column.

10.4.3 Case Histories Dealing with Load Distribution

Chapters 2 and 3 deal with mathematical models to determine transverse load distribution in various components of bridge decks. With the help of several examples, it is shown in this Subsection that the mathematical models are sometimes different than the actual bridge structures.

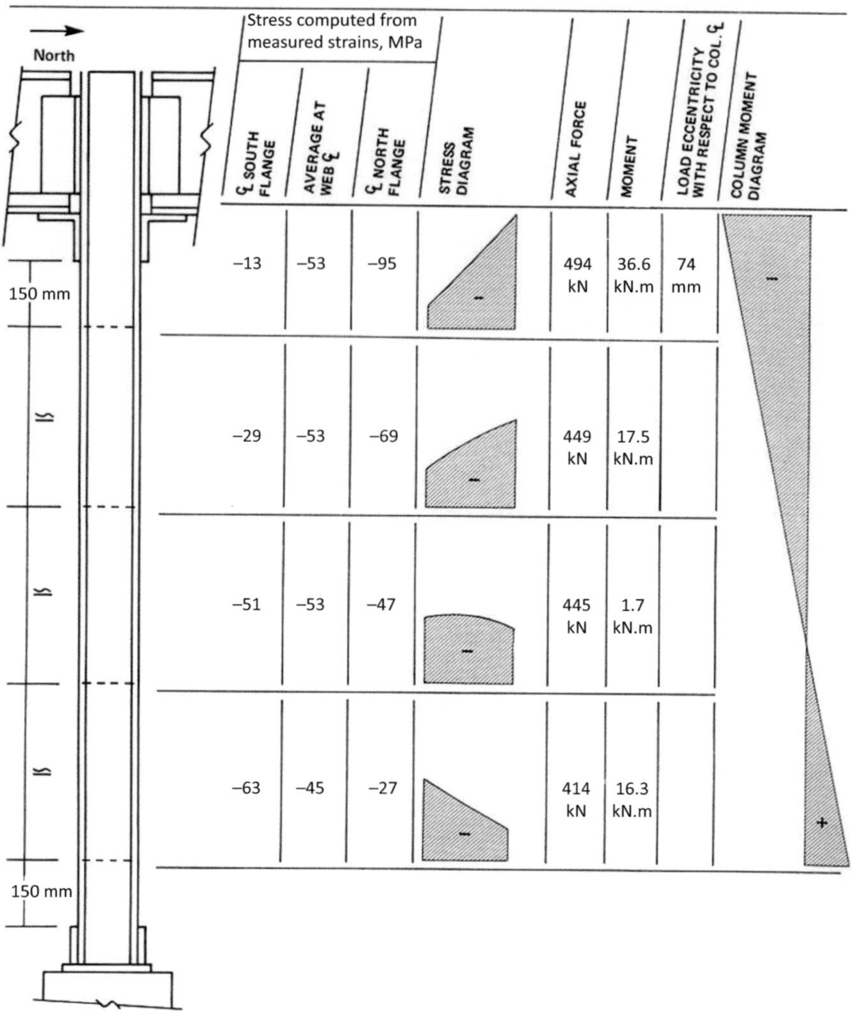

Fig. 10.37 Data computed from measured strains in a bridge column (Compressive stresses shown as –ve)

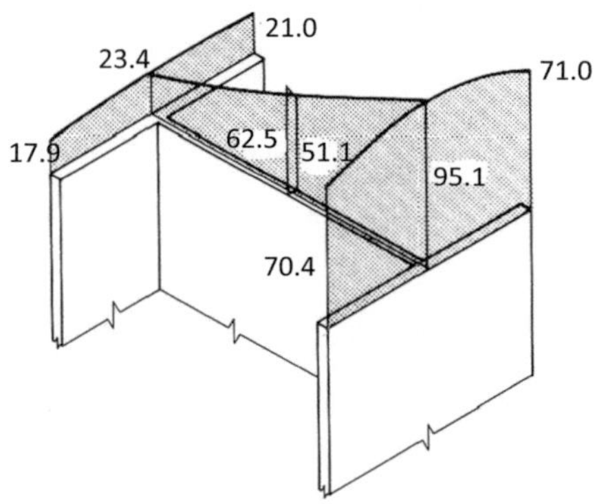

Fig. 10.38 Live load stress (in MPa) distribution at a section near the top of a column due to a truck weighing 1,414 kN

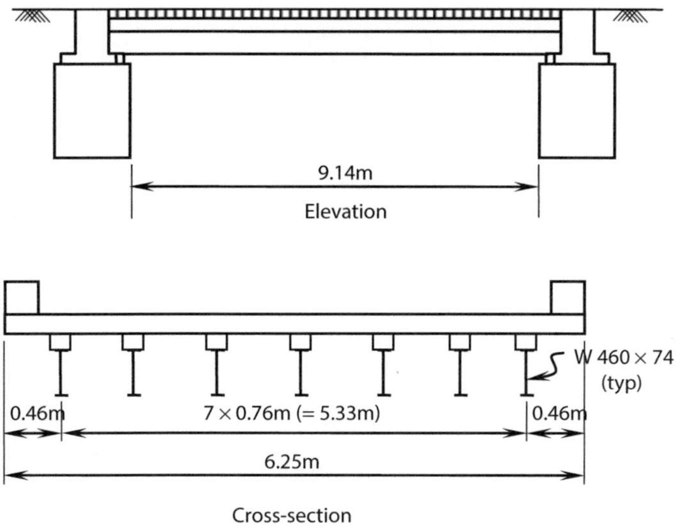

Fig. 10.39 Details of the Lord's Bridge: (**a**) elevation, (**b**) cross-section

Bridge with Timber Decking The example presented is that of the rolled steel girder Lord's Bridge with nail-laminated timber decking in which the wood laminates are laid transversely. As described by Bakht and Mufti (Bakht and Mufti 1992a), the bridge is 6.25 m wide and has a single span that is apparently simply-supported. The girders are 10.2 m long, with a bearing length of 0.53 m at each end, and rest directly on timber crib abutments. There are no mechanical devices to transfer the interface shear between the girders and the timber decking, although there are 100 × 200 mm nailing strips bolted to the top flanges of the girders; the decking is nailed to these strips. Details of this bridge are shown in Fig. 10.39.

The Lord's bridge was tested with a test vehicle under several load levels and different longitudinal and transverse positions. Even up to the highest load level, the girders responded in a linear elastic manner. For two of the load cases, the longitudinal position of the vehicle was the same but the eccentric transverse positions were the mirror images of each other. For these two load cases at the highest test load level, the distribution factors for mid-span deflections are plotted in Fig. 10.40 by viewing the cross-section of the bridge from two different ends so that the two transverse distribution profiles overlap each other for easy comparison. It is recalled that the distribution factor for deflection is the ratio of the actual and average girder deflections at the transverse section under consideration.

If the geometrically symmetrical bridge were also symmetrical with respect to its structural response, the distribution factors for the two mirror-image load cases, noted above, would have led to transverse distribution profiles that lie exactly on top of each other. As can be seen in Fig. 10.40, the two profiles are fairly close to each other but are not exactly the same, thus indicating that the two transverse

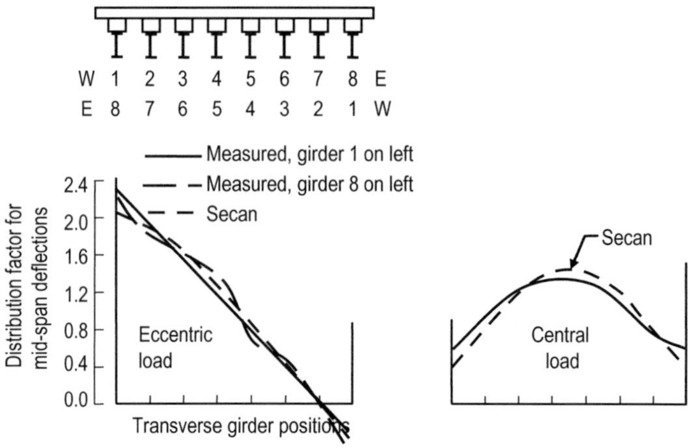

Fig. 10.40 Distribution coefficients for mid-span deflections

Fig. 10.41 Mid-span deflections

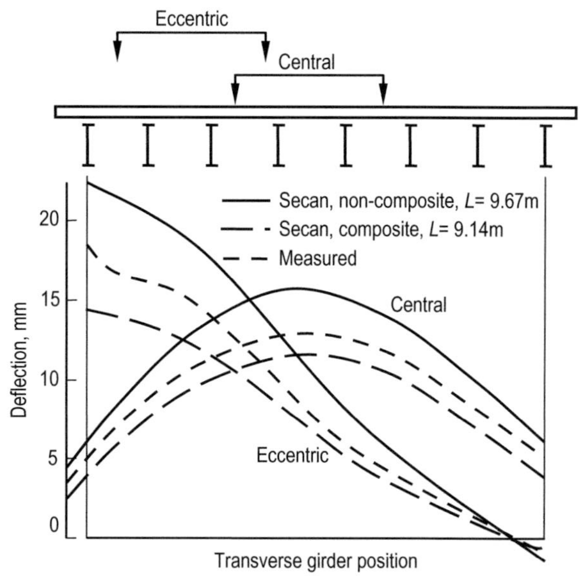

halves of the bridge do not respond in an exactly similar manner to corresponding loads. The two sets of distribution factors obtained from measured deflections are compared in Fig. 10.40 with those obtained by the semi-continuum analysis; it can be seen that the observed distribution factors are very close to their analytical counterparts. This confirms that for the bridge under consideration, the semi-continuum method used for analysis is able to predict the pattern of transverse distribution of load fairly accurately.

The same accuracy of prediction, however, cannot be claimed in the case of the absolute values of girder deflections, which are plotted in Fig. 10.41. This is because of uncertainty in quantifying the parameters discussed below.

As noted earlier, the girders for the Lord's Bridge are 10.2 m long and have an unusually long bearing length of 0.53 m at each end. It is customary to assume that the nominal point-support for a girder lies midway along the bearing length, in which case the nominal span of each girder would be 9.67 m. It can be demonstrated that for the case under consideration, the vertical pressure under the supported length of a girder, should have its peak away from the midway point of the bearing length and towards the free edge of the abutment. Determination of the exact location of this peak requires detailed knowledge of the modulus of subgrade reaction of the timber crib abutment. Clearly, this factor is not easily quantifiable thus making the task of determining the effective span very difficult. The clear span of the girder, being 9.14 m, is clearly the lower-bound of the effective span of the girder.

The transverse modulus of elasticity of the wood deck, which is operative in the longitudinal direction of the bridge, is very small as compared to the longitudinal modulus. Even if the transverse laminated deck were made composite with the girders, the contribution of the deck to the strength and stiffness of the composite section would usually be expected to be so small as to be negligible. Consequently, no attempt is usually made to provide shear connectors in such bridges. There are some holding down devices, however, to connect the deck to the girders through the nailing strips; these devices, by transferring some interface shear, do make the girders partially composite with the nailing strips and the decking. From the measured girder strains, it was discovered that despite the absence of shear connectors, the decking and the nailing strips of the Lord's Bridge were partially composite with the girders. The degree of composite action was found to vary from girder to girder, and clearly was not quantifiable.

The Lord's Bridge was analysed using two different sets of idealizations. In one idealization, the girders were assumed to be non-composite and with a simply-supported span of 9.67 m. In the other idealization, full composite action was assumed between the girders and the timber components, being the nailing strips and the decking; the girders were assumed to have the lower bound span of 9.14 m. As can be seen in Fig. 10.41, the measured deflections for the same load case for which the distribution factors are plotted in Fig. 10.40 are bracketed entirely with very large margins by the analytical results corresponding to the two idealizations. It is tempting to believe that the actual condition of the bridge lies somewhere between the two sets of conditions assumed in these idealizations and consequently, errors in analysis are related only to the uncertainties of span length and degree of composite action. However, there is at least one other complicating factor, namely bearing restraint, which was not accounted for in these idealizations and which can have a significant influence on bridge response; this factor is discussed later.

Two-Girder Bridge with Floor Beams The Adair Bridge is a single-span, single-lane structure with a clear span of 12.8 m, as shown in Fig. 10.42. As is also shown in this figure, the bridge comprises a concrete deck slab supported by two outer longitudinal steel girders and five inner longitudinal steel stringers, with the latter spanning between the abutments but also supported within the span by two

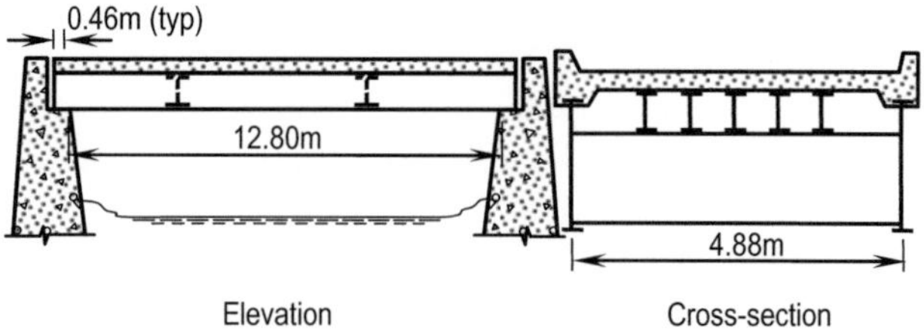

Fig. 10.42 Details of the Adair Bridge (not to scale)

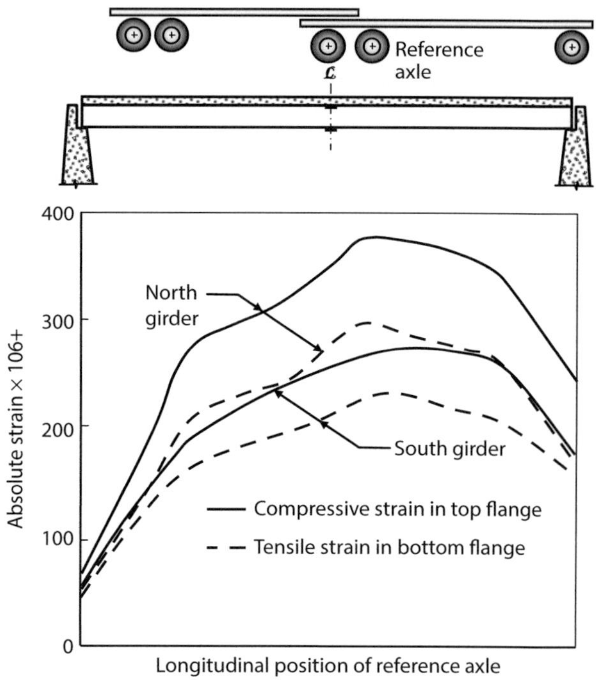

Fig. 10.43 Girder strains at mid-span in the Adair Bridge

transverse floor beams that frame into the two girders. A proof test on this bridge is described by Bakht and Mufti (1992b).

Mid-span strains in the top and bottom flanges of the two girders due to one load case are plotted in Fig. 10.43 against the longitudinal position of the test vehicle. It can be seen in this figure that the magnitude of strains in the top flanges are always much higher than the magnitude of corresponding strains in the bottom flanges. This observation confirms the presence of fairly large bearing restraint forces. Large compressive strains in the bottom flanges of the girders near their supports re-confirm the presence of significant bearing restraint, which again cannot be practically quantified for inclusion in the mathematical model for analysis.

Fig. 10.44 The Stoney Creek Bridge being tested to failure

Much larger magnitudes of strains in the top flanges of the girders also indicate the possible lack of composite action between the girders and the deck slab; this bridge did not have any mechanical shear connection between the girders and the deck slab. Because of the lack of composite action, the top flanges of the girders, getting little relief from bearing restraint at the bottom flanges, govern the load carrying capacity of the girders.

It is interesting to note that, unlike the case in the Lord's Bridge and other bridges discussed later, bearing restraint does not provide any significant reserve of strength in the Adair Bridge.

The uncertain nature of the composite action in slab-on-girder bridges without mechanical shear connection is underlined by the observation that, in the same Adair Bridge, the inner stringers are able to develop full composite action with the deck slab despite the lack of mechanical shear connectors.

Because of the composite action, the stringers had become considerably stiffer, thus relieving the non-composite girders of a much greater share of the applied loading than would have been the case if they were also non-composite. It can be appreciated that analysis cannot be very effective without knowledge of the degree of composite action in the various beams; such knowledge is practically impossible to obtain without a test.

Load Distribution at Ultimate Limit State An ultimate load test on a single-span, right, i.e. skew-less, slab-on-girder bridge, called the Stoney Creek Bridge, is described by Bakht and Jaeger (1992). The bridge, which had a clear span of 13.26 m, was loaded to failure in 1978 with concrete blocks piled in six layers. A view of the bridge during testing can be seen in Fig. 10.44, which also shows a temporary wooden structure designed to prevent a catastrophic collapse of the bridge. A crane at one end of the bridge carried the concrete blocks, and a crane at the other end had two buckets for persons who manoeuvred the blocks into place. Longitudinal girder strains at the mid-span were recorded after each layer of blocks had been placed on the bridge.

To check the validity of the recorded data, the mid-span moments taken by the girders and the associated portions of the deck slab, computed from measured

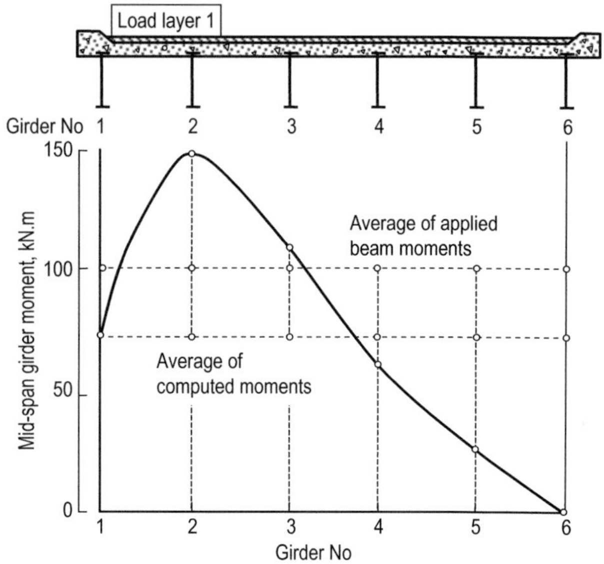

Fig. 10.45 Girder moments in the Stoney Creek Bridge computed from observed data by ignoring bearing restraint

strains, were compared with the total applied moments. It is recalled that in a right, simply supported bridge, the total moment across any transverse section is obtained by simple beam analysis and is statically determinate. When it was found that the moments computed from measured strains were up to 30 % smaller than the applied moments, the accuracy of the measured data was initially questioned. An example of the comparison of moments thus computed from measured strains and average applied moments is presented in Fig. 10.45 for load due to one layer of concrete blocks, under which loading the girder strains were well within the limit of computed elastic strains.

The initial computations of moments from measured strains were made by assuming that the girders were free from any horizontal restraint at the bearings. The bearing restraint forces were not initially entertained as the possible cause for the moment discrepancies mentioned above, because bearing restraint forces of the magnitude needed to reduce the applied moments by up to 30 %, were believed to be unlikely to develop in practice.

Subsequent tests, some of which are discussed earlier in this chapter, confirmed the presence of significant bearing restraint forces in similar slab-on-girder bridges in which girders rest upon steel bearing plates. The presence of these forces invalidates the assumption of simple supports and the computation of moments obtained from measured strains on the basis of no external forces. In light of the knowledge gained from the other tests, the data from the test on the Stoney Creek Bridge were re-analysed about 10 years after the test by back-calculating the bearing restraint forces that might have occurred. From these revised computations, it was found that the bearing restraint reduced the applied moment by up to 18 %, rather than 30 % that had been wrongly deduced by previous calculations.

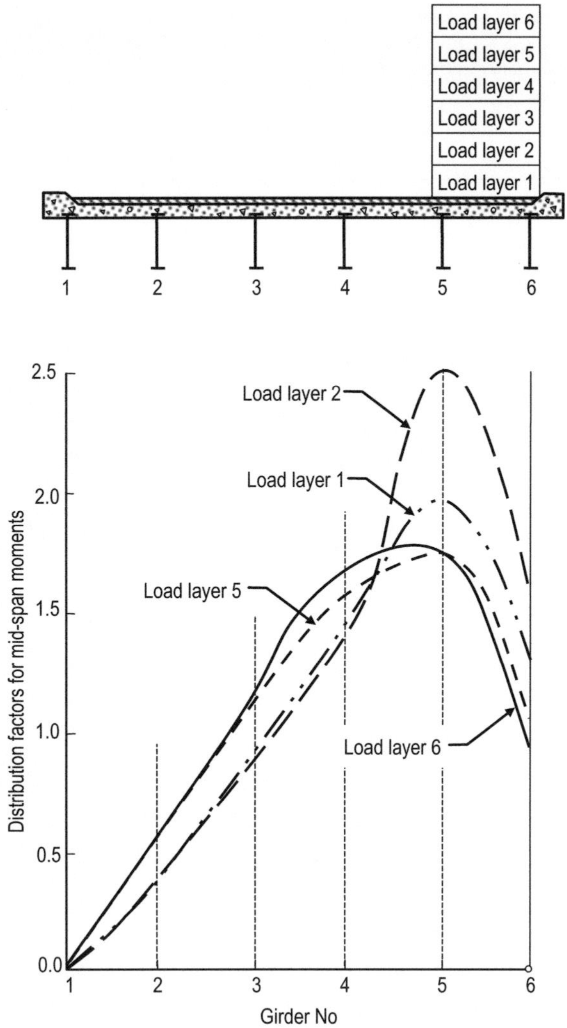

Fig. 10.46 Distribution factors for girder moments in the Stoney Creek Bridge due to load at different levels

Distribution factors for mid-span moments taken by the girders and the associated portion of the deck slab are plotted in Fig. 10.46 for loads at different levels. It is interesting to note that the transverse distribution pattern of the bridge does not change very significantly as the load approaches the ultimate sixth layer. As the failure of the bridge approaches, the load gets redistributed only slightly among the most heavily loaded girders. The girder most remote from the applied loading, receiving little load at the early stages of loading, continues to receive low levels of load even when the load approaches the failure load of the bridge.

An important outcome of the test was the observation that in the absence of mechanical shear connection, the composite action between a girder and the deck slab that may exist at low levels of load, breaks down completely as the load approaches the failure load for the girder.

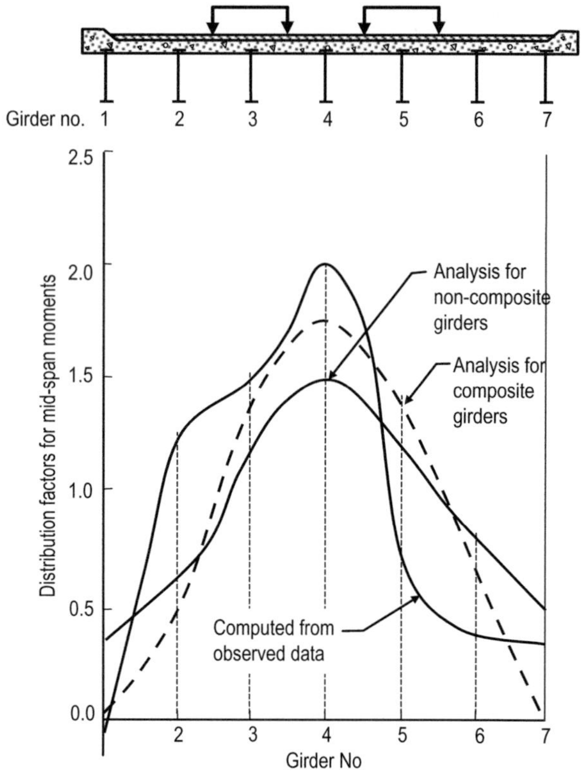

Fig. 10.47 Distribution factors for mid-span moments in the Belle River Bridge

A Non-composite Slab-On-Girder Bridge The un-quantifiable and random nature of the bearing restraint forces, and of the degree of composite action in the absence of mechanical shear connection, is illustrated by the results obtained from a test on the Belle River Bridge (Bakht, 1988). The Belle River Bridge is a slab-on-girder bridge with steel girders and an apparently non-composite concrete deck slab. The nominal span of the bridge is 16.3 m and the width 9.1 m.

As indicated earlier, the transverse load distribution analysis of slab-on-girder bridges without mechanical shear connectors between the girders and the deck slab is made difficult, to the point of becoming impossible, by the uncertain degree of the composite action. One is tempted to believe that the actual load distribution pattern of such bridges could be bracketed by two sets of analyses, one corresponding to full composite action and the other to no composite action at all, with the former analysis always leading to safe-side estimates of the maximum load effects in the girders. In reality, a deterministic analysis, no matter how advanced, might fail completely to predict safely such maximum load effects. The assertion is illustrated below with the help of the results from the test on the Belle River Bridge.

Transverse profiles of the distribution factors for mid-span girder moments in the bridge under consideration are plotted in Fig. 10.47 for a transversely symmetrical

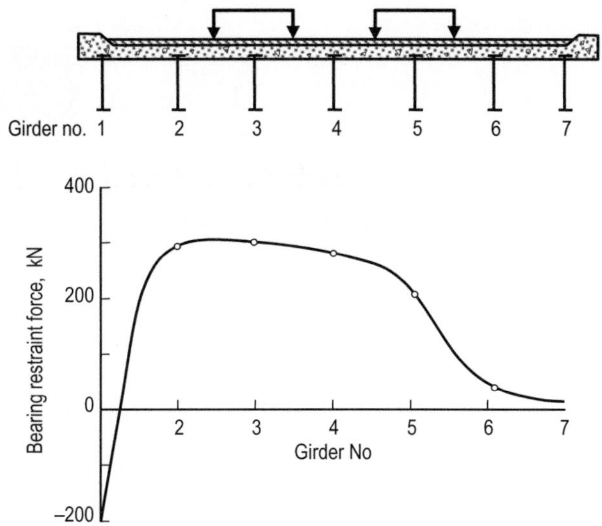

Fig. 10.48 Bearing restraint forces in the Belle River Bridge

load case. One of these profiles corresponds to moments computed from observed girders strains both at the mid-span and near the abutments, with the latter providing information regarding the bearing restraint forces. The other two transverse profiles are obtained from the results of the semi-continuum method of analysis, which is discussed in Chap. 12, for the two bounds of the composite action. It is noted that the bearing restraint forces were not considered in these analyses.

It can be seen in Fig. 10.47 that the pattern of transverse distribution of actual moments is similar, but only in a general way, to the two analytical patterns. It is also quite irregular. Unlike the analytical patterns, the actual pattern is far from being symmetrical. In fact, the actual distribution factor for maximum girder moments is about 10 % larger than the corresponding analytical factor for the fully non-composite bridge. It can be appreciated that the occurrence of the very high distribution factor and significant departure from symmetry are probably caused by the middle girder becoming accidentally much stiffer through composite action by bond than the adjoining girders. In light of the results plotted in Fig. 10.47 there can be little doubt that, for the kind of bridge under consideration, even the most rigorous deterministic analysis is at best only a fairly close approximation.

Bearing restraint forces in the girders of the Belle River Bridge were computed from observed girder strains near the abutments. From these bearing restraint forces and approximately calculated girder reactions at the supports, it was concluded that the effective coefficient of friction varied between 0.66 and 0.95; the former limit relates to loading by single vehicles and the latter to two side-by-side vehicles. Such effective coefficients of friction may be on the high side but are not uncommon in bridges in which the girders rest directly on highly rusted steel bearing plates.

Bearing restraint forces computed from measured girder strains are plotted in Fig. 10.48 for the same load case for which the distribution factors for mid-span

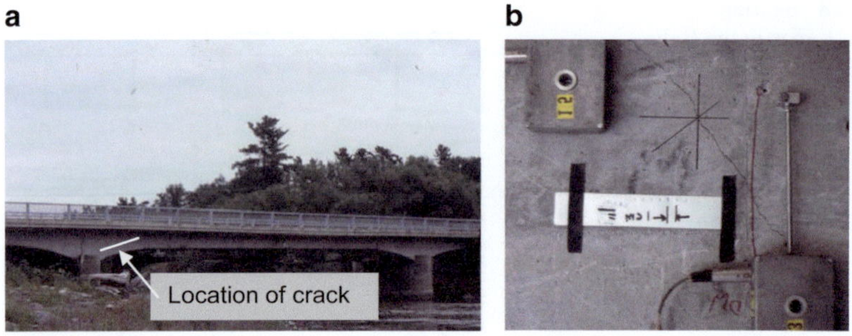

Fig. 10.49 The Galleta Bridge: (**a**) location of cracks, (**b**) *vertical* strains along crack being measured with strain transducers

girder moments are plotted in Fig. 10.47. The bearing restraint forces are shown as positive when they tend to push the abutment away from the girders.

It can be seen in Fig. 10.48 that the bearing restraint forces, in all the girders except one, are positive. At the location of the left hand outer girder, the bearing restraint force was found to be not only negative but also fairly large in magnitude. It was postulated that this unusual response is the result of a relatively soft pocket in the backfill behind the abutment in the vicinity of the left hand outer girder.

In light of the uncertainties discussed above, it can be seen that for the kind of bridge under consideration, no deterministic analysis can be expected to predict the actual behaviour of the bridge.

A Concrete Girder Bridge with Diagonal Cracks An important aspect of the interpretation of SHM data should be the determination of the effect of temperature on the behaviour of a structure. While most structures do not change their pattern of load response with temperature, a few do. An example is given in the following to illustrate this point.

The Galleta Bridge in Ontario, Canada, is a two-lane three-span continuous bridge of reinforced concrete T-beam construction. The main span is about 28 m long, and the side spans are each about 7 m. The superstructure of the bridge is cast monolithically with piers and abutments. All three girders of the bridge developed inclined cracks in the middle span (Fig. 10.49a) presumably due to the substantial restraint offered by the box-like side spans to thermally-induced movements. With these cracks running right through the beams, the load carrying capacity of the bridge could not be determined analytically. The bridge was load tested in the summer of 1981, with strain transducers measuring vertical strains across the crack (Fig. 10.49). By plotting the vertical strains across the crack, it was possible to obtain the smallest value of the vertical strain, which was assumed to correspond to the position of the vertical leg of a stirrup. It was assumed that the smallest measured strain was nearly the same as the strain in the stirrup.

When it was found that the live load strains in stirrups increased with a drop in the temperature, it was decided to re-test the bridge in winter (February 1982). The proof load on the bridge comprised two test vehicles, each with a gross weight of 87 t, under which load the bridge remained fully elastic. The bridge was declared safe for the 10 t posting that it carried.

10.5 Summary

Several significant "surprises" encountered during bridge testing by the authors have been presented in this chapter. They are given mainly to introduce the reader to the field of bridge testing, where the instruments sometimes seem to lie. It is tempting to disregard such readings as being the result of instrument malfunction. In most cases, however, it was found that the unexpected readings from the instruments, instead of resulting from instrument malfunction, were caused by unexpected bridge behaviour.

The surprises given in this chapter also underline the fact that some aspect of bridge behaviour, because of never entering into design considerations, can escape the attention of even the most experienced bridge designers and analysts. Some of the surprises found in bridge testing may have a significant effect on the load carrying capacity of a bridge, while others may have a minor effect. From the examples given in this section, it is clear that in most cases, the load carrying capacities of bridges are higher than those obtained from the usual calculations. However, there are some cases in which the load carrying capacity of a bridge can be lower than expected. A carefully planned and executed bridge test is invaluable in identifying the strengths and weaknesses of an existing bridge.

References

AASHTO (1994) AASHTO LRFD bridge design specifications. American Association of State Highway and Transportation Officials, Washington, DC

Bakht B (1988) Testing of an old short span slab-on-girder bridge, Structures research report SRR-88-01. Ministry of Transportation of Ontario, Downsview

Bakht B, Csagoly PF (1977) Testing of Perley Bridge, Research report 207. Research and Development Division, Ministry of Transportation of Ontario, Downsview

Bakht B, Csagoly PF (1979) Bridge testing, Structures research report 79-SRR-10. Ministry of Transportation of Ontario, Downsview

Bakht B, Csagoly PF (1980) Diagnostic testing of a bridge. ASCE J Struct Eng 106(ST7): 1515–1529

Bakht B, Jaeger LG (1985) Computer simulation of failure in timber bridges. In: Second international conference on computing in civil engineering, Elsevier, Hangchou\Amsterdam, 5–9 June 1985, pp 484-494

Bakht B, Jaeger LG (1987) Behaviour and evaluation of pin-connected steel truss bridges. Can J Civ Eng 14(3):327–335

Bakht B, Jaeger LG (1990a) Observed behaviour of a new medium span slab-on-girder bridge. J Inst Eng India 70:164–170

Bakht B, Jaeger LG (1990b) Bridge testing – a surprise every time. ASCE J Struct Eng 116(5):1370–1383

Bakht B, Jaeger LG (1992) Ultimate load test on a slab-on-girder bridge. ASCE J Struct Eng 118(6):1608–1624

Bakht B, Mufti AA (1992a) Behaviour of a steel girder bridge with timber decking. Structures Research Report SRR-92-02. Ministry of Transportation of Ontario, Downsview, Ontario

Bakht B, Mufti AA (1992b) Evaluation by testing of a bridge with girders, floor beams and stringers, Structures research report SRR-91-05. Ministry of Transportation of Ontario, Toronto

Bakht B, Jaeger LG, Mufti AA (2002) Use of in-service performance of a bridge to modify its safety assessment: an introduction. In: Proceedings of the first international conference on structural health monitoring, held in Winnipeg

CHBDC (2000) Canadian highway bridge design code CAN/CSA S6.00. Canadian Standards Association, Toronto

CHBDC (2006) Canadian highway bridge design code CAN/CSA S6.06. Canadian Standards Association, Toronto

Csagoly PF, Bakht B (1978a) A computer program to determine the collapse load of steel trusses, Research report 218. Research and Development Division, Ministry of Transportation of Ontario, Downsview

Csagoly PF, Bakht B (1978b) In-plane buckling of steel trusses. Can J Civ Eng 5(4):533–541

Csagoly PF, Bakht B, Ma A (1975) Local buckling of pony truss bridges, Research report 199. Research and Development Division, Ministry of Transportation of Ontario, Downsview

Melchers RE (1987) Structural reliability and analysis prediction. Ellis Horwood Ltd., (Division of John Wiley & Sons), Chichester, USA

Mufti AA (2001) Guidelines for structural health monitoring, Design manual no. 2, ISIS Canada research network. University of Manitoba, Winnipeg

Mufti AA (2003) Integration of sensing in civil engineering structures: development of the new discipline of civionics. In: Proceedings for the first international conference on Structural Health Monitoring and Intelligent Infrastructure (SHMII-1), Tokyo

Mufti A, Bakht B (2015) Reformatting of a program for analyzing lateral buckling of pony trusses. To be published

Nowak AS, Collins KR (2000) Reliability of structures. McGraw-Hill, New York, OHBDC, *Ontario Highway Bridge Design Code*. 1979. Ontario, Canada

Tharmabala T (1986) Performance studies of Keewatin channel bridge, structures research report SRR-85-05. Research and Development Branch, Ministry of Transportation of Ontario, Downsview

Chapter 11
Bridge Weighing-in-Motion

11.1 Introduction

In a realistic bridge design code, the design live loads should be based on actual traffic loads, as discussed in Chap. 1. In the past, the information about the actual traffic loads on highway bridges was obtained from truck surveys, in which the trucks are stopped for measurement and weighing on ‘static’ weighing scales. During the past few decades, however, the information about the truck loads is collected while the trucks are moving at normal speeds. Some of the weighing-in-motion (WIM) scales are installed in the pavement. In other WIM systems, a highway bridge is used for the weighing-in-motion of the trucks; this latter system is referred to herein as the bridge WIM (BWIM) system, and the system installed in the pavement as simply the WIM system. The WIM systems are significantly more expensive and are expected to be more accurate than the BWIM systems.

During the past few decades, a number of BWIM systems, with varying degrees of accuracy in predicting the axle and gross vehicle weights (GVW) of the trucks, have been introduced. An international society for weighing-in-motion (ISWIM) is already in existence (http://iswim.free.fr), it being noted that this society deals with both the WIM and BWIM systems. The main purpose of this chapter is to introduce four new BWIM systems for determining the gross vehicle weights (GVWs), developed by authors and their research associates; the accuracies of the four methods have been examined through a number of calibration tests on a bridge in Manitoba (Helmi et al. 2014).

11.2 State-of-the-Art

Some of the various BWIM systems introduced during the past few decades are summarized in this section.

B. Bakht, A. Mufti, *Bridges*, DOI 10.1007/978-3-319-17843-1_11

11.2.1 Ohio Method

Moses (1979) appears to be the first researcher to introduce a BWIM system; his method, which is referred to herein as the Ohio method, determines the truck configuration, number and spacing of axles, and the vehicle speed through tape switches that are installed on the top surface of the bridge. The vehicle speed, which is assumed to be constant, enables the transformation of the length scale on the response diagram into the time scale and vice versa, thus leaving the axle weights to be determined through the analysis of the observed bridge responses. The unknown weights of N axles of a truck can be determined by N different measurements of the bridge response for N different positions of the truck along the bridge. However, since the bridge response is measured almost continuously, there is an abundance of 'redundant' data. By using the influence line of the static response of the bridge at the instrumented section and the truck speed, the total static response of a girder under a truck load is predicted in terms of the unknown axle weights and a scalar error function obtained by summing the squares of the differences in the predicted and calculated responses at different intervals. Minimizing the error function leads to N simultaneous equations to give N axle weights.

The Ohio method was adopted by numerous researchers, and several enhancements have been introduced to the algorithm that was developed by Moses. Quilligan et al. (2002), for example extended the algorithm to be two dimensional, i.e. to be able to detect the lane position of the truck in addition to its weight. However the major drawback for the method is the need for the identification of the number of axles and the speed of the vehicle using an external method like tape switches (Znidaric et al. 2002).

Although Znidaric et al. 2002, developed an axle detector free WIM algorithm by using the peaks responses of multiple strain gauges to estimate the number of axles, the speed of the truck is estimated by using the time delay in the peak responses of several sensors. The axle spacing is estimated using the time delay in the peak response of the same sensor. This method is only applicable to short span concrete bridges.

11.2.2 Ontario Method

After experimenting with two other methods, a method was developed in Ontario in early 1980s for obtaining truck weights through BWIM (Harman 1984). The method used for actual vehicle weight surveys in Ontario, idealized the axle weights of a truck as point loads, and the set of point loads in turn were represented as an equivalent uniformly distributed load (UDL) having the same total load W as that of the set of point loads, but having an equivalent base length B_m, which is calculated by Eq. (1.1), given in Chap. 1. The significance of the equivalent UDL is that the bending moments and shear forces induced by this load, in beams of any span length in the region of maximum responses, are nearly the same as those induced by the set of point loads, and as discussed in Chap. 1, the W-B_m space is very useful in representing a very large number of truck loads as they affect the longitudinal responses of bridges.

In the Ontario method, theoretical longitudinal strains at the mid-span in the bottom flange of the girders of the instrumented simply supported bridge, idealized as a simply supported beam, were calculated for B_m varying between 1.22 and 30.5 m at intervals of 0.61 m; the distances were converted into the time scale by using the speed of the vehicle, which was computed from the response of tape switches installed on the top surface of the bridge. The time-strain plots due to different theoretical UDLs were then used to find the best fit for the sum of observed girder strains at the instrumented transverse section of the bridge, which was always at the mid-span. The equivalent base length, B_m, for which the discrepancy between the theoretical and observed responses was the smallest was adopted as the assumed value of B_m, and the corresponding value of W as the GVW of the truck (Harman 1984).

11.2.3 Australian Method

The determination of actual axle weights requires the identification of strain peaks in the observed strain-time plots, which is made difficult by the fact that the influence lines for longitudinal moments at transverse sections within the middle-third length of the bridge are relatively insensitive to individual axle weights because of which the time-strain plots do not have distinct peaks associated with the axle weights.

The problem of an influence line without well-defined peaks corresponding to axle weights was solved by a BWIM method developed in Australia, called CULWAY; this method, which was introduced in late 1980s and early 1990s, measures axle weights of trucks by using small span concrete box culverts, instrumented at the bottom surfaces of the top slabs of the boxes by means of strain transducers (e.g. Scott 1987; Heywood 1991). The method depends upon the knowledge of the transverse position of the truck on the culvert, and cannot measure spacing between axles without the help of tape switches or other similar devices. It is also important to note that the strain transducers used in the CULWAY system are subject to large drifts, thus requiring frequent calibration tests.

11.2.4 Japanese Reaction Force Method

A novel BWIM system was introduced in Japan by Yamada and Ojio (2003), in which the problem of the strain-time curves with ill-defined peaks was overcome by a method that the authors call a reaction force method; in this method, the stiffeners in a support cross-frame at one end of a simply supported girder bridge are instrumented to measure vertical strains, corresponding to the reactions at the support. The influence lines for support reactions do indeed have distinct peaks corresponding to each axle load; however, as discussed in the Sect. 11.2.8, the accuracy of the predicted axle weights by the Japanese method is not very good.

11.2.5 A Variation of the Reaction Force Method

To test the efficiency of the Japanese reaction force method, Bakht et al. (2006) instrumented a support diaphragm of a slab-on-girder bridge in Manitoba, Canada. The instruments consisted of strain gauges placed on the stiffeners of the diaphragm. The strain-time plots for the diaphragms had distinct peaks, and permitted the calculation of axle loads of moving trucks; however, the strains were found to be highly sensitive to the transverse position of the vehicle on the bridge, and thus not very accurate in calculating the axle weights for trucks in random transverse positions.

11.2.6 Connecticut Method

A BWIM method was recently developed in Connecticut, USA, to predict GVWs, axle weights and axle spacings of trucks through strains measured near the mid-span of a simply supported steel girder bridge with a concrete deck slab (Wall et al. 2009). The unique aspect of the Connecticut method is that, it determines the speed of a vehicle from only girder strains, i.e. without resorting to 'intrusive' devices such as tape switches on riding surface of the bridge.

For calculating vehicle speeds, Wall et al. use the second derivative of recorded strains with respect to time. For an idealized beam, these second derivatives are pure spikes, each corresponding to an axle load, which is idealized as a point load. Positive spikes correspond to the time when each axle load enters and leaves the beam, and the negative spikes represent the time when each axle load is over the instrumented section, which was at the mid-span for the bridge dealt with by Wall et al. (2009). Unlike the case of the idealized beam, the spikes in second derivative of strains in an actual bridge are not so well defined, because of which the calculation of speed by the Wall et al. method is subject to numerical errors that are not small. For their calibration tests, they had used a test truck which was installed with an accurate speed measuring device. The differences between the speeds measured with device and by the second derivative method range between +7.8 and −13.5 %. Since the calculated speed of a vehicle has a direct influence on the accuracy of prediction of the GVW, it is not surprising that the GVWs of the test truck predicted by this method are subject to large errors.

11.2.7 Other Methods

The above methods are all based on the static response of the bridge. Several researchers have attempted different methods for using the dynamic response of the bridge for BWIM. However most of these methods are computationally complicated and deal only with a limited number of variables e.g. only two axle loads and were based on simulations. The only case where such a method seems to have been applied was in the case of the Vransko Bridge in Slovenia (Rowley et al. 2009), in which a BWIM algorithm based on the moving load identification theory developed by González et al. (2008)) was used. However this method still needed extra axle detecting equipment to be installed.

The methods described so far, or their variations, have also been used in Europe since early 1990s; the ISWIM website cited earlier provides several references to the European applications of BWIM. Some of the important references to the European WIM practice include Arroyo et al. (2010), and Dowling et al. (2010).

11.2.8 Accuracy

To the authors' knowledge, all BWIM systems rely in their calculations on one of the methods described above, or their variations. Two important factors in deciding upon the selection of an appropriate BWIM system are its cost and the accuracy of its prediction. The pavement WIM systems, invariably proprietary and more expensive than the BWIM systems, are usually assumed to be more accurate than the BWIM systems. The American Society for Testing and Materials (ASTM 2002) in its document E 1318–02 deals mainly with three kinds of pavement WIM systems, all 'capable of accommodating vehicles moving at speeds from 16 to 130 km/h'; these WIM systems are regarded by Jacob et al. (2002) as high-speed WIM systems. The most rigorous system (Type I) is required by ASTM to be capable of determining 14 characteristics of the truck, ranging from weights of wheels, axles and axle groups to the 'violation code'. Type II system should determine all 14 characteristics of the truck, except wheel loads, and Type III system is exempt from determining 'vehicle class', 'wheel base' and 'violation code'. The ASTM standard requires the following tolerances for the different types of pavement WIM systems, each conforming to 95 % probability. For axle loads, the tolerance is ±20, ±30 and ±15 % for Type I, II and III systems, respectively, and that for GVW is ±10, ±15 and ±6 %, respectively. The ASTM standard also gives specifications for a Type IV system, which could be used for law-enforcement; in this system, the vehicle speeds are between 3 and 16 km/h, so the Type IV system could be regarded as the low-speed WIM system. It is interesting to note that even for a system that could be used for law enforcement, the tolerance for accuracy of prediction of axle weights and GVW is about ±4 %. Recognizing that the sensors in the pavement WIM systems are subject to drift, the

ASTM standard requires that all WIM systems be calibrated at least once every year.

All BWIM systems rely on measuring strains in some bridge components. Agarwal and Bakht (1988) have examined the effect of small inaccuracies in strain measurement on the accuracy of the prediction of the Ohio and Ontario BWIM methods; they concluded that while the Ohio system was generally more accurate, a combination of the Ohio method for axle loads and the Ontario method for GVW might constitute an ideal system. The accuracy of the Australian BWIM system is compromised by the fact that the strain transducers that it employs are subject to very large drifts, because of which the system without on-site correction and frequent calibrations is not suitable for long-term monitoring of loads. Yamada and Ojio (2003) calibrated the Japanese BWIM system with the help of three pre-weighed 3-axle trucks; from the data that they have provided in it was found that their BWIM system could predict axle weights and GVW with accuracies of ±75 % and ±15 %, respectively. It is noted that the accuracy of prediction of this BWIM system might have been poor because only one vertical member of the cross-frame was instrumented by Yamada and Ojio (2003).

11.3 Manitoba Methods

Four methods were developed in the Province of Manitoba, Canada, by the authors and their research associates to determine the GVW of trucks from bridge responses; these methods are described in this section with the help of specific examples.

11.3.1 Asymmetry Coefficient Method

Similar to the Ontario method, this method uses UDLs as being representative of all the axle loads of a truck, which are usually regarded as point loads in theoretical calculations. In the proposed method, it is assumed that the total load of a truck W is distributed uniformly over a length nL, where L is the span of the bridge and n is a fraction, so that the intensity of the UDL is W/nL per unit length.

Selection of a Bridge for BWIM The proposed method is based on the assumption that the BWIM system is applied to a right simply supported slab-on-girder bridge, having a span length greater than the length of most trucks, which is assumed to be a maximum of about 20 m. All girders, assumed to be parallel to each other, should preferably be of steel.

Basis of the Asymmetry Coefficient Method Consider a simply supported beam with span L, with an instrumented section at a distance aL from the left-hand support, where a is a fraction of the span length, as shown in Fig. 11.1. The instruments at the section measure strains, from which the bending moments in

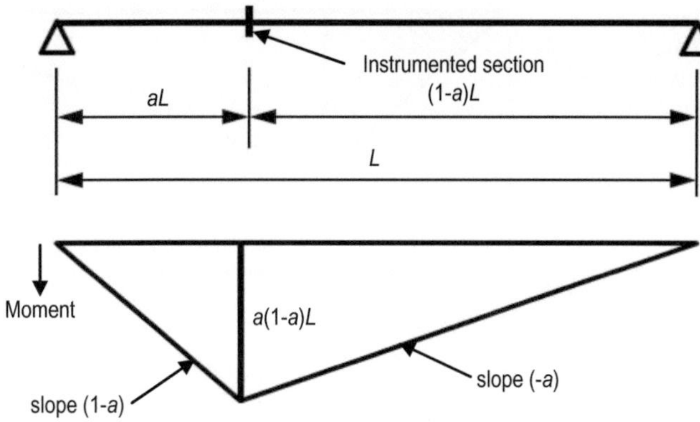

Fig. 11.1 Influence line for bending moments at the instrumented section in a simply supported beam

the beam can be calculated directly through calibration constants, which are discussed later. As shown in Fig. 11.1, the influence line for bending moments at the instrumented section is the familiar triangle, and the area under the influence line is $a(1-a)L^2/2$. From the records of the instrumented section, the objective is to determine n, from which W can be determined, also as explained later. Assuming that the UDL travels from left to right, the complete record for a given truck has the length $(1+n)L$, as it covers the leading edge of the UDL coming to the left-hand support and leaving the beam when the trailing edge of the load just leaves the right-hand support. Figure 11.1 also shows the slope of the influence line on both sides of its peak.

When only a portion of the total UDL W enters the beam a distance x from the left-hand support, which is smaller than aL, the moment M at the instrumented section is given by the following equation (Fig. 11.2).

$$M = \frac{1}{2}\frac{W}{nL}x^2(1-a) \tag{11.1}$$

As shown in Fig. 11.3, the maximum moment at the instrumented section, M_{max}, occurs when the leading and trailing edges of the UDL have the same height ordinate on the influence line, so that a fraction a of the UDL is to the left of the instrumented section and a fraction $(1-a)$ to the right. This closely parallels the situation for point loads for which the equality of $a/(1-a)$ is reversed as a point load gets on the instrumented section and gives the maximum moment due to point loads. M_{max} for the UDL is given by the following equation.

$$M_{max} = WLa(1-a)(1-n/2) \tag{11.2}$$

Figure 11.4 illustrates the situation when the UDL leaves the right-hand support of the beam. Consider a small portion of the load, say w_1, passing over the beam; it

Fig. 11.2 Bending moments due to load advancing a small distance from the left-hand support

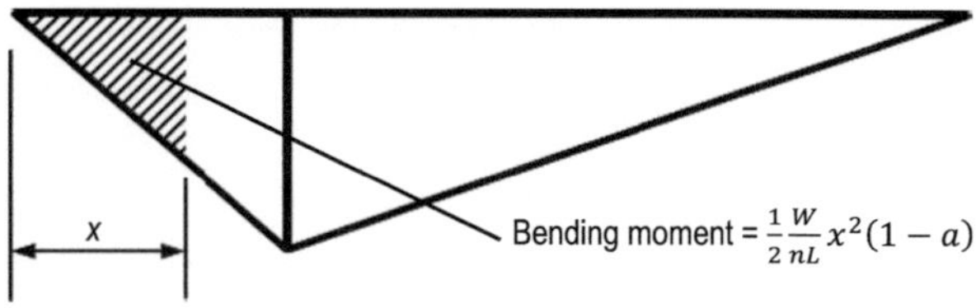

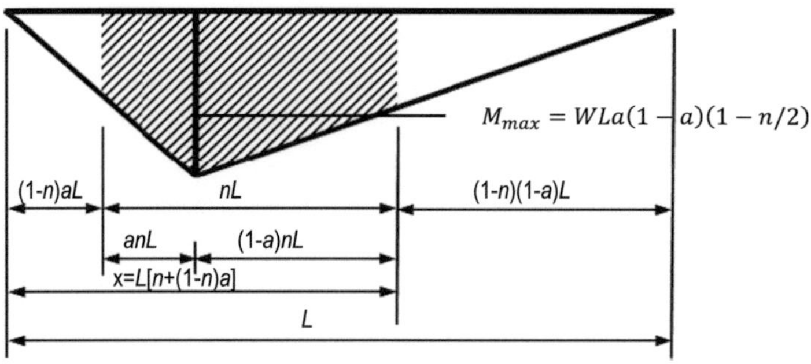

Fig. 11.3 Maximum bending moment position

passes over every point on the influence line, and the area under the moment-distance curve at the instrumented section is $w_1a(1-a)L^2/2$. The same observation is true for any other contribution of the total load W. By adding the contributions of all the loads together, a very important conclusion is reached: When a total load W crosses over the simply supported beam, the area under the moment-space curve at the instrumented section is $Wa(1-a)L^2/2$, which is independent of the way W is distributed.

The above observation is illustrated in Fig. 11.5, in which the fraction (bending moment/WL) at the mid-span of a simply supported beam with span L is plotted over the length x due to the passage of a load W, distributed uniformly over lengths of zero, $L/5$, $2L/5$ and $3L/5$, respectively. It is noted that the load distributed over length zero is indeed a point load.

The fraction (bending moment/WL) is plotted in Fig. 11.6 for the reference section at ¼ span over the length x. The moment-distance plots of Figs. 11.5 and 11.6 not only show that the area under the various plots is independent of the way that the total load W is distributed, but they also show graphically that the distribution of the beam moments is governed by only two factors: (a) the peak value of the moment, which decreases as the length of load distribution increases, i.e. n increases, and (b) the distance of the peak moment from the reference section. The first factor is quantified by the fraction n that is equal to the ratio of the length of the UDL and L; when W is a point load, $n =$ zero, and when W covers the entire span length, $n = 1.0$. The second factor defines the distribution of W by the ratio α, which is called the asymmetry coefficient.

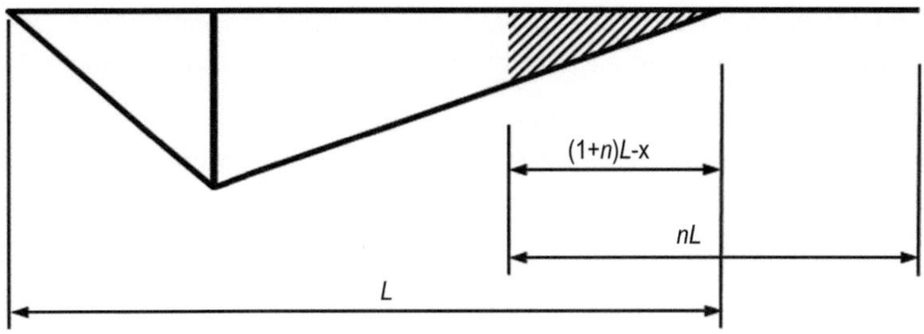

Fig. 11.4 The UDL leaving the beam

Fig. 11.5 Moments at mid-span due to W moving from left to right

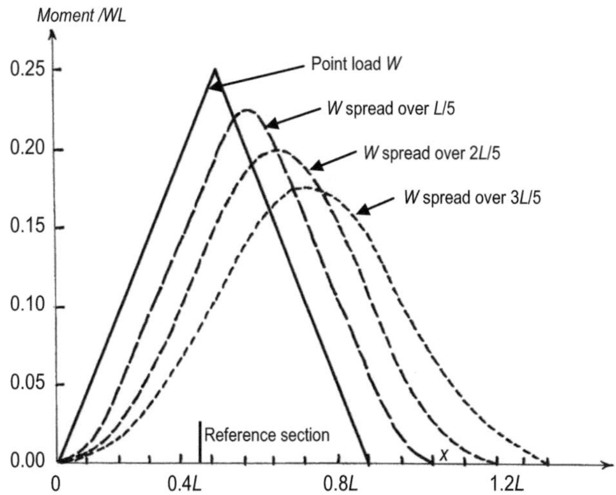

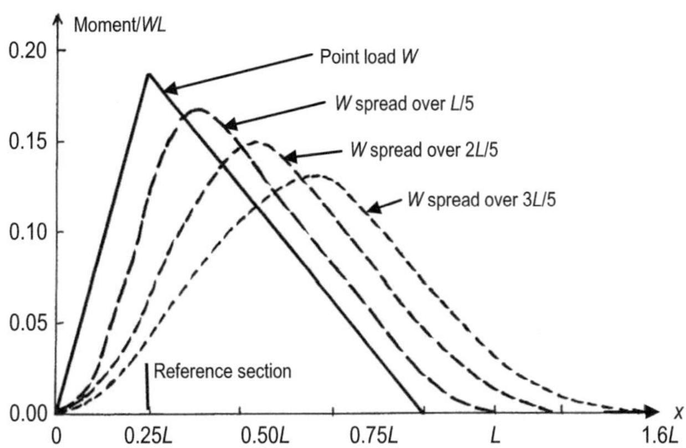

Fig. 11.6 Moments at ¼ span due to W moving from left to right

The maximum bending moment, M_{max}, in a simply supported beam at a reference section aL from the left support due to a load W distributed uniformly over a length nL is given by Eq. (11.2).

An Ideal Beam An ideal beam is considered first, in which the time of the UDL at entering and leaving the beam is known exactly. When the leading edge of the UDL comes on the left-hand support of the beam, its distance from the left-hand support, x_1, is equal to zero, and when its tailing end leaves the right-hand support, its distance from the left-hand support is x_3, given by:

$$x_3 = (1 - n)L \tag{11.3}$$

For the maximum moment, the leading edge of the UDL is a distance x_2 from the left-hand support; this distance is given by:

$$x_2 = L\{n + a(1 - n)\} \tag{11.4}$$

The asymmetry coefficient α is really the ratio of the distance of the front end of the UDL inducing M_{max} from the left-hand support and the length of the moment-distance plot, so that:

$$\alpha = \frac{x_2 - x_1}{x_3 - x_1} = \frac{n + (1 - n)a}{(1 + n)} \tag{11.5}$$

The bridge response collected for BWIM systems is usually strains plotted over time, which can be converted into the distance scale only if speed of vehicle is known. In the absence of the knowledge of the vehicle speed, the asymmetry coefficient α can be calculated from the time records as follows.

$$\alpha = \frac{t_2 - t_1}{t_3 - t_1} \tag{11.6}$$

where t_1 is the time when the leading edge of the load just comes on the beam, t_2 is the time at which M_{max} occurs, and t_3 is the time when the trailing edge of the load just leaves the bridge. Equation (11.7) gives directly the value of n in terms of α (alpha) and a:

$$n = \frac{\alpha - a}{1 - a - \alpha} \tag{11.7}$$

Modification to Ideal Beam Figure 11.7 shows longitudinal strains in a bridge girder due to the passage of a vehicle plotted against time. The plot has been smoothed to eliminate the dynamic effects, as discussed later.

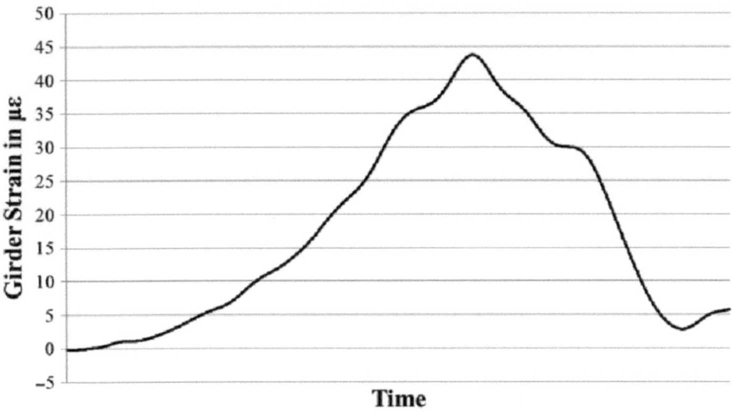

Fig. 11.7 Longitudinal girder strains under a moving truck

It can be seen in Fig. 11.7 that from this strain-time plot, time t_3 cannot be established for the truck under consideration because the strain at the end of the plot does not cross the zero strain line, possibly due to another vehicle following closely. Similarly, there can be difficulty in estimating t_1 in a multi-span bridge even if the spans are simply supported, because the girders in the instrumented span start vibrating even before a truck comes on it. For overcoming the difficulty in estimating t_1 and t_3, the baseline of the strain-time or moment-distance plot is moved up by an arbitrary 2 % of the value of the maximum strain or moment. The distances corresponding to t_1, t_2 and t_3 with respect to the new baseline are designated as x_1, x_2 and x_3, respectively.

Figure 11.8 shows the moments in a bridge calculated from observed strains at an instrumented section. It can be seen in this figure that the initial and final portions of the moment curve are parabolic, so that the initial portion of the curve is to the left of x_1 and the final portion is to the right of x_3; from this observation, it follows that the areas under the leading and trailing edges of the curve are very small as compared to the total area under the moment-time curve, thus leading to negligible errors in the estimation of the total area.

To find x_1 for the moment-distance plot truncated at 2 %, M according to Eq. (11.1) is equated to $0.1M_{max}$, where M_{max} is given by Eq. (11.2):

$$\frac{1}{2}\frac{W}{nL}x_1^2(1-a) = 0.02WLa(1-a)(1-n/2) \tag{11.8}$$

From Eq. (11.8), the following equation is obtained for x_1.

$$x_1 = L\sqrt{0.04an(1-n/2)} \tag{11.9}$$

Since M_{max} is not affected by the 2 % truncation, x_2 is still obtained by Eq. (11.4), and similarly to x_1, x_3 is obtained from:

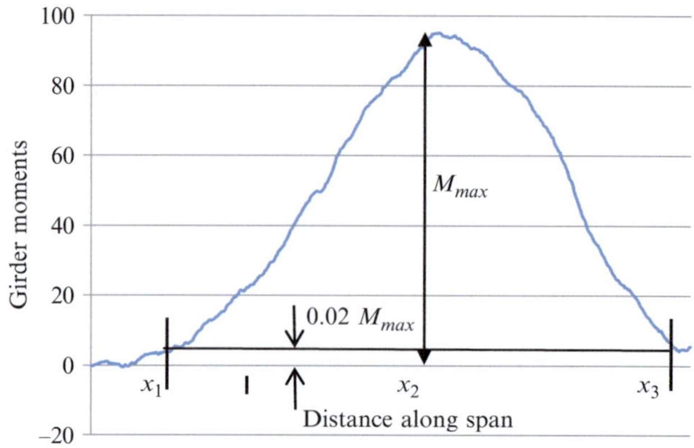

Fig. 11.8 The girder response diagram for the instrumented section showing curtailment at 2 % of the maximum moment level

$$x_3 = L\left[(1+n) - \sqrt{0.04n(1-a)(1-n/2)}\right] \tag{11.10}$$

The asymmetry coefficient can now be calculated as follows.

$$\begin{aligned}\alpha &= \frac{x_2 - x_1}{x_3 - x_1} \\ &= \frac{n + (1-n)a - \sqrt{0.04an(1-n/2)}}{(1-n) - \sqrt{0.04an(1-n/2)} - \sqrt{0.04n(1-a)(1-n/2)}}\end{aligned} \tag{11.11}$$

From a known value of α, Eq. (11.11) can give the corresponding value of n, but the calculation requires a time-consuming iterative process. It is convenient to either tabulate values of α and corresponding values of n, or to store the n-α relationship in a graphical format as discussed later. As also shown later, the curtailment of the response diagram at the 2 % level has negligible effect on the value the asymmetry coefficient.

Calculation of GVW The proposed BWIM system requires four pieces of data for each truck event, these being M_{max}, x_1, x_2 and x_3, which are identified in Fig. 11.8.

The equivalent length of the UDL representing a weighed truck is equal to nL, where n is then obtained with the help of Eq. (11.11).

Having found M_{max} and n for a truck event at the instrumented transverse section, the value of W, the GVW of the truck, is calculated from the following equation, which is a rewrite of Eq. (11.2).

Fig. 11.9 A very long truck

$$W = \frac{M_{max}}{aL(1-a)(1-n/2)} \tag{11.12}$$

11.3.2 Area Method

As discussed later with the help of a specific example, the accuracy of prediction of the asymmetry coefficient method depends upon the base length of the truck, with accuracy deteriorating with the increase in the base length. There are some very long trucks on the road, such as the one shown in Fig. 11.9.

It is quite possible that such long trucks are not permitted in many jurisdictions, but very long trucks do exist elsewhere and the proposed methods should be able to handle them. The area method is proposed to weigh very long trucks through BWIM.

Premise of the Area Method The premise of the proposed method can be explained with the help of Fig. 11.10, which shows bending moments at a reference point in a simply supported beam due to a set of three moving point loads.

It is well known that the influence line for each of the point loads has a triangular shape with maximum moment being $PaL(L\text{-}aL)/L$, where P is magnitude of the load, aL is the distance of the reference section from the left support, and L is the span of the beam. Although the moments due to P_2 and P_3 are offset from left support by x_1 and $(x_1 + x_2)$, respectively, the length of the bending moment diagram remains L for each load. It can be shown that the total area, A, of the bending moment diagram under the three moving loads is given by:

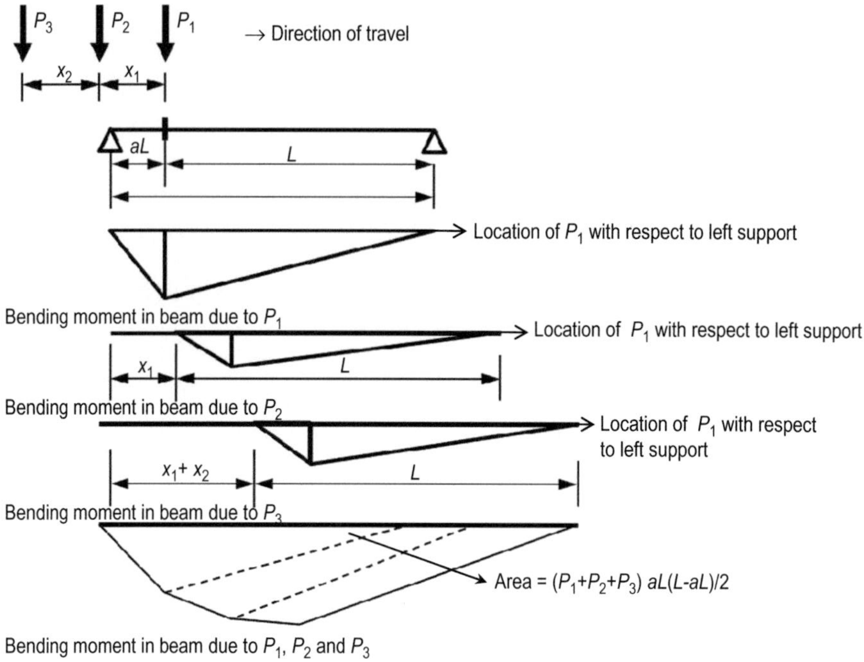

Fig. 11.10 Bending moments at the reference section in a simply supported beam due to a set of three moving point loads

$$A = (P_1 + P_2 + P_3)\ \frac{aL(L - aL)}{L}\frac{L}{2} = W\frac{aL(L - aL)}{2} \tag{11.13}$$

where, W is the sum of all three loads. Equation (11.13) shows that the total area under the response diagram plotted along the longitudinal direction of the beam depends only upon the sum of all loads; this important observation is borne out by field observations on the South Perimeter Bridge, which are discussed in detail in a later section.

For example, Fig. 11.11 shows the sum of girder strains at Section CC of the South Perimeter Bridge due to Truck B, as discussed later, travelling Eastbound in the same transverse position at three different speeds, being nearly 50, 75 and 95 km/h. Truck B had $W = 389$kN, and $B_m = 14.22$ m, and Section CC was 18.8 m away from the east support of the simply supported span with a length of 28.8 m. In Fig. 11.11, the horizontal time scale for the three curves was converted into the length scale by assuming that beginning and ending of all the curves, i.e. the zero strain points, are at the same locations. While the peak values of strains for three strain plots in Fig. 11.11 have a spread of about ± 4 %, it can be visualized that the total areas under each of three curves are likely to have a much smaller spread.

Figure 11.12 shows the sum of strain plots at Section CC due to Truck B travelling Westbound at speeds of nearly 50, 75 and 90 km/h. It can also be seen

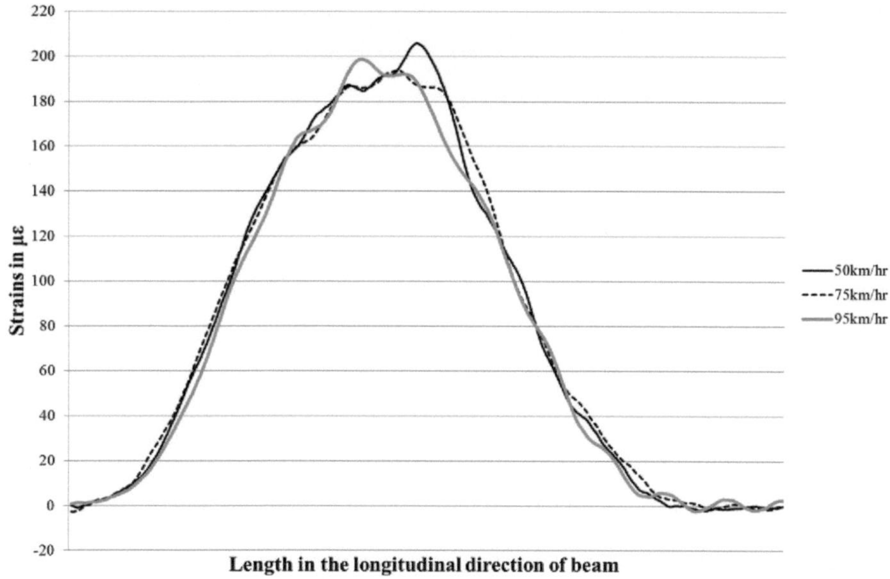

Fig. 11.11 Sum of girder strains at Section CC due to Truck B travelling Eastbound

in this figure that the three strain curves due to the same truck travelling at three different speeds are sitting virtually on top of each other, so that the areas under the three curves are also likely to be virtually the same. Similar observations can also be made about the sum of strain plots for Section DD, which was 5.6 m away from support, due to Truck B travelling Eastbound and Westbound, as shown in Fig. 11.13.

Details of the Area Method The total area A under a response diagram can be easily obtained by numerical integration, as explained later with reference to the particular example of the South Perimeter Bridge. The advantage of using numerical integration is that A becomes relatively insensitive to the smoothing process. By using Eq. (11.13), W is given by the following equation.

$$W = \frac{2A}{aL(L - aL)} \tag{11.14}$$

It is noted that the response diagrams of Figs. 11.11, 11.12, 11.13, and 11.14 correspond to the sum of girder strains. These diagrams can be converted into diagrams related to the sum of girder moments by multiplying the sum of strains by the constant C obtained from the calibration tests.

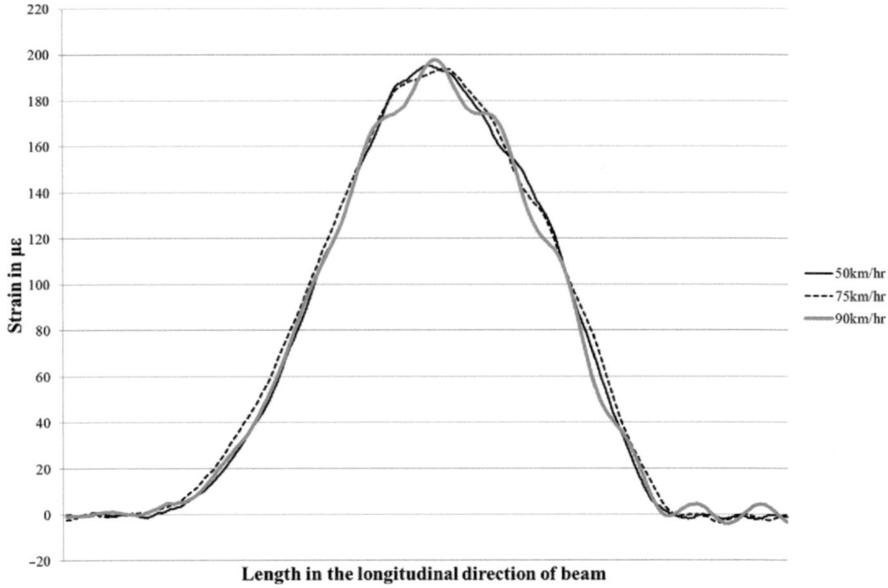

Fig. 11.12 Sum of girder strains at Section CC due to Truck B travelling Westbound

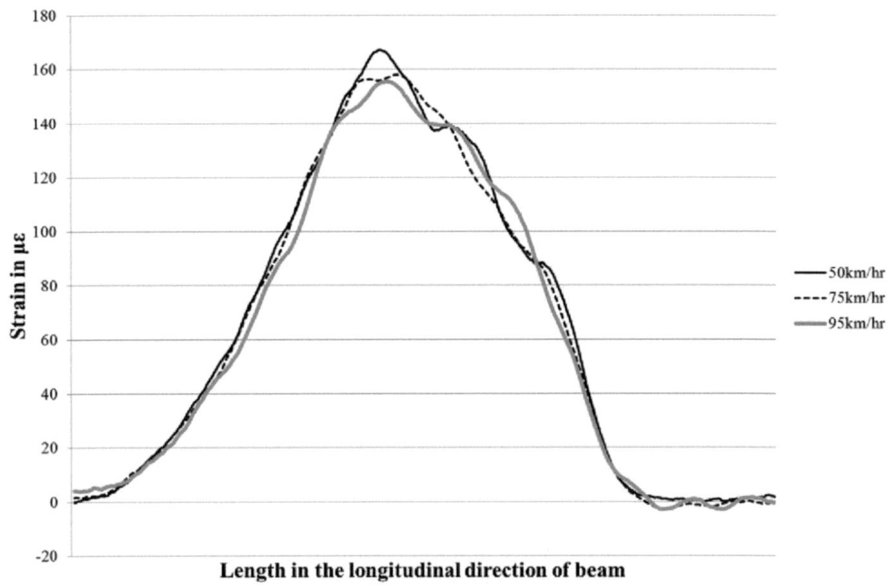

Fig. 11.13 Sum of girder strains at Section DD due to Truck B travelling Eastbound

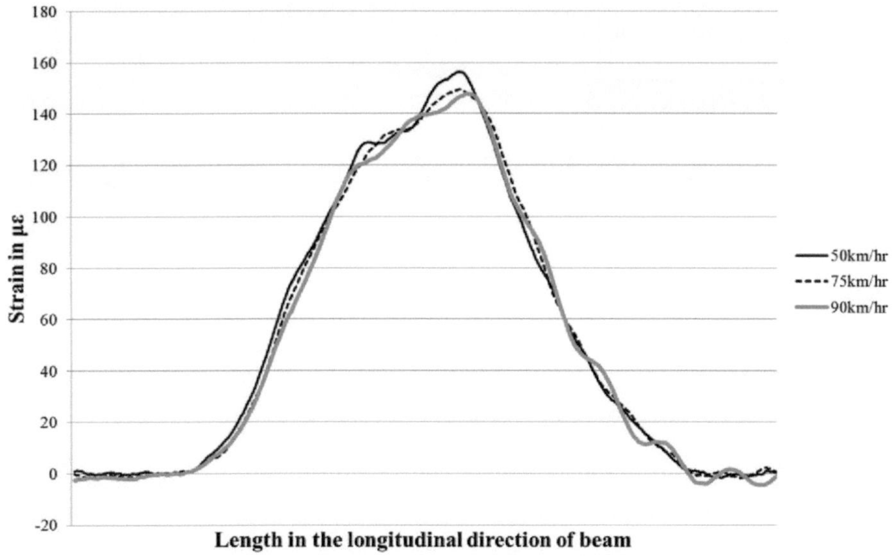

Fig. 11.14 Sum of girder strains at Section DD due to Truck B travelling Westbound

11.3.3 Two Stations Method

In the asymmetry coefficient and area methods, the estimate of W is obtained from girder responses at only one transverse section of the bridge. In the two-station method, the estimate of W is obtained from girder responses at two transverse sections, which as shown in Fig. 11.15 are designated as DD and CC, with distances of a_1L and a_2L, respectively from the left hand support. Also as shown in this figure, the distance between DD and CC is denoted as rL, where L is the span length. The total truck axle load, W, is again represented as a uniformly distributed load over a length nL.

The total time interval for the leading edge UDL to enter the span and for the trailing edge to leave the span is denoted as T. If the load is travelling at speed V, T is given by the following equation.

$$T = \frac{(1+n)L}{V} \tag{11.15}$$

The maximum response at an instrumented section occurs when the leading edge of the load has advanced a distance x from the left support:

$$x = L[n + (1-n)a] \tag{11.16}$$

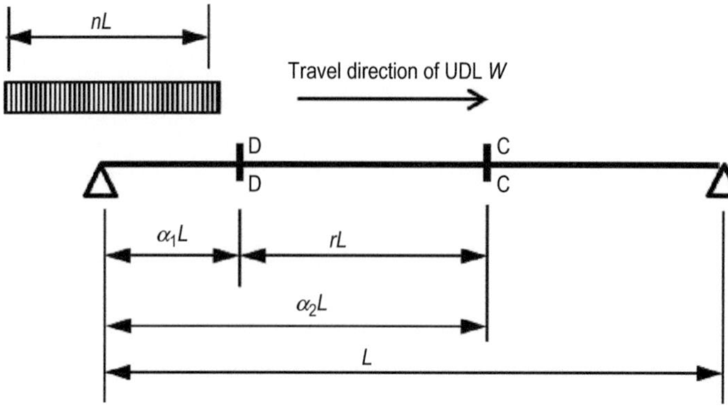

Fig. 11.15 Notation for the two-station method

where aL is the distance of the instrumented section from the left support. Using x_1 and x_2 for sections DD and CC, respectively, Eq. (11.16) can be written as follows for the two sections.

$$x_1 = L[n + (1 - n)a_1] \tag{11.17}$$

$$x_2 = L[n + (1 - n)a_2] \tag{11.18}$$

By subtracting Eq. (11.7) from Eq. (11.8), and denoting $(a_1 - a_2)$ as r, the following equation can be written.

$$x_2 - x_1 = (1 - n)rL \tag{11.19}$$

Ideally, the responses at each of sections DD and CC due to the moving UDL will have peaks at different times; the time interval between the two peak responses is denoted as t. Since $(x_2 - x_1) = Vt$, Eq. (11.19) can be written as:

$$Vt = (1 - n)rL \tag{11.20}$$

Denoting t/T as s, and dividing Eq. (11.15) by Eq. (11.20), the following equation is obtained.

$$n = \frac{r - s}{r + s} \tag{11.21}$$

After determining n from Eq. (11.21), W can be found using Eq. (11.12).

11.3.4 Beta Method

It can be seen from Eq. 11.14 that the relation between the GVW of a truck and the moment area is linear, and could be represented by the following equation.

$$GVW = constant \times A \tag{11.22}$$

Christenson et al. (2011) note that the GVW of a test truck of known weight GVW_k and that of a truck of unknown weight GVW_u are related by the following equation.

$$\frac{A_k}{GVW_k} = \frac{A_u}{GVW_u} \tag{11.23}$$

where A_k and A_u are the areas of the strain-time curves for the truck of known and unknown weights, respectively; a typical strain-time plot, showing strains from the beginning to the end of a vehicle pass, is shown in Fig. 11.16.

The ratio GVW_k and A_k is defined as the calibration constant β, which is given by:

$$\beta = \frac{GVW_k}{A_k} \tag{11.24}$$

Although referred to as a 'constant', β is really a variable that depends upon: (a) the speed of the vehicle, (b) the direction of travel, (c) the location of the instrumented transverse section, and (d) sometimes on the transverse position of the vehicle. It is noted for calculating the area under bridge response curve, Wall et al. (2009) and Christenson et al. (2011) convert the time scale into length scale by multiplying time with the calculated speed, so that the units of GVW_k and GVW_u are kN/με.m. For the exercise at hand, the time scale is not changed so that the units of GVW_k and GVW_u are kN/με.s.

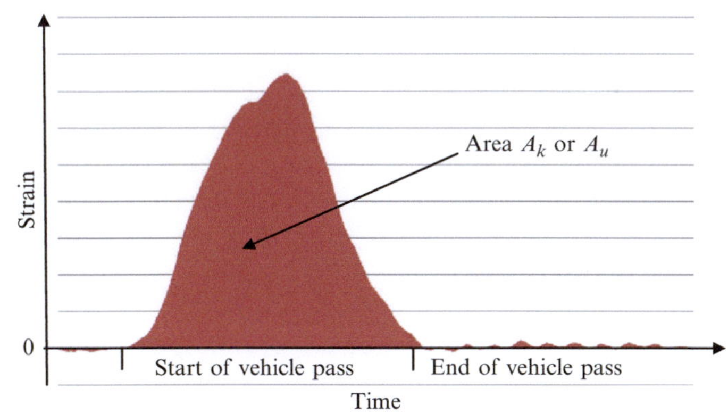

Fig. 11.16 A strain-time plot

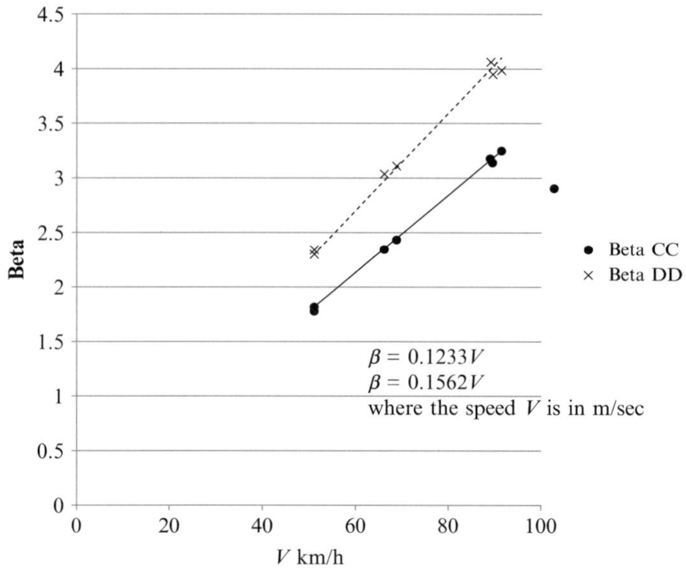

Fig. 11.17 A chart for calculating β for East-bound trucks by using V calculated from strap strains

As shown in Sect. 11.4, the transverse position of the vehicle does not change the calibration constants C and β for the instrumented bridge. The calculated speed of the vehicle, V, is used to obtain the values of the latter constant from the kind of chart given in Fig. 11.17, which was obtained from the results of the calibration tests, described in 11.4. The chart shows that each set of discrete β-V points can be represented by straight lines quite accurately. The values of β for instrumented transverse sections CC and DD are given by the equations given in Fig. 11.17 with a goodness fit that is very close to 1.00.

11.4 A Case History

In order to apply the proposed BWIM system, a bridge in Manitoba, Canada, was instrumented and tested for calibration under two vehicles of known configurations. This section describing the bridge and the tests, also provides calculations for the parameters that will be used in estimating the GVW for the test trucks described in the following.

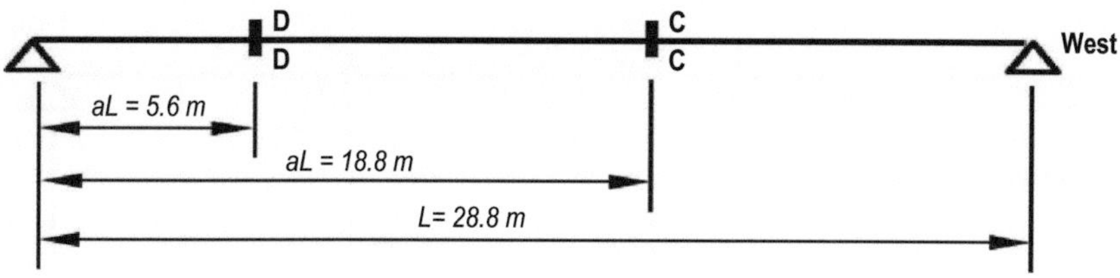

Fig. 11.18 The simply supported span instrumented for BWIM

11.4.1 Details of Bridge and Calibration Trucks

The South Perimeter Bridge in Winnipeg, Manitoba, is a bridge with seven spans, of which four are continuous and the remaining three simply supported, as shown in Fig. 10.30 in Chap. 10, and which shows that the simply supported span No. 2 was instrumented at three transverse sections, DD, BB and CC, and at the transverse section AA over the third pier from the western end. The bridge has both Westbound and Eastbound traffic. For the sake of convenience, it is assumed that the end, from which a vehicle enters the instrumented span is always the left support. The notation for the instrumented span for the Westbound traffic is shown in Fig. 11.18, from which it can be concluded that a for Sections DD and CC are 0.194 and 0.652, respectively.

Similarly, as shown in Fig. 11.19, a for Sections DD and CC for the eastbound traffic are 0.806 and 0.347, respectively.

As shown in Fig. 10.31 in Chap. 10, the bridge comprises an externally restrained concrete deck slab on eight steel girders. At section DD, which is closer to the East support of bridge, electrical resistance strain gauges, measuring longitudinal strains were installed on the upper side of the bottom flanges of all girders. At Section CC, the girders also included electrical resistance strain gauges on the underside of the top flanges, as well as strain gauges on the external steel straps, restraining the deck slab. The steel straps of the externally restrained deck slab of the bridge were instrumented at Sections AA and BB. The instrumented sections were calibrated for BWIM under two trucks, one of which had a single steering axle and two 2-axle groups; this truck, designated as Truck A, is shown in Fig. 11.20a, b.

The other truck, designated as Truck B, had a single steering axle, followed by 2-axle, 3-axle, and 2-axle groups. Details of Truck B are shown in Fig. 11.21a, b. Figures 11.20 and 11.21 also show W and B_m values for the two test trucks. As noted earlier, B_m is the length of the equivalent uniformly distributed load that induces nearly the same maximum moments and shears in a beam as the set of idealized point loads. In the context of the asymmetry coefficient method, $B_m = nL$.

The calibration tests were conducted under two sets of loads. In the first set, Truck A was made to travel at a crawling speed in 14 different transverse positions, which are identified in Fig. 11.22 as crawling speed load cases (CSLC). The purpose of this series of tests was to determine the value of the constant C that relates the sum of theoretical peak girder moments M to the corresponding sum of observed girder strains, ε, according to the following equation.

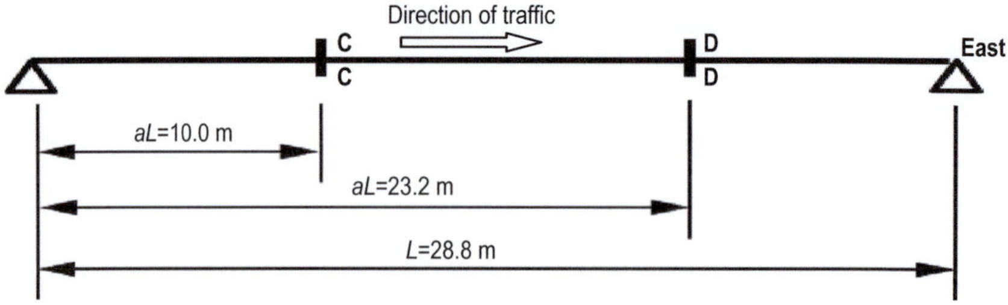

Fig. 11.19 The simply supported instrumented span under Eastbound traffic

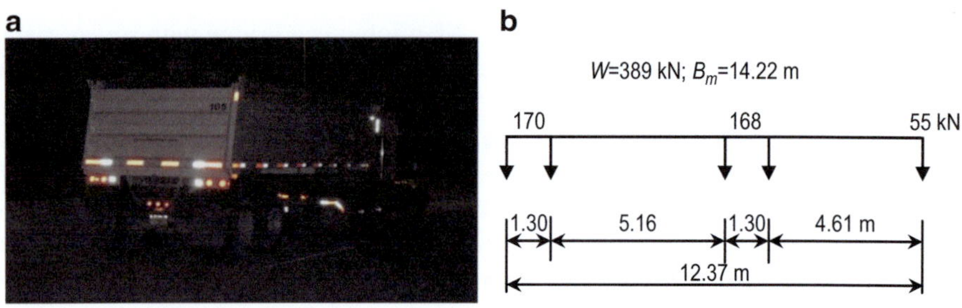

Fig. 11.20 Details of Truck A: (**a**) photo, (**b**) axle weights and configurations

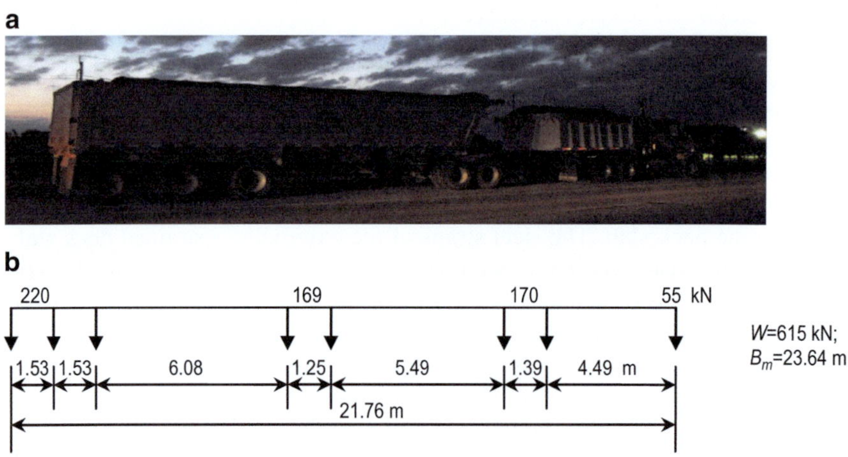

Fig. 11.21 Details of Truck B: (**a**) photo, (**b**) axle weights and configurations

$$M = \varepsilon C \tag{11.25}$$

Clearly, C has the same units as M, and relates to girder moments without dynamic amplification.

For the other series of tests, both Trucks A and B were made to travel at different traffic speeds in the normal and passing lanes; the transverse positions of trucks

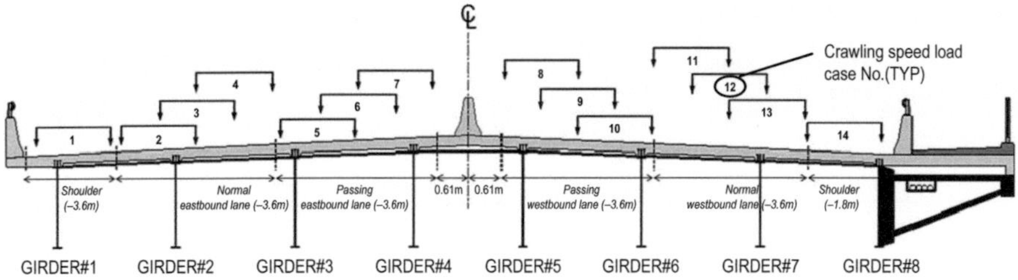

Fig. 11.22 Transverse positions of Truck A, moving at crawling speed for calibration tests

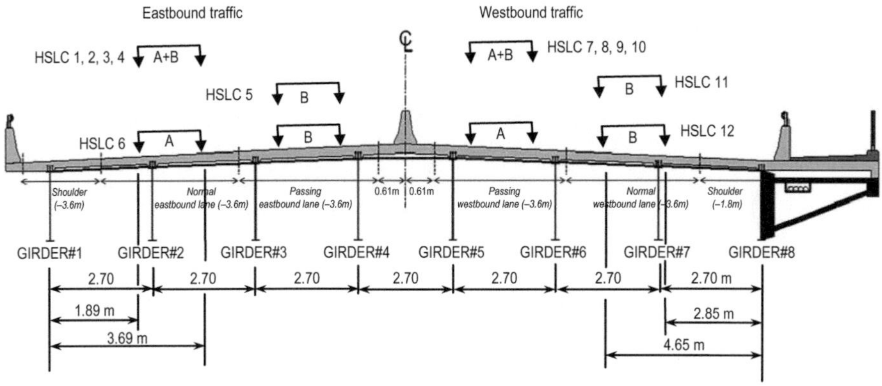

Fig. 11.23 Transverse positions of test trucks, moving at fast speeds for calibration tests

during these tests are shown in Fig. 11.23, along with the high speed load case (HSLC) numbers. As can be seen in this figure, all except two load cases comprised Truck A followed by Truck B at a 'safe' distance. When the trucks travelled at speeds of 50 km/h., or greater, the distance between the two trucks was large enough so as not to interfere with the responses due to one in the instrumented span with those due to the other truck. However, at speeds of nearly 25 km/h, the distance between the two trucks became too small to isolate clearly the responses of one truck from those of the other.

In Fig. 11.23, 'A' and 'B' refer to Test Trucks A and B, respectively; and (b) 'A + B' refers to the load case when Test Truck B followed behind Truck A at a safe distance. The purpose of tests of the second series was to determine the accuracy of the various BWIM methods described earlier. The load case Nos., speeds of the trucks and other information are provided in Table 11.1, along with the titles of the files containing the raw numerical data.

Table 11.1 Details of tests with vehicles running at high speeds

Direction	High speed load case no.	Description	Approximate speed, km/h.	Title of numerical file
Eastbound	1	Truck A followed by Truck B	25	test_B_01
	2	Truck A followed by Truck B	50	test_B_02
	3	Truck A followed by Truck B	75	test_B_03
	4	Truck A followed by Truck B	95	test_B_04
	5	Truck A followed by Truck B	95	test_B_05
	6	Truck A beside Truck B	95	test_B_06
Westbound	7	Truck A followed by Truck B	25	test_B_07
	8	Truck A followed by Truck B	50	test_B_08
	9	Truck A followed by Truck B	75	test_B_09
	10	Truck A followed by Truck B	90	test_B_10
	11	Truck A followed by Truck B	90	test_B_11
	12	Truck A beside Truck B	90	test_B_12

11.4.2 Calculation of Bridge Constant C

By using the observed data, smoothed using a low pass filter, under crawling speed load cases 1 through 14, shown in Fig. 11.22, and Eq. (11.25), values of constant C were calculated for both Sections CC and DD for Truck A travelling in each of eastbound and westbound directions. The data was smoothed by a low pass Butterworth filter. By idealizing Span 2 as a simply supported beam it was found that Truck A travelling in the eastbound direction induces maximum moments of 1303 and 1928 kN.m at Sections DD and CC, respectively. For the truck moving in the westbound direction, the corresponding maximum moments are 1377 and 1994 kN.m, respectively.

The percentage difference form the average in the calculated values of C are plotted in Fig. 11.24 against the transverse positions of the centre of Truck A. It can be seen in this figure that the magnitude of C is little affected by the transverse position of the vehicle. The average value of C for Section CC was found to be 11.307×10^{12} N.mm for the east bound direction and 12.077×10^{12} N.mm for the west bound direction, with a maximum variation of 1.5 %, and the average value of

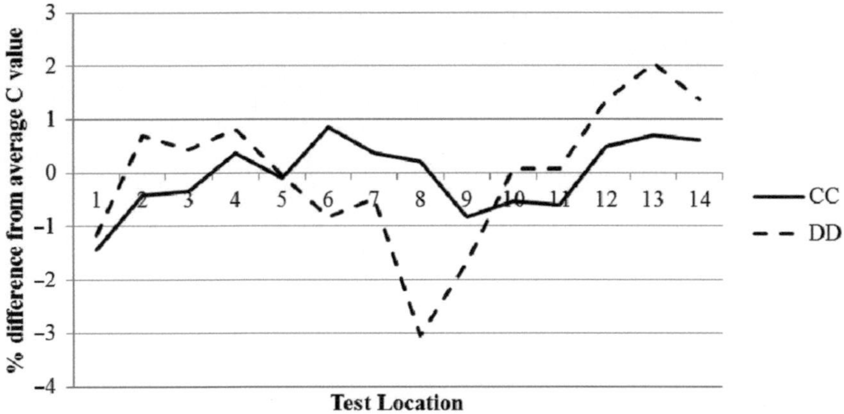

Fig. 11.24 Percentage difference from average *C* value

C for Section DD was found to be 9.836×10^{12} N.mm for the east bound direction and 10.239×10^{12} N.mm for the west bound direction, with a maximum variation of 3 %,. The average values of *C* was used for converting the BWIM data into truck weights.

11.4.3 Calculation of Vehicle Speed

The speed of the vehicle is calculated from the peak strains in the straps at Sections AA and BB. Figure 11.25 presents the strain-time plots for Strap No. 8 (at - Section AA) and Strap No. 2 (at Section BB) due to Trucks A and B travelling in high speed test No. 3; the plots were smoothed by a low pass filter at a cut off frequency of 10 Hz. It can be seen in this figure that the peak strains in both the straps are well defined. Because of its proximity to the support diaphragm, the peak strains in Strap No. 8 are considerably smaller than the corresponding peak strains in Strap No. 2.

Dividing the distance between Sections AA and BB by the time interval between corresponding peaks at the two sections directly gives the vehicle velocity *V*. Table 11.2 compares the calculated velocity by the strap method with those reported by the truck driver. It is noted that in the absence of accurate measuring device, the reported speeds can only be used as guides. The fact that for each test, the velocities of both Trucks A and B are very nearly the same confirms the validity of the method. The speed calculations are further validated by the accuracy of prediction of GVWs of the two test trucks.

Fig. 11.25 Strains in straps due to Trucks A and B during high speed load case No. 3

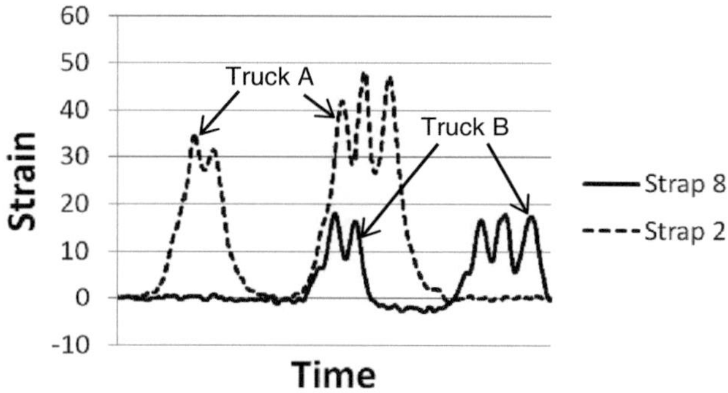

Table 11.2 Reported and calculated speeds of test trucks in km/h by the strap method

	Test truck	East-bound traffic				West-bound traffic			
		High speed test no.				High speed test no.			
		2	3	4	5	8	9	10	11
Reported speed (approximate)	A	50	75	95	95	50	75	90	90
	B	50	75	95	95	50	75	90	90
Calculated speed	A	45.2	71.2	83.1	87.8	50.9	68.7	89.4	91.4
	B	45.2	72.1	83.8	85.9	51.0	66.1	89.1	89.6

11.4.4 Observed Transverse Load Distribution

The transverse load distribution characteristics of a girder bridge are identified by non-dimensional load distribution factors (DFs) of longitudinal responses of the girders, such as strains, moments or deflections. A DF can either be ratio of the actual and average longitudinal responses, or the ratio of the actual and total longitudinal responses. In the former case, the sum of DFs for the girders is equal to the number of girders (8 in the bridge under consideration), and for the latter case, which is used here, the sum is equal to 1.0.

The numerical values of DFs for peak longitudinal girder strains near the bottom flanges corresponding to all the 14 crawling speed test cases, shown in Fig. 11.22, are listed in Table 11.3 it being noted that the strains due to the crawling speed tests were without dynamic magnification, and therefore did not require filtering other than that needed to remove the electric noise.

During the crawl speed test Nos. 3 and 12, the test vehicles traversed the bridge in the centre of normal travelling lanes in the eastbound and westbound directions, respectively. The corresponding high speed test numbers for the test vehicle travelling in the eastbound directions were 1, 2, 3 and 4, and for those in the westbound directions were 7, 8, 9 and 10, as shown in Fig. 11.23. The DFs for girder strains due to crawling speed test Nos. 3 and 12 are plotted against girder Nos. at Section DD in Fig. 11.26a, b, respectively. The normal travelling lanes for

Table 11.3 Distribution factors for longitudinal strains near the bottom of girders due to crawling speed tests identified in Fig. 11.22

Crawling speed test no.	Distribution factors for girder no. 1	2	3	4	5	6	7	8	Sum of DFs
1	0.46	0.29	0.14	0.06	0.03	0.02	0.01	0.00	1.00
2	0.31	0.32	0.21	0.09	0.05	0.02	0.01	0.00	1.00
3	0.23	0.30	0.24	0.11	0.06	0.03	0.02	0.00	1.00
4	0.19	0.27	0.28	0.13	0.07	0.04	0.01	0.00	1.00
5	0.08	0.15	0.31	0.23	0.14	0.07	0.02	0.00	1.00
6	0.07	0.13	0.28	0.29	0.15	0.06	0.03	0.00	1.00
7	0.05	0.11	0.23	0.30	0.18	0.08	0.04	0.00	1.00
8	0.00	0.03	0.09	0.17	0.32	0.23	0.11	0.05	1.00
9	0.00	0.02	0.07	0.12	0.28	0.29	0.16	0.07	1.00
10	0.00	0.02	0.04	0.07	0.15	0.26	0.29	0.17	1.00
11	0.00	0.01	0.04	0.07	0.15	0.27	0.30	0.17	1.00
12	0.00	0.01	0.03	0.05	0.12	0.22	0.34	0.24	1.00
13	0.00	0.01	0.02	0.04	0.09	0.16	0.36	0.33	1.00
14	0.00	0.00	0.01	0.02	0.06	0.13	0.31	0.47	1.00

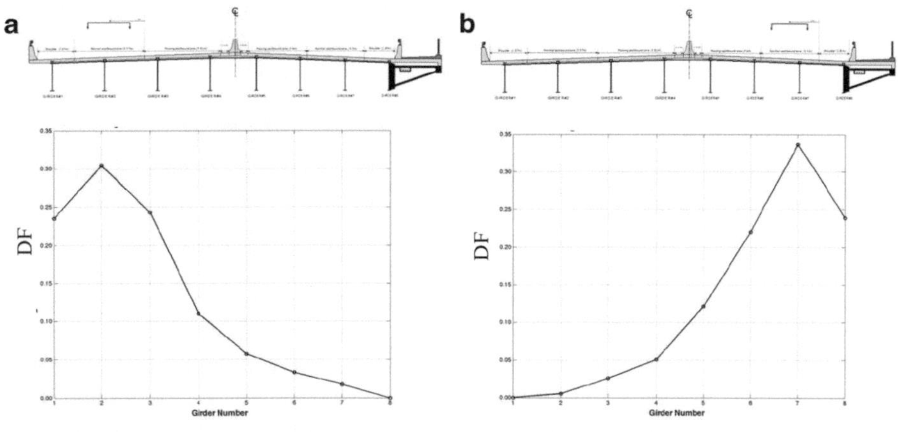

Fig. 11.26 Distribution factors for girder strains plotted against girder numbers at Section DD: (**a**) for crawling speed test No. 3, (**b**) for crawling speed test No. 12

east- and west bound traffic are almost mirror images of each other. Accordingly, the transverse load distribution curves are also almost mirror images of each other.

An interesting aspect of the transverse load distribution characteristics of the South Perimeter Bridge is that when a vehicle is travelling in the east-bound lanes (over Girder Nos. 1, 2, 3 and 4), the sum of the DFs for strains in Girder Nos. 5, 6, 7 and 8, designated as $DF_{5,6,7,8}$ is relatively small as compared to the sum of the DFs for Girder Nos. 1, 2, 3 and 4, designated as $DF_{1,2,3,4}$. Similarly, for the vehicle travelling in the west-bound lanes, $DF_{1,2,3,4}$ is relatively small as compared to $DF_{5,6,7,8}$; this aspect can be observed in Table 11.4.

Another interesting and useful, although predictable, aspect of the load distribution characteristics of the bridge is that the maximum girder strains occur in the girders directly under the vehicles; this aspect, as explained later, can be used to identify if the observed girder strains are due to a single- or a multiple-truck event. Table 11.5 identifies the girder/s which experience maximum strains under the various load case, it being noted that the strains in two girders are assumed to be 'maximum' when they are within 5 % of each other; this table also shows the distances of the centre of the test truck with respect to the outer-most girders, designated as x_E for the truck in east-bound lanes and x_W for the truck in west-bound lanes, as shown in Fig. 11.27.

Values of $DF_{5,6,7,8}$ and $DF_{1,2,3,4}$ are plotted in Fig. 11.28 against x_E and x_W, respectively. The curves of this figure together with the Girder Nos. with maximum strains, listed in Table 11.5 can be used with advantage to identify single-truck events and the transverse position of the truck.

It is important to note that the calibration tests discussed in this subsection were conducted under vehicles, in which the distance between the centrelines of their two lines of wheels was 1.80 m. Accordingly, many of the conclusions drawn above are applicable to similar normal vehicles. For wider vehicles, in which there are more than two lines of wheels or the distance between the centrelines of two lines of wheels is greater than 1.80 m, calculating the distribution factors must be made using mathematical models, i.e. numerical methods of analysis.

11.4.5 Smoothing of Raw Strains

After they have been converted from observed voltages from the electrical resistance strain gauges, the strain-time curves include 'electrical noise' as well as effects of dynamic amplification due to the interaction of the dynamic systems of the vehicle/s and the bridge. To obtain strain-time curves which are free of the electrical noise and dynamic amplification, the data has to be smoothed by a suitable scheme. Initially, the smoothing was done by a method using moving averages. The moving average method was abandoned in favor of smoothing by the Butterworth filter at a cut-off frequency of about 3.9063 Hz, which corresponds to the natural frequency of the bridge. Smoothing by the Butterworth filter provided better accuracy in calculating GVW.

The Butterworth filter is a type of signal processing filter designed to have as flat a frequency response as possible in the passband, i.e. the frequencies that are included. It is also referred to as a maximally flat magnitude filter (Bianchi and Sorrentino 2007).

The validity of the smoothing process is provided by a comparison of the smoothed signals from high speed tests with the un-smoothed signals from

Table 11.4 Sum of distribution factors for strains in girders remote from the vehicle at Section DD

	East-bound lanes							West-bound lanes						
Crawling speed load test no.	1	2	3	4	5	6	7	8	9	10	11	12	13	14
$DF_{5,6,7,8}$	0.05	0.07	0.11	0.13	0.23	0.24	0.30	Not shown						
$DF_{1,2,3,4}$	Not shown							0.29	0.21	0.13	0.11	0.08	0.07	0.03

Table 11.5 Girders with maximum strains under various load cases at Section DD

Crawling speed load case no.	East-bound lanes							West-bound lanes						
	1	2	3	4	5	6	7	8	9	10	11	12	13	14
Girder no./nos. with maximum strains	1	2	2	2,3	3	3,4	4	5	5	5,6	6,7	7	7,8	8

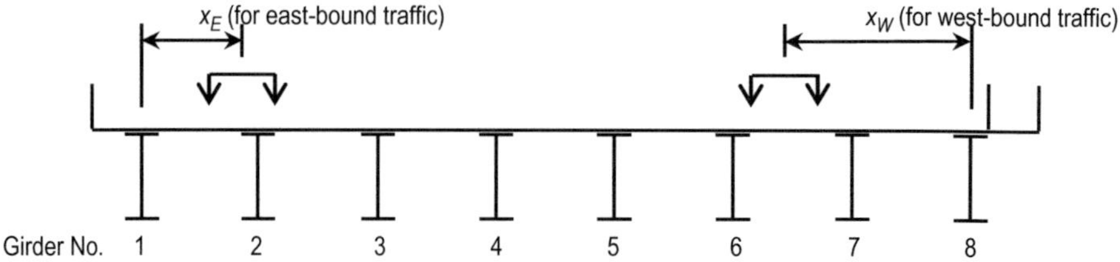

Fig. 11.27 Notation for transverse distance of vehicles

Fig. 11.28 Sum of distribution factors for strains in girders remote from the truck, plotted against transverse truck position (*Note*: For clarity, actual points are shown only for vehicles in East-bound lanes)

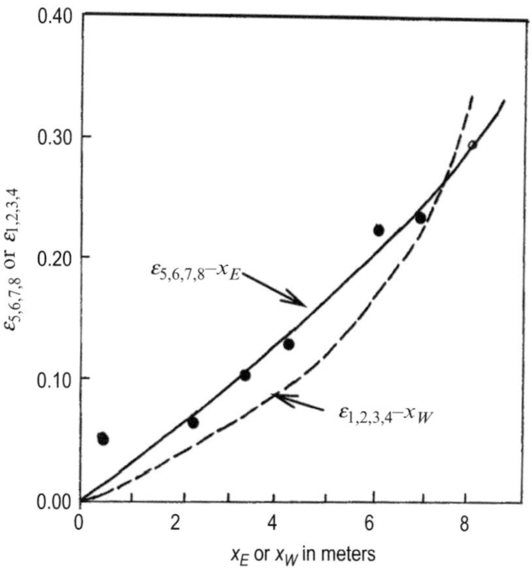

corresponding crawl speed tests. All these comparisons are provided in Fig. 11.29 for tests in east-bound lanes and in Fig. 11.30 for tests in west-bound lanes. It is noted that the charts in both these figure relate to Section DD.

It is interesting to note in Figs. 11.29 and 11.30 that the smoothed curves for the high speed tests are virtually the same as the corresponding un-smoothed curves for the crawling speed tests. The slight difference between the two can be attributed to the fact that the transverse positions of trucks travelling at high speed are unlikely to be exactly the same as those for the corresponding crawling speed tests.

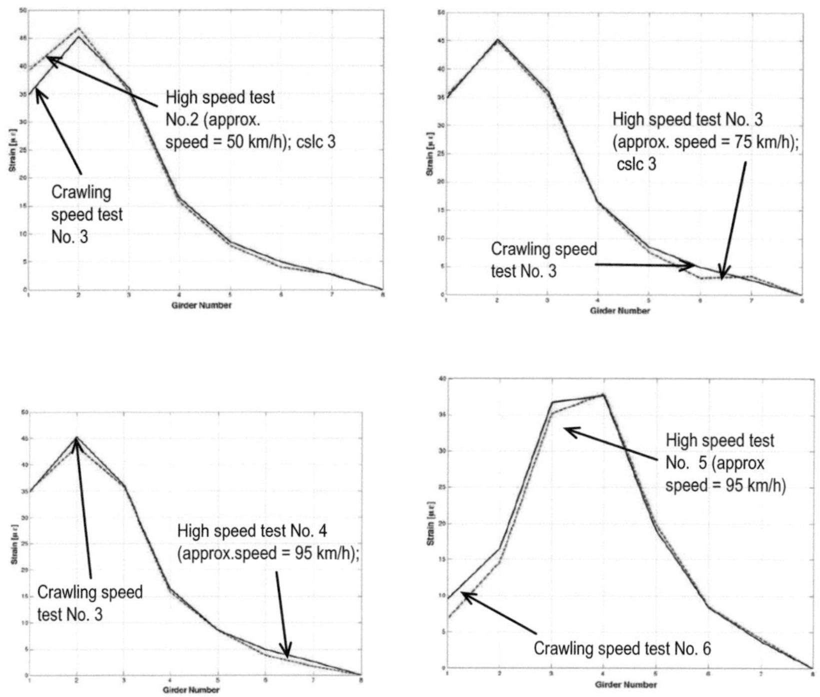

Fig. 11.29 Comparisons of smoothed signals from high speed tests and un-smoothed signals from corresponding crawl speed tests, at Section D-D in East-bound lanes

The comparisons of Figs. 11.29 and 11.30 confirm that the smoothing process through the Butterworth filter does give quite accurate results.

11.4.6 Analysis for Load Distribution

The analytical method selecting for analyzing the South Perimeter Bridge was the semi-continuum method (Jaeger and Bakht 1989), which is discussed in Chap. 3. The steel plate girders of the South Perimeter Bridge, although having nearly constant depth of construction along their length, have additional bottom plates that change the value of I as shown in Fig. 11.31 for Girder No. 1; this figure also shows the notation which is used to calculate the value of the equivalent moment of inertia I_e for purposes of load distribution analysis (Jaeger and Bakht 1989). As can be seen in Fig. 11.31, the moment of inertia of Girder No. 1 in the middle region is 0.064 m^4, and that in the end regions is 0.050 m^4. The moment of inertia in the regions between the middle and end regions is 0.059 m^4. According to method of Jaeger and Bakht, the equivalent I_e of the girder with variable symmetrical stepped moments of inertia is given by the following equation.

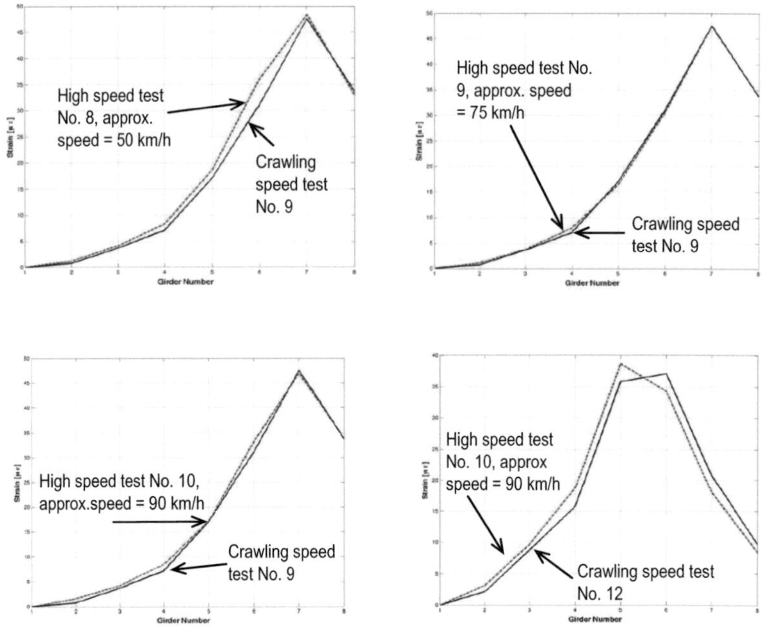

Fig. 11.30 Comparisons of smoothed signals from high speed tests and un-smoothed signals from corresponding crawl speed tests, at Section DD in West-bound lanes

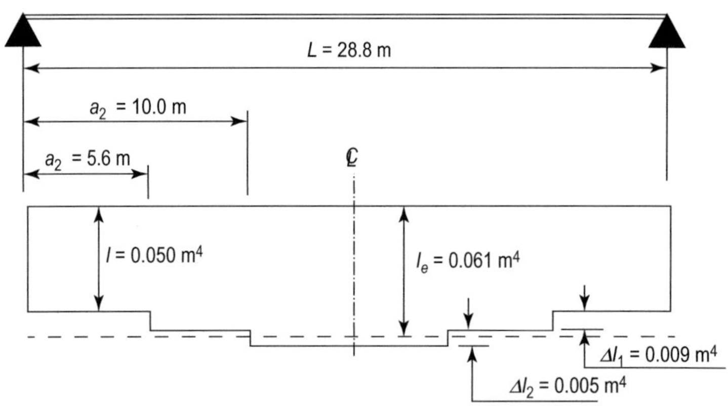

Fig. 11.31 Calculated values of moments of inertia for Girder No. 1

$$I_e = I + \sum_{r=1}^{r=n} \Delta I_r cos \frac{\pi a_r}{L} \tag{11.26}$$

where n is the No. of stepped moments of inertia, being 2 in the case under consideration, and other notation is as shown in Fig. 11.31. It is noted that π in

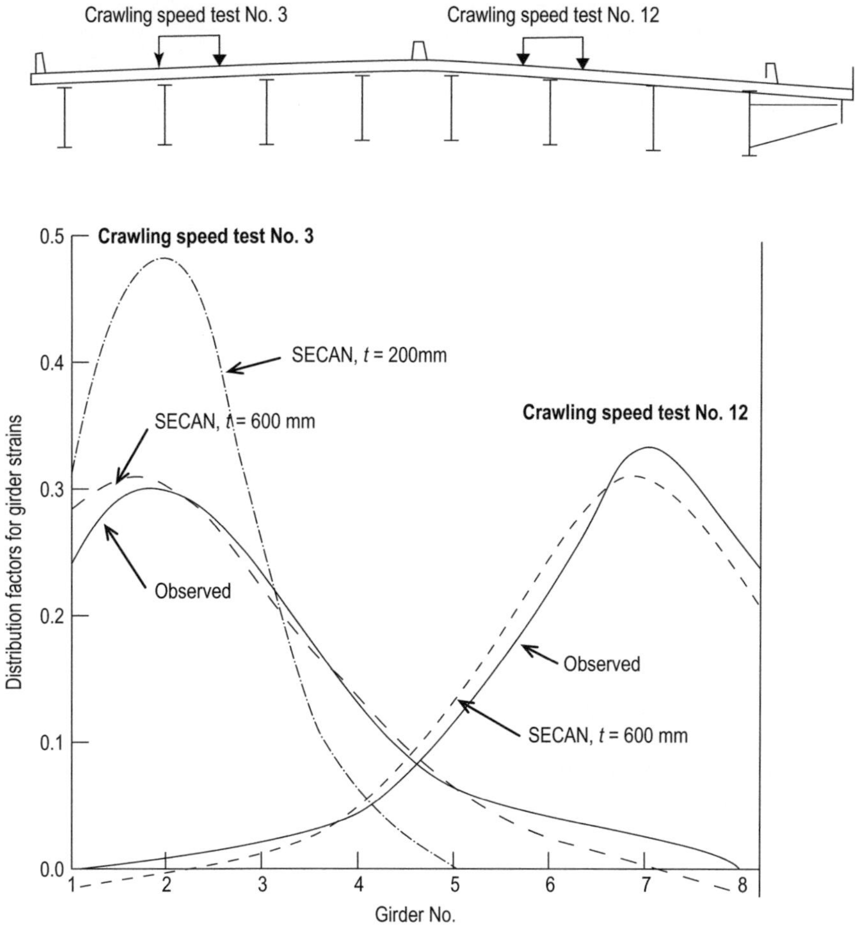

Fig. 11.32 Comparison of observed and analytical DFs for girder strains near the bottom flanges at Section DD

the above equation is in radians. By using Eq. (11.26), I_e for Girder No. 1 was found to be 0.061 m^4; this value is also applicable to Girder Nos. 4 and 5. Similarly, for Girder Nos. 2, 3, 6 and 7 I_e was found to be 0.059 m^4, and that for Girder No. 8 was 0.067 m^4.

The above values of moments of inertia are in terms of the units of steel, for which the modulus of elasticity E was assumed to be 200,000 MPa. By using these values of EI, the superstructure of the South Perimeter Bridge was analyzed under Test Truck A in the same transverse position as in crawling speed test No. 3, in which the truck is in the middle of the normal travelling east-bound lane; longitudinally, the third axle of the truck was placed directly above Section DD. It is recalled that slight changes in the longitudinal location of the truck has little effect on the transverse load distribution pattern. Initially, the deck slab thickness, t, was

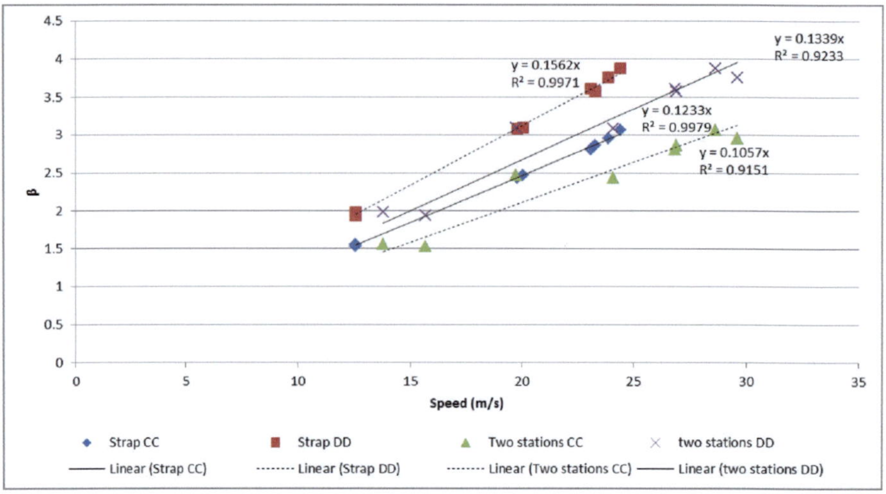

Fig. 11.33 Calibration of Beta for East bound direction

taken as 200 mm. The transverse pattern of DFs for girder moments thus obtained was found to be much different than that of the observed girder strains, as can be seen in Fig. 11.33. This outcome is not surprising because (a) the transverse steel straps of an externally restrained deck slab are well away from the underside of the slab, making its effective thickness to be much greater than the actual thickness of the concrete slab, and (b) the analysis did not take into account the frequently spaced transverse diaphragms.

The SECAN analysis was repeated several times by gradually increasing the effective thickness of the deck slab, until it was found that the analytical and observed patterns of transverse distributions were fairly close to each other. As shown in Fig. 11.32, the operative effective thickness of the deck slab was found to be 600 mm.

The superstructure of the bridge was also analyzed for the crawling speed test No. 12, in which the truck travelled in the middle of the west-bound normal lane. It can be seen in Fig. 11.32 that for $t = 600$ mm, the analytical transverse load distribution again matches well with the observed pattern. While it is not exact, the comparison should be considered reasonable in light of uncertainties related in exactly ascertaining the properties of the composite girders and the degree of bearing restraint that nullifies the assumption of simply supported ends.

It will be noticed that while the observed DFs were calculated from girder strains, their analytical counterparts were obtained from girder moments. Admittedly, strains in girders with different moments of inertia are not directly proportional to the moments. However, it was found that the moments of inertia of different girders of the South Perimeter Bridge and distances of their neutral axes from the instrumented locations were so close to each other that the errors incurred by ignoring the differences had negligible effect on the outcome.

11.4.7 Calculation of n for Asymmetry Method

As shown in Fig. 11.18, the span of the bridge is 28.8 m, and the instrumented section DD is 5.6 m from the centre of the left-hand support, so that a for this section is 5.6/28.8 = 0.19444. Similarly, a for section CC is 18.8/28.8 = 0.65278. The value of n for the ideal beam can be calculated for any value of α from Eq. (11.7) for both Sections DD and CC. However, for the modified beam, it is convenient to use Eq. (11.11) to calculate α for several values of n.

11.5 GVW Estimation for High Speed Tests

The four methods discussed above were implemented on the data that was obtained during high speed tests numbers 2, 3, 4, 5, 8, 9, 10 and 11 for both trucks A and B. In tests 1 and 7 the two trucks were too closely spaced to isolate the signals for each truck for processing. Tests 6 and 12 included both of the trucks traveling side by side. The purpose of these cases was to test multi presence algorithms in the future.

The following is a summary of the results of each method.

11.5.1 The Asymmetry Method

Table 11.6 presents the summary of the predicted GVW of the truck and the percentage of error in calculating the GVW, using the asymmetry method while taking the start and end times as 2 % of the maximum sum of strains for all girders.

It can be seen from the results that this method produces large errors in calculating the GVW. In the case of test 2 for truck B, the value of α was calculated as 0.59, which produced a value of n larger than 1 giving the negative value. Although in many cases when taking the average GVW for sections CC and DD the error is relatively small.

11.5.2 The Two Stations Method

Table 11.7 presents the results of calculating the GVW using the two statins method. The 2 % and 0 % present the threshold value for determining the start time t_1 and end time t_3 as a percentage of the maximum sum of strains.

The results show that the threshold value had little effect on the overall results of the method; however the results using a 2 % threshold seem to be more consistent. This may be due to the fact that at lower strains the signal to noise ratio and the effect of vibration is larger and which could lead to a shift in the position of the

Table 11.6 GVW results calculated using the asymmetry method

Travel direction		East bound				West bound			
Test number		Test 2	Test 3	Test 4	Test 5	Test 8	Test 9	Test 10	Test 11
Truck A	GVW CC	462.02	693.55	666.69	853.47	366.96	351.14	358.93	327.31
	GVW DD	350.38	386.25	373.46	344.94	440.24	441.10	438.47	487.73
	GVW Av.	406.20	539.90	520.07	599.20	403.60	396.12	398.70	407.52
	% error CC	18.89	78.46	71.55	119.61	−5.57	−9.65	−7.64	−15.78
	% error DD	−9.84	−0.61	−3.90	−11.24	13.28	13.50	12.83	25.50
	% error Av.	4.52	38.93	33.83	54.19	3.86	1.93	2.59	4.86
Truck B	GVW CC	−1706.6	547.18	787.68	1072.72	673.28	413.06	423.49	424.92
	GVW DD	577.11	509.29	613.51	766.75	764.92	829.09	768.12	796.51
	GVW Av	−564.75	528.23	700.60	919.73	719.10	621.08	595.80	610.72
	% error CC	−377.34	−11.08	28.01	74.33	9.42	−32.87	−31.18	−30.95
	% error DD	−6.21	−17.24	−0.30	24.61	24.31	34.74	24.83	29.44
	% error av	−191.78	−14.16	13.86	49.47	16.86	0.93	−3.17	−0.75

Table 11.7 GVW results using the two stations method

Threshold value	Travel direction		East bound				West bound			
	Test no.		Test 2	Test 3	Test 4	Test 5	Test 8	Test 9	Test 10	Test 11
Threshold value as percentage from max strain = 2 %	Truck A	n	0.42	0.57	0.51	0.50	0.45	0.48	0.50	0.49
		V (m/s)	13.77	24.07	26.84	28.63	15.71	21.29	27.48	27.50
		GVW CC	383.62	416.84	393.35	380.93	392.10	389.32	389.12	378.78
		GVW DD	368.64	395.64	387.09	382.15	399.69	390.17	398.64	422.19
		GVW (av.)	376.13	406.24	390.22	381.54	395.90	389.74	393.88	400.49
		error CC	−1.29	7.26	1.22	−1.98	0.89	0.18	0.13	−2.53
		error DD	−5.14	1.81	−0.39	−1.66	2.85	0.40	2.58	8.64
		error av.	−3.21	4.53	0.41	−1.82	1.87	0.29	1.35	3.05
	Truck B	n	0.99	0.57	0.97	0.96	0.99	0.65	0.53	0.54
		V (m/s)	15.67	19.72	26.88	29.60	17.16	19.48	24.58	24.12
		GVW CC	744.87	498.36	754.68	725.80	713.33	541.09	544.80	554.73
		GVW DD	751.11	502.40	731.14	709.07	739.14	535.31	531.95	555.71
		GVW av.	747.99	500.38	742.91	717.44	726.23	538.20	538.37	555.22
		error CC	21.05	−19.01	22.64	17.95	15.92	−12.07	−11.46	−9.85
		error DD	22.06	−18.35	18.82	15.23	20.12	−13.01	−13.55	−9.69
		error av.	21.56	−18.68	20.73	16.59	18.02	−12.54	−12.51	−9.77

(continued)

Table 11.7 (continued)

	Travel direction		East bound				West bound			
Threshold value	Test no.		Test 2	Test 3	Test 4	Test 5	Test 8	Test 9	Test 10	Test 11
Threshold value as percentage from max strain = 0 %	Truck A	n	0.47	0.64	0.55	0.54	0.47	0.51	0.57	0.53
		V (m/s)	12.52	20.17	24.82	26.68	15.07	20.06	23.45	25.52
		GVW CC	396.90	438.13	403.32	389.79	397.81	397.09	409.10	388.23
		GVW DD	381.39	415.84	396.90	391.04	405.51	397.95	419.11	432.72
		GVW av.	389.15	426.99	400.11	390.42	401.66	397.52	414.11	410.47
		error CC	2.13	12.74	3.78	0.30	2.36	2.18	5.27	−0.10
		error DD	−1.86	7.01	2.13	0.62	4.35	2.40	7.85	11.35
		error av.	0.14	9.87	2.96	0.46	3.36	2.29	6.56	5.62
	Truck B	n	0.99	0.59	0.97	0.96	0.99	0.67	0.56	0.56
		V (m/s)	13.94	18.61	23.65	27.64	16.30	18.02	23.17	23.17
		GVW CC	745.36	507.02	757.38	727.86	713.79	551.93	554.90	561.64
		GVW DD	751.60	511.14	733.75	711.08	739.62	546.03	541.82	562.63
		GVW av.	748.48	509.08	745.56	719.47	726.70	548.98	548.36	562.13
		error CC	21.13	−17.60	23.08	18.29	16.00	−10.31	−9.82	−8.73
		error DD	22.14	−16.93	19.24	15.56	20.20	−11.26	−11.95	−8.57
		error av.	21.64	−17.27	21.16	16.92	18.10	−10.78	−10.89	−8.65

beginning and end of truck load event. The results also show that the errors in calculating the GVW using section CC is less than that of section DD. This may be because section CC has higher bending moment and therefore the noise to signal ratio is smaller.

The errors in calculating the GVW for truck A using the averages from sections CC and DD had a maximum of 6 %, while for truck B the error ranged from 9 % to 22 %. It therefore could be concluded that error in calculating the GVW using the two stations method increases with the increase of the total truck length.

The test speed reported was provided by the truck driver from the speedometer reading. The speed calculated using the two stations method therefore could not be verified accurately. However as will be shown in the following section the predictions of the GVW using speeds obtained from the two stations method had relatively high errors, and it could be concluded that the speed also had higher errors, especially for truck B.

11.5.3 The Area Method

Table 11.8 presents the results of calculating GVW using the area method. The results show that the error in calculating the GVW for truck A ranged from 7 % to 20 % when using the speeds calculated using the two stations method, while the errors ranged from 0.9 % to 2.8 % when using the speeds calculated from the straps.

The results also show that for truck B the error ranged from 2.3 % to 24 % when using the speeds calculated using the two stations method, while the errors ranged from 0.4 % to 2.7 % when using the speeds calculated from the straps.

11.5.4 The Beta Method

Figures 11.33 and 11.34 present the calibration of beta for the eastbound and west bound directions respectively. It could be shown that the speeds from straps give smaller scatter for beta than the speeds calculated from the two stations method, thus providing more accurate calibration for the parameter beta. This could be demonstrated in Table 11.9 which present the error for the GVW calculated using the beta parameters obtained from Figs. 11.34 and 11.35.

The results show that the error in calculating the GVW for truck A ranged from 0.23 % to 8.9 % when using the speeds calculated using the two stations method, while the errors ranged from 0.36 % to 4.4 % when using the speeds calculated from the straps.

The results also show that for truck B the error ranged from 0.19 % to 18.5 % when using the speeds calculated using the two stations method, while the errors ranged from 0.6 % to 4.4 % when using the speeds calculated from the straps.

Table 11.8 GVW results using the area method

Travel direction			East bound				West bound			
Test no.			Test 2	Test 3	Test 4	Test 5	Test 8	Test 9	Test 10	Test 11
V m/s (straps)			12.55	19.79	23.09	24.41	14.15	19.07	24.71	25.38
Truck A	Speed two stations	GVW CC	419.48	469.85	452.69	442.53	423.90	430.23	423.55	416.23
		GVW DD	412.52	465.04	442.79	440.45	414.57	421.42	416.97	424.43
		GVW av.	416.00	467.44	447.74	441.49	419.23	425.82	420.26	420.33
		% error CC	7.94	20.90	16.49	13.87	9.08	10.71	8.99	7.10
		%error DD	6.15	19.67	13.94	13.34	6.68	8.44	7.30	9.21
		%error av	7.05	20.28	15.21	13.61	7.88	9.57	8.14	8.16
	Speed straps	GVW CC	382.41	386.27	389.42	377.27	381.86	385.42	380.91	384.08
		GVW DD	376.06	382.33	380.90	375.50	373.44	377.52	375.00	391.65
		GVW av.	379.23	384.30	385.16	376.38	377.65	381.47	377.96	387.86
		% error CC	−1.60	−0.60	0.21	−2.92	−1.74	−0.82	−1.98	−1.17
		%error DD	−3.23	−1.62	−1.99	−3.38	−3.91	−2.86	−3.50	0.78
		%error av	−2.42	−1.11	−0.89	−3.15	−2.82	−1.84	−2.74	−0.19
Truck B	speed two stations	GVW CC	769.34	599.78	705.79	751.59	747.73	644.20	604.46	596.15
		GVW DD	762.52	602.12	710.50	743.69	730.69	627.01	589.92	595.25
		GVW av.	765.93	600.95	708.15	747.64	739.21	635.61	597.19	595.70
		% error CC	25.03	−2.53	14.70	22.14	21.52	4.69	−1.77	−3.12
		%error DD	23.92	−2.15	15.46	20.86	18.75	1.90	−4.13	−3.26
		%error av	24.47	−2.34	15.08	21.50	20.13	3.29	−2.95	−3.19
	Speed straps	GVW CC	616.60	609.47	611.05	605.89	616.94	606.93	608.69	614.88
		GVW DD	611.14	611.84	615.12	599.52	602.88	590.74	594.04	613.95
		GVW av.	613.87	610.65	613.08	602.71	609.91	598.84	601.37	614.42
		% error CC	0.20	−0.95	−0.70	−1.54	0.26	−1.37	−1.08	−0.07
		%error DD	−0.68	−0.57	−0.04	−2.57	−2.03	−4.00	−3.46	−0.23
		%error av	−0.24	−0.76	−0.37	−2.05	−0.88	−2.68	−2.27	−0.15

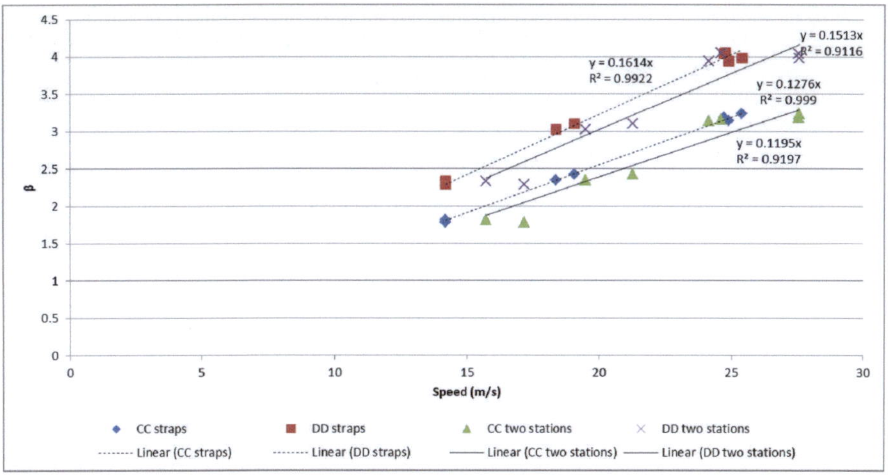

Fig. 11.34 Calibration of Beta for West bound direction

11.6 BWIM: A Tool for Bridge Management

The case history described in Sects. 11.4 and 11.5 was commissioned by Manitoba Infrastructure and Transportation (MIT) of the province of Manitoba, Canada, to gain knowledge about the weights of the vehicles that travel on its roads. The ISIS Canada Resource Centre monitors and processes the BWIM data and provides MIT with periodical reports about the exceptionally heavy vehicles. The GVW data are provided in the form of histograms of GVWs, a sample of which is provided in Fig. 11.35 for a certain 1 month period.

It can be seen in Fig. 11.35 that the bridge experiences vehicles of exceptionally high GVWs. The BWIM system described above is also complemented with a video camera that provides photographs of all vehicles. An example of a photograph is provided in Fig. 11.36.

11.7 Concluding Remarks

Based on analysis of the test data and the results of calculating the GVW using the four proposed methods the following conclusions could be made.

- The location of a truck could be accurately determined using the strain distribution of the bridge girders.
- The asymmetry method produces large errors and is not reliable.
- The two stations method provide good prediction for the GVW for normal 5 axle trucks, however the method produces large errors for longer trucks.

Table 11.9 Errors in GVW using Beta method

Direction		East bound				West bound			
Test no		Test 2	Test 3	Test 4	Test 5	Test 8	Test 9	Test 10	Test 11
Truck A	Speed (two stations)	−6.87	4.65	0.23	−1.16	2.70	−7.71	−8.91	−8.89
	Speed (straps)	−0.97	0.36	0.58	−1.71	−1.27	−3.55	−4.44	−1.92
Truck B	Speed (two stations)	8.29	−15.03	0.12	5.70	1.19	−13.00	−18.26	−18.45
	Speed (straps)	1.24	0.72	1.12	−0.60	−2.61	−4.38	−3.98	−1.89

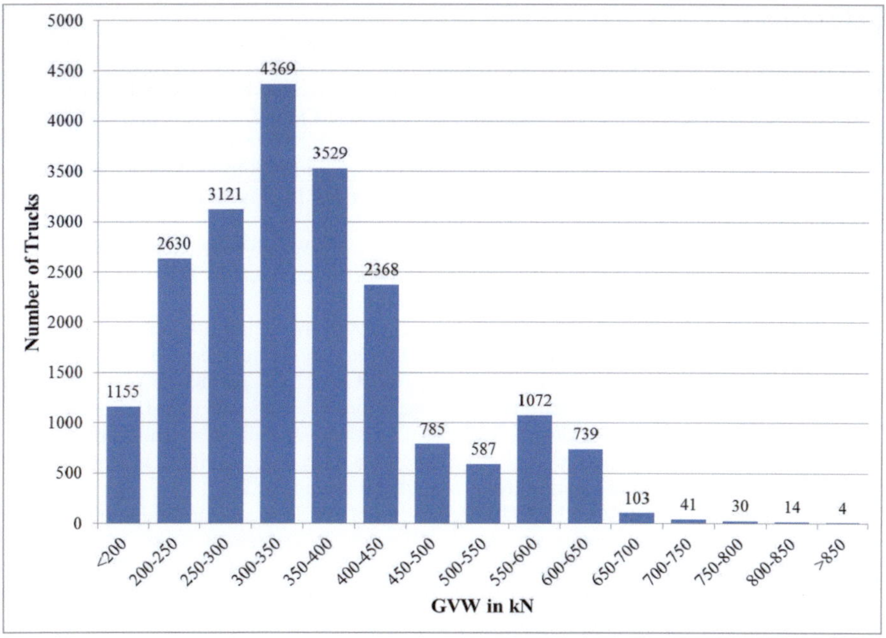

Fig. 11.35 Histogram of GVW of vehicles over a 1 month period

Fig. 11.36 Photograph of a vehicle weighed by the BWIM system

- The area and the beta method provide predictions with errors less than 5 % of the GVW when calculated using speeds from the straps.
- The vehicle speeds could be accurately determined using data from straps.
- The Beta method gives the most accurate assessment of the GVW, its accuracy being ±5 % with a confidence limit of 90 %.

References

Agarwal AC, Bakht B (1988) Error analysis of weighing in motion by highway bridges. Transportation Forum, Roads and Transportation Association of Canada, vol 3(3). pp 31–40, Ottawa

Arroyo M, Hannachi M, Siegert D, Jacob B (2010) Extreme traffic load effects on medium span bridges. In: Clemente P, DeStefano A (eds) WIM (Weigh-in-Motion) load capacity and bridge performance. International Society for Health Monitoring of Intelligent Infrastrures, Politico di Torino, pp 9–14, Italy

ASTM E1318-02 (2002) Standard specification for highway Weigh-in-Motion (WIM) systems with user requirements and test methods. ASTM International, West Conshohocken, Pennsylvania

Bakht B, Mufti AA, Tadros G, Eden R, Mourant G (2006) Weighing-in-motion of truck axle weights by Japanese reaction force method. In: 3rd international conference on bridge maintenance, safety and management, held in Porto, 16–19 July 2006, pp 981–982

Bianchi G, Sorrentino R (2007) Electronic filter simulation & design. McGraw-Hill, New York

Christenson R, Bakulski S, McDonnell AMH (2011) Establishment of a dual-purpose bridge health monitoring and weigh-in-motion system for a steel girder bridge. In transportation research board 90th annual meeting (No. 11–1598). Washington, DC

Dowling J, Gonzalez A, Obrien EJ (2010) Monitoring of traffic loads and bridge performance using a bridge weigh-in-motion system. In: Clemente P, DeStefano A (eds) International society for health monitoring of intelligent infrastrures. Politico di Torino, pp 107–113, Italy

González A, Rowley C, OBrien EJ (2008) A general solution to the identification of moving vehicle forces on a bridge. Int J Numer Methods Eng 75(3):335–354

Harman DJ (1984) A study of commercial vehicles and their multiple presence, OJTCRP project no. 23141. University of Western Ontario, Faculty of Engineering Science, London, Ontario

Helmi K, Bakht B, Mufti A (2014) Accurate measurement of gross vehicle weight through bridge weighing-in-motion. A case study. J Civ Struct Heal Monit 4:195–208

Heywood RJ (1991) Bridge applications of WIM data. In: Heywood RJ (ed) Proceedings of the AUSTROADS bridges conference. Austroads, Haymarket, pp 139–150

Jacob B, O'Brien E, Jehaes S (2002) COST 323, weigh-in-motion of road vehicles. Laboratoire Central des Ponts et Chausées, Paris

Jaeger LG, Bakht B (1989) Bridge analysis by microcomputer. McGraw Hill, New York

Moses F (1979) Weigh-in-motion system using instrumented bridges. ASCE Transp Eng J 106(TE3):233–249

Quilligan M, Karoumi R, O'Brien EJ (2002) Development and testing of a 2-dimensional multi-vehicle Bridge-WIM algorithm. In: Third international conference on Weigh-in-Motion (ICWIM3), Florida, USA

Rowley CW et al (2009) Experimental testing of a moving force identification bridge weigh-in-motion algorithm. Exp Mech 49(5):743–746

Scott G (1987) Weighing-in-motion technology – status CULWAY in Australia. Australian research board seminar on road traffic data collection using weighing-in-motion, Melbourne

Wall CJ, Christenson RE, McDonnell AMH, Jamalipour A (2009) A non-intrusive bridge weigh-in-motion system for a single span steel girder bridge using only stain measurements. Report no. CT-2251-03-09-5. Connecticut Department of Transportation, Bureau of engineering and construction, office of research and materials

Yamada K, Ojio T (2003) Bridge weigh-in-motion system using reaction force method. In: Proceedings of the international workshop on structural health monitoring of bridges/colloquium on bridge vibration'03, Japan Society of Civil Engineers, pp 269–276

Znidaric A, Kalin J, Lavric I (2002) Bridge weigh-in-motion measurements on short slab bridges without axle detectors. In: Third international conference on Weigh-in-Motion (ICWIM3), Florida, USA

Chapter 12
Bridge Aesthetics

12.1 Introduction

The word "aesthetics" is derived from the Greek word "aisthetike" denoting sensory perception; it has been linked strongly to the perception of beauty.

Classical Greek education recognised five divisions of knowledge as the five pillars of education:

- Logic – examining reason;
- Metaphysics – examining existence;
- Ethics – examining morals;
- Epistemology – examining knowledge; and
- Aesthetics – examining beauty.

12.2 Theory of Numbers

The Greeks and other ancient cultures recognized that the study of beauty was an essential part of the education of young minds. They also recognized that there were numbers connected to geometry that were in turn related to equations. As shown in Fig. 12.1, they perveivd numbers as either squares such as 1, 4, 9, 16, 25 or triangles such as 1, 3, 6, 12, and 15.

Relationships between numbers, geometry and equations, essential to the conceptualization and quantification of aesthetics are discussed in the following.

B. Bakht, A. Mufti, *Bridges*, DOI 10.1007/978-3-319-17843-1_12

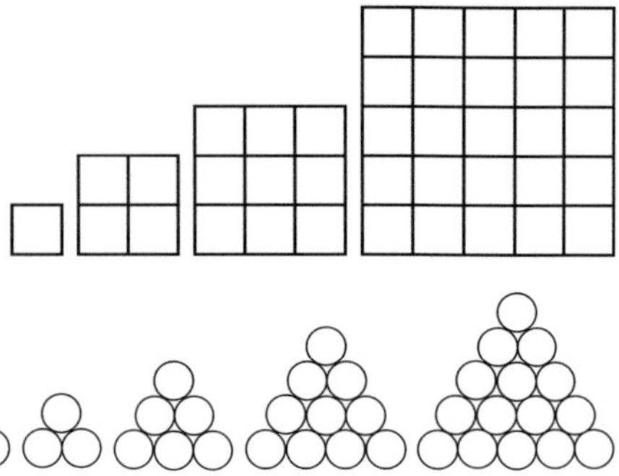

Fig. 12.1 Square and triangular numbers

12.3 Pythagorean Theory

To connect the idea of numbers to geometry, it is instructive to review the Pythagorean Theorem, which states that the number of squares on the hypotenuse of a right-angled triangle is equal to the sum of the squares on the other two sides. Pictorially, this can be expressed by drawing squares along Side C on the hypotenuse of a triangle, as well as along Sides B and A, as shown in Fig. 12.2. The equation which evolves is:

$$C^2 = A^2 + B^2 \tag{12.1}$$

If these squares were placed 5×5 on Side c, 3×3 on Side b, and 4×4 on Side a, it becomes apparent that 5^2 is 25, 4^2 is 16 and 3^2 is 9, as shown in Fig. 12.2; and if the right-hand sides (a and b) are added, they equal 25, which equals the squares on side c. Clearly, the numbers are connected with the geometry and the geometry is very much connected with the equation.

The ancients also recognized that there are numbers such as 'zero' and 'infinity' and they again related these numbers to geometry by looking at a pentagram, shown in Fig. 12.3. If the pentagram is connected by straight lines from one vertex to the vertices opposite, a new shape emerges in the form of another pentagram. In the centre of the pentagram is another pentagram exactly the same as the outer pentagram but scaled down and rotated by 180°. If this division process is continued, the central shape is eventually reduced to zero. Conversely, if the starting point is the smallest pentagram shape 'zero' an infinite number of shapes can be created to return to the very large pentagram figure.

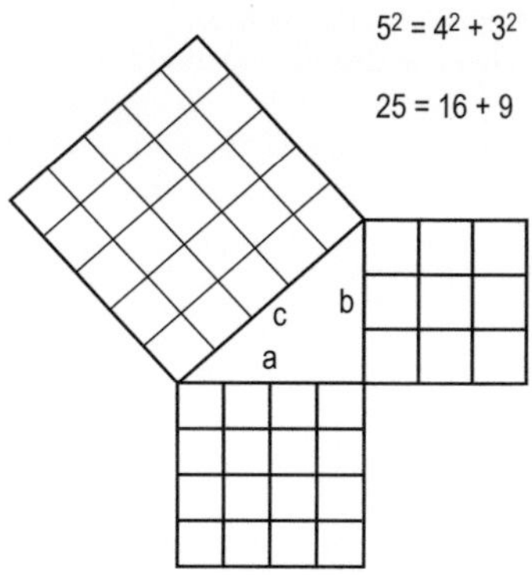

Fig. 12.2 Pythagorean theory

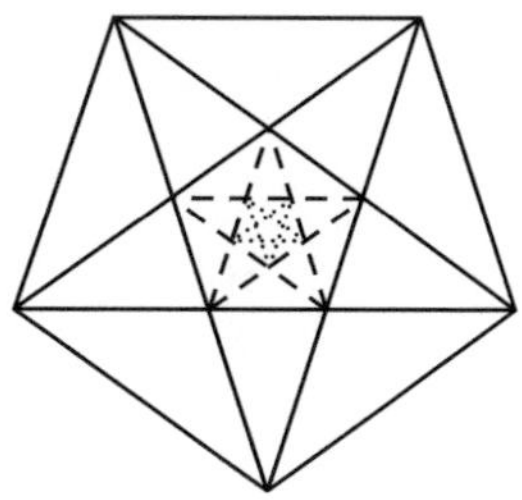

Fig. 12.3 Pentagram and zero

12.4 The Golden Mean

The other number which the ancients found to be extremely useful was the 'Golden Mean', illustrated in Fig. 12.4. This number was used to great advantage by the Greeks in the design and construction of temples and other prominent buildings. The number that emerges as the Golden Mean or Ratio is $(1+\sqrt{5})/2$ i.e., 1.618. To determine this ratio, draw a rectangle, as is shown in Fig. 12.4 with a height of 1.618 and a width of 1. A rectangle created in this manner forms one of the most beautifully balanced shapes in existence. Divide this rectangle into one equal-sided square (1×1) and the area remaining will create a new rectangle with a height of 0.618 resulting in a ratio of 0.618–1, thereby creating the Golden Ratio, i.e., 1.618. This process can be continued until eventually arriving at a beautiful cascading subdivision of numbers, which really follow an infinite pattern. For example, the Parthenon in Greece, shown in Fig. 12.5, was built using the Golden ratio.

The diagram of Fig. 12.6a is constructed in the following successive steps:

(a) Consider two 1×1 squares, each with the aspect ratio (ratio of the two sides) $= 1.0$
(b) Add a 2×2 square to form a rectangle having an aspect ratio of $3/2 = 1.5$
(c) Add a 3×3 square to form a rectangle having an aspect ratio of $5/3 = 1.667$
(d) Add a 5×5 square to form a rectangle having an aspect ratio of $8/5 = 1.6$

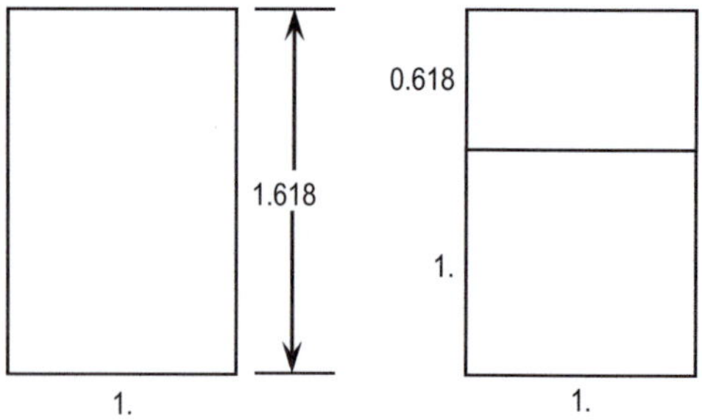

Fig. 12.4 Golden Mean (Greek aesthetic evolution)

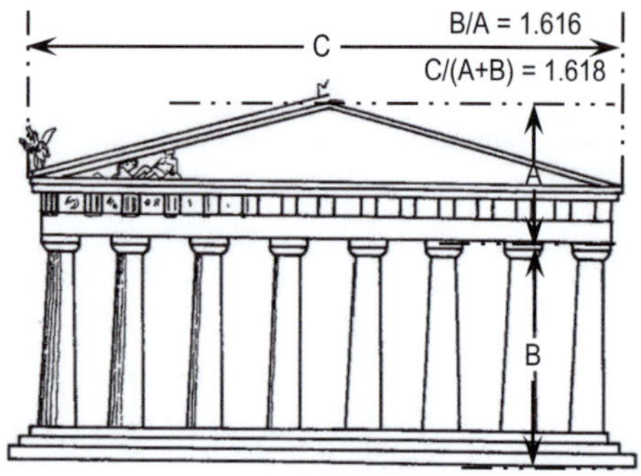

Fig. 12.5 The Parthenon

(e) Add an 8×8 square to form a rectangle having an aspect ratio of $13/8 = 1.625$
(f) Add a 13×13 square to form a rectangle having an aspect ratio of $21/13 = 1.615$

If the process of adding squares continues similarly, it will be found that the aspect ratios of the resulting rectangles will converge to the same ratio of 1.618, which is referred to as the *Golden Ratio*.

If tangents are drawn through the corners of successive rectangles, a spiral emerges, as shown in Fig. 12.6b. It is interesting to note that such a spiral also occurs in nature, for example in the shell of a snail, shown in Fig. 12.6c.

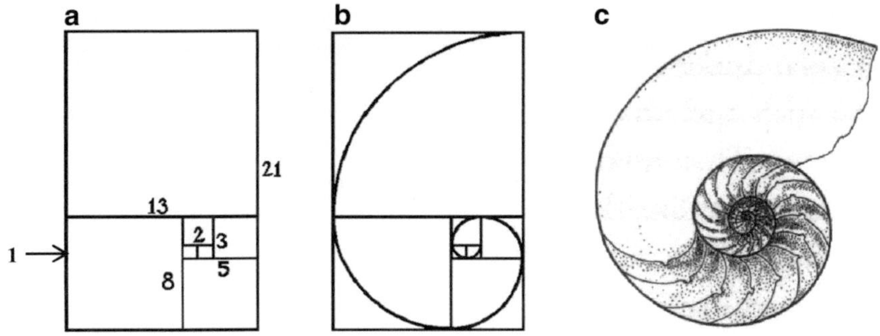

Fig. 12.6 The Golden ratio (Mean) (**a**) rectangular shapes, (**b**) spiral through tangents, and (**c**) the shell of a snail

Given this background, it can be summarized that numbers are connected to geometry, and geometry is connected to equations. When viewing the geometry of a statue or a structure, the engineer's brain is determining the beauty of the numbers and the equations that resulted in its creation. Clearly, there must be a link between numbers, geometry and equations. Further, if an engineer or someone with a similar mindset knows that a certain structure has historical connections to events, which occurred in the past and which human beings either liked or disliked, then it influences their decision as to whether or not they find a structure aesthetically pleasing.

The Parthenon, shown in Fig. 12.5, makes use of several extremely important motifs. Not only does the structure follow a geometry based on the Golden Ratio, it demonstrates clearly how the forces flow from the roof, through the beams, through the columns and thence to the foundation. The third quality, which is very interesting in the Parthenon, is that it also incorporates the motif articulated by Pheidias, the architect of the structure, who wrote in a letter to the elected leader of the City of Athens, "It is my dream to see Athens, the supreme centre of beauty as well as philosophy and science. And for generations to remember, I want to crown Acropolis with Parthenon in memory of Athene Parthenos." This is very interesting because it enunciates the idea that beautiful structures are associated with the cultural and sociological events of the time in which they were constructed; therefore, when viewing a structure, its beauty may be evident, but it also brings to the viewer's mind the motifs incorporated into the structure, including the historical period and the human events that took place during its construction as well as the uses to which the structure was put after it was built.

Developing this line of thought further, one is led to looking at the aesthetic issue from the standpoint of Kayser's coupling, which states that aesthetic issues can be viewed qualitatively or quantitatively depending upon the viewer. Kayser suggests that, generally, an artist looks at a structure or object in qualitative terms, whereas a scientist or engineer looks at the same structure and sees its characteristics in quantitative terms. For example to an artist, geometry, ratio, and shape will be qualitatively defined as the tone; an engineer or a scientist will see those same characteristics as dimensions, as for example in the rectangle and the Golden Ratio.

An artist will look at an object or structure and say that it has a beautiful tone, and an engineer will assess the same object or structure and say it has beautiful dimensions. The second coupling suggested by Kayser concerns perception versus logic. An artist may perceive that a structure has a certain way of taking the forces acting upon it whether these are its own weight or the applied weights of any other forces which may come from nature such as earthquakes, floods, wind, rain or any other hazards. This perception to an engineer will be viewed in terms of logic. Engineers use logical and scientific methods to determine how forces acting upon a structure will affect it. Kayser's third coupling involves feelings. An artist will feel that a structure may be extremely beautiful. An engineer will look at the same structure to determine how much knowledge is available about that structure and what can be learned from it. For example, the Parthenon is a structure of great beauty that follows excellent geometrical ratios and is exposed in such a way that the forces traveling from the top to the foundation can be easily visualized. However, there is something else about the Parthenon that affects the viewer and that is the knowledge of its history. Democracy started in Greece during the time of the Parthenon and it was very much an external symbol of what human beings were thinking at the time it was built. If the Parthenon is compared to a similar structure, which is a little older, the Temple of Luxor in Egypt, the viewer will find that the Temple evokes a vastly different set of feelings. The Temple was constructed under the rule of one absolute ruler, Pharaoh, and its construction is believed to have involved the use of slave labour. These issues seem to affect how a structure is judged in terms of history, and may affect whether it is considered beautiful or aesthetically pleasing. The Parthenon, shown in Fig. 12.7a, and the Temple of Luxor, shown in Fig. 12.7b, are very good examples of this coupling because their geometry or ratios are very similar; their flow of forces is also similar; but the knowledge about these two structures is quite different and hence the perception, or feelings, they evoke eventually seem to supersede their physical form when deciding which structure is more aesthetically pleasing.

12.5 Harmonizing Beauty, Utility and the Environment

The Craigellachie Bridge, shown in Fig. 12.8, is a cast iron bridge over the Severn River near Tewkesbury, England, designed by Thomas Telford (1757–1834). It can be seen that the designer of this bridge was consciously aware of the value of aesthetics, because this bridge has a beautiful structural component. It was designed and constructed at the start of a period when the use of metal in bridge building was just beginning. To fully utilize this material and show it to advantage, Thomas Telford designed a bridge that complements its surroundings while serving a very functional purpose, thus proving that aesthetics and utility can be successfully linked together. The beauty of the arch, the deck and the strut members, which are very slim in the Craigellachie Bridge, still retain their fascination even today, proving that this structure has withstood the test of time both in aesthetics and

Fig. 12.7 (a) The Parthenon and (b) the Temple at Luxor

Fig. 12.8 The Craigellachie Bridge

Fig. 12.9 A bridge that cannot be regarded as beautiful

utility. On the other hand, if the Mythe Bridge is contrasted with the bridge shown in Fig. 12.9, it becomes obvious that the designer did not take into account aesthetic values either in the design and construction of this bridge or in blending the structure with its environment. Further, it is unlikely that any economy was realized because the ugly structure would have required a similar amount of labour and materials as that necessary in a more aesthetically pleasing structure, which would have had the added benefit of becoming a work of art that could be enjoyed by citizens and visitors of the area for the generations that followed its construction.

The bridge in Fig. 12.9 is extremely heavy and does not seem to indicate how the flow-of-forces travel through it. As well, its geometry is extremely confusing, and of course, since it obviously does not attract a caring component from society, its history is not recorded for posterity. As a result, there is little knowledge beyond perhaps the most basic technical details about this bridge compared to Telford's Craigellachie Bridge, which is recognized as one of the first metal bridges in the world incorporating the new shapes of the deck arch and the connecting members.

Differing Visions of What Is Aesthetically Pleasing Put two different people in a room and it is likely that they will have two different views on what they consider aesthetically pleasing. For example, Louis Riel is an historic personage in the Province of Manitoba. Early settlers from the European continent that came to Manitoba considered Riel a treasonous rebel. However, to the indigenous peoples of the area, particularly the Métis, he is considered to be the 'Father of Manitoba.' Times change and people change, and now that tempers have cooled and events can be judged from an historical perspective, the elected leaders in the Province of Manitoba as well as many of its citizens, decided that a debt is owed to this man who fought for his people and suffered great personal trials and tribulations as a result. To honour Louis Riel, the Province of Manitoba commissioned a statue, which was to be placed on the grounds of the Legislative Buildings. However, when the statue was completed and situated in a place of honour, the hue and cry from the public including aboriginal, Métis and non-aboriginal alike was overwhelming in their condemnation of the structure, which can be seen in Fig. 12.10a.

Although meant to convey the image of a tortured man who suffered much for his people but came through strong and still struggling, the statue evoked, for some people, the negative image of a man who went suffering to his grave, and not that of a person who was a shining light, a statesman who stood up for them when no one else would. It was this latter image that they held in their hearts and minds; and it was in this manner that they wanted to see Louis Riel portrayed. Clearly, it was the

Fig. 12.10 Two statues that pay tribute to Louis Riel (**a**) an abstraction, and (**b**) a life-like statue

perception and the knowledge, which the people had of Riel that made them decide which statue was beautiful and which statue they considered to be ugly. Bowing to public pressure, another statue of Louis Riel was commissioned, which is shown in Fig. 12.10b; this new statue depicting Riel in a European dress looking like a lawmaker replaced the first statue on the grounds of the Manitoba Legislature and the first statue was moved to l'École St. Boniface, a college in the French quarter of the City of Winnipeg. With the passage of time, both versions of the tribute by two different artists to this very important historical figure to the people of Manitoba have come to be accepted. In fact, both are artistically very well done. However, many are still divided over which version is aesthetically pleasing. It should be mentioned, however, that the more conservative image, which was more acceptable to the vast majority, blends in much better in its current environment, a site on the grounds of the Legislature, which is home to statues of other historical figures significant to the province such as Queen Victoria, Lord Selkirk, Major General James Wolfe, etc.

12.6 Artists Who Work in 3-D Forms

Artists who work in 3-D forms include structural engineers, architects and sculptors. Professor David Billington has written an excellent treatise on great structural engineers who were also artists; he has also written several other books on this subject. In his book, *The Tower and the Bridge* (1983), he has highlighted several contemporary structural engineers whom he identifies as artists in the same category as architects and sculptors. Four structural engineers, who have really focused their efforts on ensuring that the structures they designed and constructed were not only aesthetically pleasing but also blended well into their environments with the overall effect that their creations have been of value to posterity are discussed in this section. The task for those of us who follow is to determine what information can be gleaned from these structures to determine how or if the structural engineers who built them conformed to the principles of the aesthetics previously outlined.

Bridges born of necessity represent an astonishing marriage of technology and art spanning 2000 years of engineering and aesthetic trends. The four structural engineers of note that will be discussed in the following sections are: Robert Maillart, Christian Menn, Toshiaki Ohta and Gamil Tadros. These professionals are but a small representation of the global engineering community who have contributed significantly to the vast state-of-the-art knowledgebase available to the practicing engineers of today.

Robert Maillart (1872–1940) was a renowned Swiss citizen very well known as an artist as well as a structural engineer. Currently, his exhibition is frequently circulated to various venues in Europe where his work on bridges is displayed in the same manner that the sculpture of Rodin such as the Thinker, shown in Fig. 12.11.

One of Maillart's most significant achievements, the Salingatobel Bridge, shown in Fig. 12.12d, was not arrived at in a simple linear fashion. It was developed over time as Robert Maillart crafted and perfected his expertise.

Maillart began with a bridge called the Zuos Bridge, shown in Fig. 12.12a and constructed in 1901; its 3-hinged arch was a new concept in bridges. The north face of the bridge is exposed to the sun in the morning, and its south face is exposed to the sun in the evening. The exposure to the sun in the morning eventually led to cracks in the portions of the beam that were near the abutments. The citizens from the little village near the bridge were concerned that there might be safety issues because of these cracks. Robert Maillart inspected the bridge and immediately determined that the cracks were temperature-related. He also confirmed that the equilibrium was very much satisfied by the forces in the arch as well as in the deck; therefore, safety was not an issue. However, there was the issue of aesthetics. Nobody likes cracks in a bridge, which create a feeling of insecurity even when unwarranted. Learning from this experience, when Maillart designed his next arch bridge, Tavanasa Bridge, shown in Fig. 12.12b, in 1905, he developed a new form in which he essentially removed the concrete that was cracking due to changes in temperature. This bridge was very beautiful because it was streamlined to remove all the material that was not needed. The next 7 years again saw the evolution of the form he had used in the Tavanasa Bridge into the Aaburg Bridge, shown in Fig. 12.12c, in which more unnecessary material was removed. This new bridge looked exactly how a well-designed arch bridge form should look. The connection between the deck and the arch was achieved by inserting cylindrical columns that provided visible geometry, which was very pleasing aesthetically; yet it also indicated how the flow-of-forces were being transmitted to the foundation. There followed an 18-year gap, while Robert Maillart was busy living his life and pursuing his career in the USSR, until he designed his final bridge called the Salginatobel Bridge, shown in Fig. 12.12d. In this bridge, Maillart used his previous experiences to lead to the most perfect geometry, because some of the material he had removed in his previous structures was put back and the approaches he designed for this new structure were further refined by proportioning the columns, height and cross-section, which fitted beautifully between the deck and the arch and enhanced the environment into which the bridge was placed.

The 18-year gap in Maillart's career is pointed out because the reader should be made aware that the artist, whether he or she is an engineer or a sculptor has a human spirit, which must also be nurtured if the person is to reach his or her full

Fig. 12.11 Rodin's sculpture the Thinker

Fig. 12.12 Bridges by Robert Maillart: (**a**) Zuos (1901); (**b**) Tavanasa (1905); (**c**) Aaburg (1912); (**d**) Salginatobel (1930)

potential and have a full life. After building the Aaburg Bridge, Robert Maillart married and with his wife left to establish his construction business in Russia, where he had access to capital and backers. When he built the Zuos and Tavanasa bridges, he was a consulting engineer. After he married, he essentially decided to change his career path. Unfortunately for him, although his business started out very successfully, the Russian Revolution changed the face of society in Russia. In the process, Robert Maillart lost all the resources and the wealth that he had accumulated and also suffered the untimely death of his wife. However, Maillart did not succumb to despair because of these tribulations. His engineer's spirit, although wounded, was not broken. He moved back to Switzerland and went on to create his most beautiful structure to date, the Salginatobel Bridge. What engineers should learn from Maillart's experiences is that no matter how much adversity or ridicule one must suffer in life, one should never lose focus of what one believes and what one wants to achieve. Maillart's strength of character and integrity allowed him to build a bridge that remains a truly great work of art, which is highly regarded and greatly admired to this day. Clearly, his example is one from which others could benefit and gain insight.

Christian Menn (b. 1927) is another Swiss engineer; and fortunately, for those of us in the engineering community, Professor Menn is still active today to give us the benefit of his knowledge and experience. Christian Menn attended the Comprehensive Secondary School in Chur, Switzerland from 1939 to 1946. He then attended ETH Zurich, Switzerland from 1946 to 1950 where he received his diploma in engineering. After graduating, to gain some experience and expertise, he worked from 1951 to 1953 in the engineering offices of private industry in Chur and Bern, Switzerland. In 1953, Dr. Menn became a research associate to Professor Pierre Lardy at ETH Zurich. It was while he was assisting Professor Lardy that Menn learned the new theory of pre-stressing. He participated in Lardy's studies of pre-stressed bridges and for the first time he came into contact with the practice of bridge engineering. Dr. Menn found this to be the greatest motivating experience of his assistantship. In 1956, he received his Doctor of Science Technology Degree from ETH Zurich. Since his graduation from the Institute, Dr. Menn has owned his own engineering company, was also a professor of structural engineering at the Institute, and has been a private consulting engineer. In 1996, he received an honorary doctorate from the University of Stuttgart. Dr. Menn has used his education to expand and explore new areas of civil engineering, and his extensive research and use of pre-stressed, concrete and reinforced concrete has changed the way bridges are built to day.

Professor Menn saw the Salginatobel Bridge and noticed that the 3-hinged arched bridge hinged at the crown and at the abutments was taking most of the forces by compression. When he designed the Ganter Bridge (1980) (Fig. 12.13), he reversed the flow-of-forces by taking the form of the Salginatobel Bridge and inverting it, so that the members in arch now became tensile members in the Ganter Bridge. Menn took these tensile forces by pre-stressing the girders and then hanging the deck from two tall columns. It was phenomenal that not only could he visualize the geometry, but he was able to visualize that the structure would remain beautiful even if it was inverted, in addition to the fact that he could see that the flow-of-forces when reversed would still be very transparent. In fact, the forces moving in the Ganter Bridge are more transparent than they are in the Salginatobel Bridge. However, both of these bridges seem to fit with great beauty and grace into their surroundings in the Swiss Alps.

Fig. 12.13 The Ganter Bridge, Switzerland

Fig. 12.14 The Sunniberg Bridge columns (Switzerland)

The question of economics is often linked to the aesthetics of structures. A few years ago, when the authors visited Professor Menn in Switzerland while he was involved in the construction of the new Sunniberg Bridge, shown in Fig. 12.14 during construction, in Chur, he said that society must and should budget from 12 % to 15 % of the capital cost of a structure to ensure that it has an aesthetically pleasing appearance. The bridge that Professor Menn was in the process of building was spectacular. The columns were done in such a way that they almost appeared to be flaring into two and their height above the deck was kept short so that the columns could match with the geometry of the deck column above it and the cables that were holding the deck, as can be seen in Fig. 12.14. The geometry of the structure was immediately apparent just by looking at exactly how the forces were flowing through the deck. The deck had been made thicker because the height of the column above it had been made shorter since more force would be going into the deck due to the shallow angle of the cables. A significant feature of the curved-in-plan bridge is that it has no expansion joints. The shortening and elongation of the

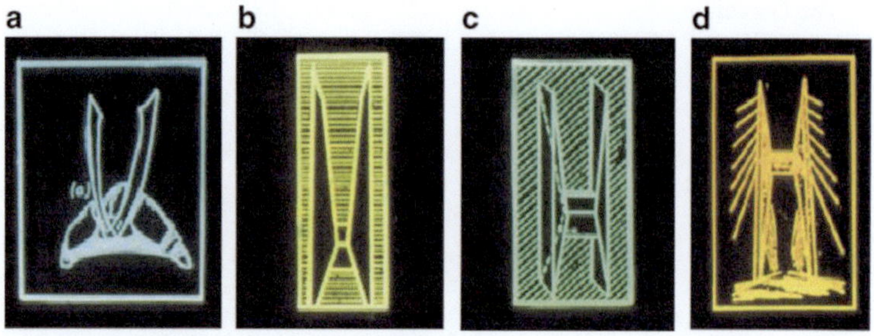

Fig. 12.15 Aesthetic design method for bridges – (**a**) Find a pleasing motif; (**b**) Abstract the essential forms from the motif; (**c**) Materialize the abstracted forms; (**d**) Produce a variation of similar forms regarding the pier, tower or abutment under the conditions required for the design

Fig. 12.16 The Hitsuishi-jima and Iwakuro-jima Bridges

deck caused by changes in temperature are accommodated by the columns, which are hinged at the bottom.

Professor Toshiaki Ohta et al. of Japan in his paper, *Aesthetic Design Method for Bridges* (1987), argues that geometry and flow-of-forces must be kept transparent – the geometry should be slender and have pleasing ratios. However, he also states that the structure should have a motif that represents the cultural, historical and the developmental aspects of the surrounding society. His argument is that bridge designers should find a desirable motif for the basic symbolic image of the structure, abstract essential forms for the motif, materialize the abstracted form, produce a variation of this form for the pier tower and abutment under the conditions required for the design, and finally choose the motif that most closely complements its environment, as can be seen in Fig. 12.15. Professor Ohta utilized this methodology of incorporating motif in the implementation of the Hitsuishi-jima and Iwakuro-jima Bridges in Osaka, Japan; a view of the bridge can be seen in Fig. 12.16. The piers are indicative of the fins on the helmets of ancient Japanese warriors and this effect is quite obvious when one looks at these bridges.

Dr. Gamil Tadros proposed that for structures with repetitive segments, each segment should be built in a similar form to the first so that the final entire structure has an aesthetically pleasing flow. He utilized this concept in the Confederation

Fig. 12.17 Confederation Bridge, New Brunswick, Canada

Bridge, shown in Fig. 12.17. The 13 km long bridge constructed from 52 modules joins the Prince Edward Island to the mainland of Canada at the Province of New Brunswick. This bridge, which is the longest bridge over ice-covered waters in the world, was completed in 1997 and is composed of precast modules designed to repeat for the full length of the structure contributing to its elegance and the flowing nature of its form. It has often been stated that the Confederation Bridge was manufactured, rather than constructed.

It is concluded, therefore, that in considering aesthetics in the art of bridge design, lightness or thinness, should be the first criterion. The flow-of-forces should be transparent. A form that fits the local environment and has a motif related to its cultural surroundings is another criterion that should be adopted when feasible. Finally, the fourth criterion is that if a structure has more than one segment, similar segments should be built in a repeating form.

12.7 Incorporation of a Cultural Motif

Several researchers writing about bridge aesthetics (Billington 1993; Leonhardt 1984; Ohta et al. 1987; Bakht and Jaeger 1983) have indicated that certain characteristics are helpful in leading to pleasing and beautiful bridges. One of these desirable characteristics is for a bridge to portray a motif that symbolizes the historical and cultural heritage of a society. Such a motif can sometimes be abstracted and then used in a variety of similar forms in piers, towers, abutments, etc.

In a computer graphics study for a skyway proposal for the City of Karachi, an Islamic motif was studied with a view to incorporating it into the structure. With the aid of computer graphics, the motif was abstracted and then materialized for embodiment into the piers of the skyway. Several forms were studied using wireframe diagrams, hidden-line-removed figures and solid modelling (Mufti and Hsiung 1991). Various vantage points of observation with respect to the skyway were also incorporated into the study, thus making it possible to assess the appearance of the bridge from different directions, lines of sight and at various speeds.

The use of computer graphics in the design of buildings has been popular and successful for some time but its application to bridge aesthetics has yet to be explored extensively (Mufti and Jaeger 1988).

12.7.1 A Skyway Proposal for Karachi

The Karachi Development Authority of Pakistan considered a number of schemes for relieving traffic congestion in the city. One possible solution, which could be used in conjunction with any of several possible schemes for mass transit, was an elevated highway, or *skyway*.

The skyway was to follow the line of the seasonal Lyari River, which contains little water during the dry season. The proposed skyway was to be approximately 18 km long and, because of its elevation, would be visible from many points throughout the city. Along the proposed 18 km route of the skyway, the river would be crossed by ten bridges, as shown in Fig. 12.18. Because of its high visibility, it was desirable that the skyway should be aesthetically pleasing, and that it should reflect traditional Muslim architecture in a significant way. With this in view, a study was made of classical arches and domes in the Muslim tradition. Figure 12.19 shows some of the domes that have been utilized on various structures in Islamic locales. For a Muslim, these domes have great historical, cultural, and even spiritual significance. Figure 12.20 depicts arches that have been utilized in Islamic architecture along with examples of arches in the Gothic style. The affinity between the Gothic and Islamic arches is quite clear.

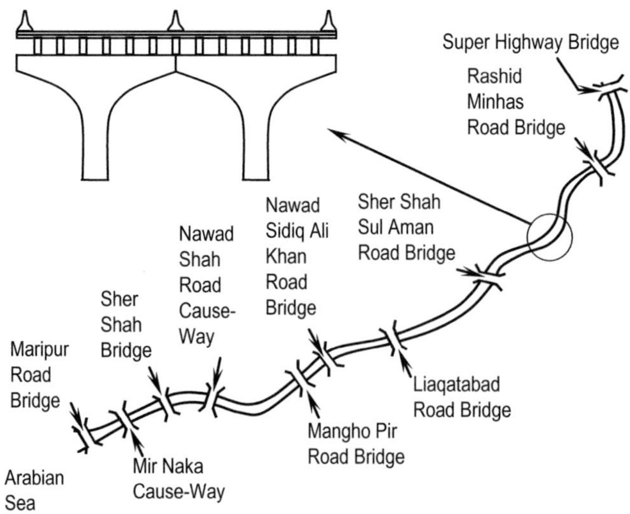

Fig. 12.18 18 km Lyari River path

Fig. 12.19 Types of domes

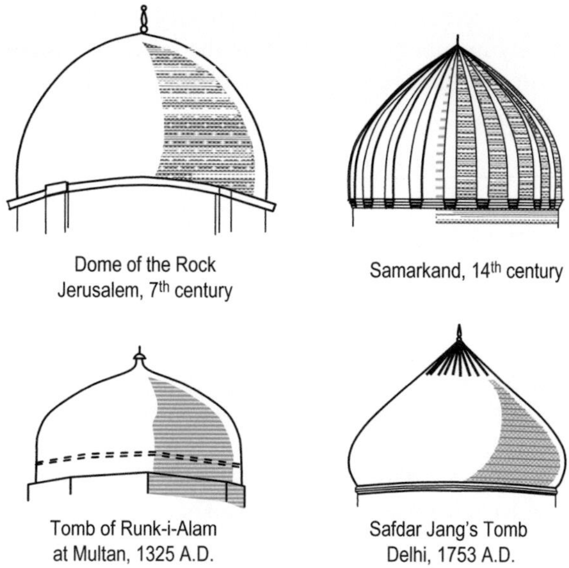

Fig. 12.20 Types of arches

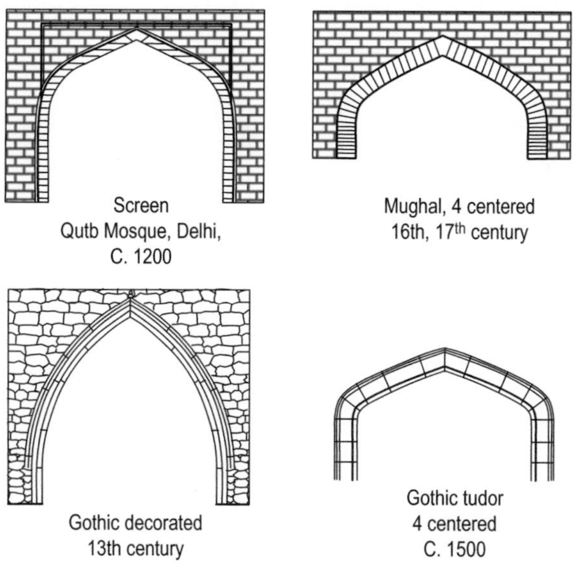

It was concluded that it was possible to distinguish between the two families of shapes, both of which exhibit an overriding property that the curvature is tightest at the point where the curve begins to take shape from the vertical column, and thereafter diminishes. One of these families is particularly prevalent for arches, whilst the other is found more frequently in domes. The advantage of having computer graphics available as a design tool is that many members of the family can be scrutinized, by systematically varying the governing parameters; and the family member best answering the needs of structural efficiency and architectural form can then be chosen. These two families will be described in the following sections.

12.7.2 Arches and Domes

Arches The curvature of an arch is constant over a certain range and then falls suddenly to zero. Thus the profile comprises a circular arc followed by a straight line. The various members of the family are distinguished by the fraction of the total that is circular arc and the fraction that is straight. In illustrating the notation, Fig. 12.21a shows arch parameters y_1, y_2, and ℓ using which general mathematical formulae are derived; these formulae can generate a large number of arch shapes that have been used in the past. The relevant formula for $0 \leq y \leq y_2$ is:

$$x = \sqrt{(2Ry - y^2)} \tag{12.2}$$

and the formula for $y_1 \leq y \leq y_2$ is as noted below

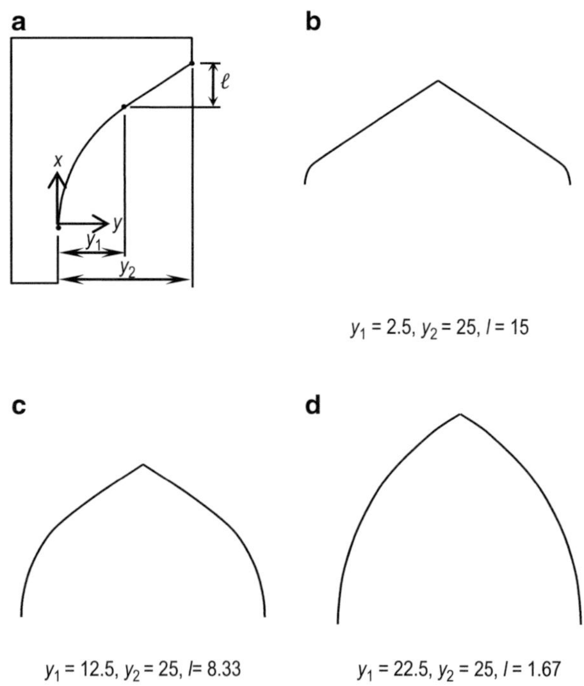

Fig. 12.21 Arch parameters (**a**) notation, (**b**) arch with low rise, (**c**) arch with medium rise, and (**d**) arch with high rise

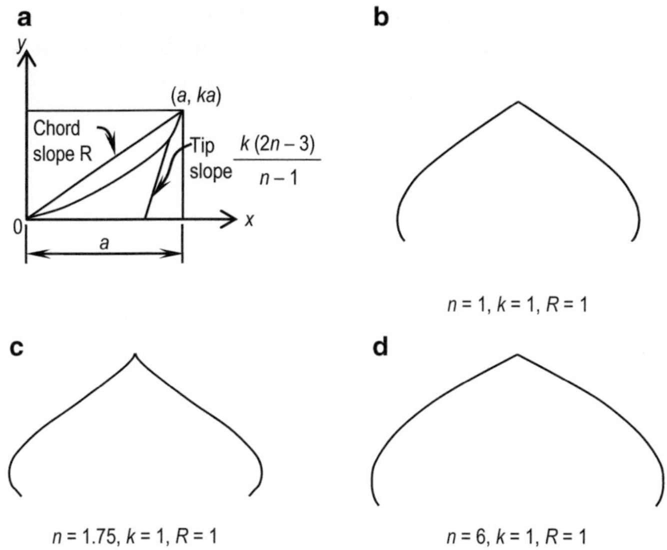

Fig. 12.22 Dome parameters (**a**) notation, (**b**) (**c**) (**d**) domes of different shapes

$$x = x_1 + \frac{l(y - y_1)}{y_2 - y_1} \tag{12.3}$$

where

$$R = \frac{y_1}{\left[1 - \frac{\ell}{\sqrt{\ell^2 + (y_2 - y_1)}}\right]} \tag{12.4}$$

$$x_1 = \frac{y_1(y_2 - y_1)}{\left[\sqrt{\ell^2 + (y_2 - y_1)^2} - \ell\right]} \tag{12.5}$$

The sequence of steps required to calculate arch shapes through computer graphics is noted below.

From the straight portion of the curve, get values of $y_1,\ y_2,\ and\ \ell$; these values will be the input.

(a) Calculate R from Eq. (12.4);
(b) Calculate x_1 from Eq. (12.5). This last equation establishes the origin;
(c) Plot $x = (2Ry - y^2)^{0.5}$ for values of y from 0 to y_1;
(d) Plot the straight line portion as usual, from $x = x_1 + l(y - y_1)/(y_2 - y_1)$ for values of y from y_1 to $(y_1 + y_2)$ to.

Figure 12.21b, c, and d show three examples of this family; in each example the left-hand half is generated as 'circular arc plus straight-line' and the right-hand half is generated by symmetry.

Domes For domes, the curvature starts at an initial value and diminishes steadily as the curve progresses. In some members of this family, the curvature can become negative, so that the curve displays a point of inflexion. Figure 12.22a defines the notation for the dome parameters. Using these parameters, the mathematical equations are derived; these equations can create a very large number of domes that have been built in the past.

$$x = at \tag{12.6}$$

$$y = \frac{R\,at^2}{(n - 1)}(n - t) \tag{12.7}$$

where

t ranges from −0.2 to 1
a scale factor
R chord slope
n range of value to define Cubic-Parabolic Curve

The sequence of calculation steps for the graphics is as follows:

(a) Select value of R; this defines the chord slope;
(b) Examine the variation of dy/dx; the slope of curve is given by:

$$\frac{dy}{dx} = \frac{dy}{dt}\frac{dt}{dx} = \frac{R}{n - 1}(2nt - 3t^2) \tag{12.8}$$

At $t = 1$, the slope is $R(2n - 3)/(n - 1)$. This tip slope is thus $(2n - 3)/(n - 1)$ of the chord slope.

(c) Examine the variation of d^2y/dx^2; the second derivative, which is directly related to curvature, is given by:

$$\frac{d^2y}{dx^2} = \frac{d}{dt}\left(\frac{dy}{dx}\right)\frac{dx}{dt} = \frac{2k}{(n - 1)^a}(n - 3t) \tag{12.9}$$

It is noted that for $n < 3$ there is a point of inflexion at $t = n/3$. The following should also be noted:

(a) For large values of n, the curve approaches a parabola.
(b) The practical lower bound of n is $n = 1.5$, this corresponding to a point of inflexion at $x = 0.5\ a$ and a "point" at $(x = a,\ \ y = ka)$ of zero angle, i.e. with $y = k\ \ at^2(3 - 2t)$.
(c) For $1.5 < n < 3$ the curve has a point of inflexion which, as n increases, moves nearer to the tip. Thus for $n = 2$ we have $y = k\ \ at^2(2 - t)$ and the point of inflexion is at $x = 2a/3,\ y = 6Ra/27$.
(d) For $n = 3$ the point of inflexion is at the tip, and $y = 0.5\ \ kat^2(3 - t)$.

(e) For $n > 3$ there is no point of inflexion; for example for $n = 6$; $y = 0.2\ kat^2(6 - t)$.

Figure 12.22b, c, and d show examples of various domes drawn by the procedure discussed above. The dome shown in Fig. 12.22c has a point of inflexion. It may be noted that it is often desirable to "back off" these curves below the normal origin as shown in Fig. 12.22c so as to provide a curvature "below the equator".

A study of classical Muslim architecture shows that the curves described in the previous sub-sections have been used for both arches and domes; however, the first family is more prevalent for arches and the second family for domes.

12.7.3 The Karachi Skyway Project

In the initial design concepts for the Lyari River Skyway, it was decided to apply the motif of arches employed in Muslim architecture to the tall 'hammer-head' piers of the structure. Each pier, comprising a central column and two cantilevers, was supposed to support four lanes of the highway. To avoid stress concentrations, the junction of the cantilever and column was made smooth by a curve. The full eight-lane highway was to be supported by two piers placed side by side. Computer graphics were used to model the resulting structure. It can be seen that the two hammer-head piers placed side by side lead to a pleasing arch of Muslim architecture, as can be seen in Fig. 12.23. By using the analytical procedure described in the sub-section on arches, it was confirmed that the arch thus formed was very close to the Mughal arches employed in the Mughal monument the Taj Mahal.

By using the standard solid modeling software package, a solid model was made of the initial design of the proposed Karachi Skyway. The two side-by-side piers supporting one span are shown in Fig. 12.23. It can be seen that the resulting arch is close in appearance to the arches of the Taj Mahal. A larger segment of the proposed structure as viewed from the ground can be seen in Fig. 12.24. It is obvious that the highly visible cluster of piers in the Skyway would not be a permanent eye-sore to the residents of Karachi.

Fig. 12.23 Two side-by-side 'hammer-head' piers supporting one span of the proposed Karachi Skyway

Fig. 12.24 A computer-generated view of the proposed Karachi Skyway

12.8 Concluding Remarks

The aesthetic characteristic that embodies a motif was studied using computer graphics. It was found that computer graphics is a versatile technology that generates families of graphical models for a bridge engineer to use in studying the aesthetics of possible design solutions. With the help of a specific example, it has been shown that the incorporation of cultural motifs in a bridge can be done without compromising the purity of structural forms, and without incurring additional expenses.

References

Bakht B, Jaeger LG (1983) Bridge aesthetics. Can J Civ Eng 12(3):408–414

Billington D (1993) The tower and the bridge, the new art of structural engineering. Princeton University Press, Princeton

Leonhardt F (1984) Bridges. MIT Press, Cambridge, MA

Mufti AA, Hsiung B (1991) Solid modelling in structural engineering. J Microcomput Civ Eng 3:58–63, Elsevier. New York

Mufti AA, Jaeger LG (1988) Use of computer graphics in bridge aesthetics. In: Proceedings of third international conference on computing in civil engineering, Vancouver

Ohta T, Takahashi N, Yamane T (1987) Aesthetic design method for bridges. Struct Eng J ASCE 113(8):1678–1687

Index

B. Bakht, A. Mufti, *Bridges*, DOI 10.1007/978-3-319-17843-1

C

D

E

F